*Plant
Evolutionary
Biology*

Plant Evolutionary Biology

Edited by

LESLIE D. GOTTLIEB*

and

SUBODH K. JAIN †

*Department of Genetics
†Department of Agronomy and Range Science
University of California, Davis
California, USA

London   New York
CHAPMAN AND HALL

First published in 1988 by Chapman and Hall Ltd
11 New Fetter Lane, London EC4P 4EE
Published in the USA by Chapman and Hall
29 West 35th Street, New York NY 10001

© 1988 Chapman and Hall Ltd

Printed in Great Britain at the
University Press, Cambridge

ISBN 0 412 29290 4 (hardback)
0 412 29300 5 (paperback)

British Library Cataloguing in Publication Data

Plant evolutionary biology
1. Plants – Evolution
I. Gottlieb, L.D. II. Jain, S.K.
581.3'8   QK980

ISBN 0-412-29290-4
ISBN 0-412-29300-5 Pbk

Library of Congress Cataloging in Publication Data

Plant evolutionary biology
edited by L.D. Gottlieb and S.K. Jain.
p.   cm.
Includes bibliographies and index.
ISBN 0-412-29290-4. ISBN 0-412-29300-5 (pbk.)
1. Plants – Evolution. I. Gottlieb, L.D. (Leslie D.),
1936– . II. Jain, S.K. (Subodh K.), 1934–
QK980.P53 1988
581.3'8 – dc19

Contents

Contents

Contributors

JANIS ANTONOVICS
Department of Botany
Duke University
Durham, North Carolina 27706,
USA

C. WILLIAM BIRKY, JR
Department of Molecular Genetics
Ohio State University
Columbus, Ohio 43210, USA

ROBERT N. BRANDON
Philosophy Department
Duke University
Durham, North Carolina 27706,
USA

CURTIS CLARK
Biological Sciences Department
California State Polytechnic
  University
Pomona, California 91768, USA

MICHAEL T. CLEGG
Department of Botany and Plant
  Sciences
University of California
Riverside, California 92521, USA

JAMES R. EHLERINGER
Department of Biology
University of Utah
Salt Lake City, Utah 84112, USA

NORMAN C. ELLSTRAND
Department of Botany and Plant
  Sciences
University of California
Riverside, California 92521, USA

BRYAN K. EPPERSON
Department of Botany and Plant
  Sciences
University of California
Riverside, California 92521, USA

J. PHILIP GRIME
Department of Botany
University of Sheffield
Sheffield, S10 2TN, UK

ROBERT L. JEFFERIES
Department of Botany
University of Toronto
Toronto, Ontario M5S 1A1, Canada

DONALD A. LEVIN
Department of Botany
University of Texas
Austin, TX 78712, USA

KARL J. NIKLAS
Section of Plant Biology
Cornell University
Ithaca, New York 14853, USA

ERAN PICHERSKY
Department of Biology
University of Michigan
Ann Arbor, Michigan 48104, USA

PETER H. RAVEN
Missouri Botanical Garden
St. Louis, Missouri 63166, USA

CHARLES M. RICK
Department of Vegetable Crops
University of California, Davis
Davis, California 95616, USA

TSVI SACHS
Department of Botany
Hebrew University
Jerusalem 91904, Israel

G. LEDYARD STEBBINS
Department of Genetics
University of California, Davis
Davis, California 95616, USA

STEVEN D. TANKSLEY
Department of Plant Breeding
Cornell University
Ithaca, New York 14853, USA

ROBERT WYATT
Department of Botany
University of Georgia
Athens, Georgia 30602, USA

Preface

There are still heroes in science. They are recognized because the issues and problems they chose to study became the issues and problems of a major field of research. They are also recognized because their insights and solutions are the ones that are tested and evaluated when new ideas and technologies become available. In the field of plant evolutionary biology, the hero is George Ledyard Stebbins. His first scientific publication appeared in 1929 and has been followed by nearly 60 magnificent years of seminal ideas, proofs, and proposals that defined much of what was worth doing in plant biosystematics, evolution and biological conservation. His energy, enthusiasm and good humor (widely shared at many congresses and symposia in the 'Singalongs with Stebbins') made him a wonderful teacher for both undergraduates and graduate students. He is the mentor of several generations of botanists, plant geneticists and evolutionists. A brief biography and publication list were included in *Topics in Plant Population Biology*, edited by Otto T. Solbrig, Subodh Jain, George Johnson and Peter Raven (Columbia University Press, 1979) which resulted from a symposium held on the occasion of Ledyard Stebbins' 70th birthday. In this volume, population biology and physiological ecology received major attention particularly in relation to plant form and function.

Now, ten years after that 1977 meeting, we again honored Professor Stebbins with a major symposium at the University of California, Davis, where he founded the Genetics Department in 1950, and taught and did research until his formal retirement to emeritus status in 1973. To the delight of all of us, he has maintained a very dynamic 'post-retirement' lifestyle and an energetic scientific career. The subject of this symposium, held in September, 1986, was plant evolutionary biology in its broadest sense, and included papers in population biology, population genetics, physiological ecology, systematics, development and molecular genetics. By juxtaposing these fields, with a greater emphasis on genetic topics, the intent was to document that scientists from now all-too-separate disciplines have much to learn from each other. The juxtaposition was also appropriate because Ledyard Stebbins has contributed important ideas in many fields and his example appeared to be worth emulating.

The need for greater interaction can be illustrated by considering how morphological or other traits are studied. Ecologists identify traits that enhance fitness, but not their developmental basis nor how they evolve with the result that they cannot account for how characters become modified to serve as adaptations. Developmental biologists and morphologists study organography and the cellular basis of changes in character expression, but fail to investigate the genetic basis of the changes and how differences in expression affect fitness. Population geneticists estimate relative fitnesses and examine the magnitudes and direction of selective forces from genotypic frequency analyses, but in many cases study the phenotype insufficiently to identify the particular traits that respond to selection or the number of gene substitutions involved in character change. Molecular biologists undertake to describe gene sequences but the relationship of the gene products to environmental factors is nearly always unknown. Clearly we are ready for more comprehensive treatments in evolutionary biology that will employ the tools and points of view of many disciplines and, in this spirit, we present this book based on the Davis symposium in honor of our friend and colleague Ledyard Stebbins.

Professor Stebbins has contributed a characteristically wide ranging and thoughtful essay. He points out that substantial new information from molecular genetics and development as well as from prokaryotic biology has recently become available, but it augments rather than replaces the neoDarwinian evolutionary synthesis in which he participated as the primary botanical representative. He foresees a more 'pluralistic' science that takes deeper account of the different lifestyles of organisms from bacteria and protists to higher plants and animals. He calls for study of developmental genetics in its broadest sense and of ecology, both likely to reveal the 'organizational' constraints that link evolutionary competence, genetic information and economy of adaptive shifts.

His paper is followed by thirteen articles contributed by symposium speakers, and five commentaries and an epilogue by the editors. Although the range of subject matter is wide and provides numerous contrasts in thoughts and technologies, many important subjects are not included. This was unavoidable. We hope that sufficient material has been provided to honor Ledyard Stebbins and to reveal the intellectual strength and excitement of plant evolutionary biology.

Many people on the Davis campus contributed to the planning and arrangements for the symposium which provided a cordial and friendly meeting on our campus. We thank the authors heartily for making the symposium meaningful and for their patience with us during the editorial process. We gratefully acknowledge the financial support of Dean Charles E. Hess, College of Agricultural and Environmental Sciences, Dean Allen G. Marr, Graduate School, Vice-Chancellor Robert M. Cello, and Professor Calvin O. Qualset, Director of the University of California Genetic Resources Conservation Program. Ms. Kimberly Strauch, Ms. Nancy Hilden and Ms. Marilee Dykstra did fine jobs handling registration, mailing and various office details. Professors Francisco Ayala, Bob Pearcy and Grady Webster helped plan the program and local arrangements. All royalties from the sale of this volume will be donated to the Botanical Society of America to establish the Stebbins Conference in Evolutionary Biology to be held in even-numbered years at the Society's national meeting. We hope that the Conference will bring together scientists of outstanding merit, thereby, honoring Ledyard Stebbins for many more years.

PART 1

An overview of evolutionary biology

CHAPTER 1

Essays in comparative evolution. The need for evolutionary comparisons

G. LEDYARD STEBBINS

1.1 INTRODUCTION

During the past 20 years, evolutionary theory has been in a turmoil. Evolutionists have been bombarded with a cascade of new facts about the nature of DNA-based genetic variation. At the same time they have discovered that the amount of genetic variation in populations is greater by several orders of magnitude than previously imagined. Some paleontologists have insisted that the concept of gradual evolutionary change should be abandoned in favor of evolution by sudden bursts, while at the same time biochemists have insisted with equal conviction that most of the proteins of which the body is constructed evolve with clock-like regularity. Meanwhile, quantitative studies of ecology have emphasized the diversity and patchiness of the environment, and hence a complexity of genotype–environment interactions far greater than that assumed by the model builders of population genetics a generation ago.

In the face of these challenges and innovations, evolutionary theories need drastic revisions. The question that now divides theorists concerns their nature. Must the synthetic theory that dominated thinking 30 to 50 years ago be completely abandoned? Can its major postulates be retained, modified and amplified to fit the newly discovered facts, or should evolutionists start thinking in entirely new directions?

Most evolutionists agree in recognizing that the new theory that will certainly be forged should, like the mid-20th century theory, be truly synthetic, rather than one in which undue emphasis is placed upon one or two factors, such as molecular genetics, neutralism, punctuation, or an

3

all-encompassing view of natural selection. New facts require modification and amplification of accepted principles and processes, rather than their rejection.

One direction in which the synthetic theory should be amplified is from the nature of new facts. Gene families and clusters as well as non-coding DNA are more prevalent and important in eukaryotes than prokaryotes; accessory genetic systems, like those of mitochondria and plastids affect only eukaryote evolution. Many evolutionary changes of enzymes and other metabolic proteins may be neutral in higher animals and plants. Nevertheless, among free-living prokaryotes, that were the only organisms existing on the earth during the entire first half of the 3.5×10^9-year time span of organismic evolution, adaptive radiation via evolution of temperature-resistant and oxygen-tolerant protein molecules are well documented. In prokaryotes, evolution of macromolecules has clearly been adaptive. In higher organisms, some proteins may well have evolved with a regularity that justifies their recognition as evolutionary clocks. Simultaneously, phenotypic characteristics that are determined by protein molecules of another kind, such as structural proteins in membranes and in skeletons, were tracking the complex and occasionally rapid changes which many environments undergo.

Many points of view, now polarized, may be resolved when enough facts become known. Many of these new facts show that the relative importance of the chief evolutionary parameters differs greatly from one major group of organisms to another. Other discoveries reveal equally drastic differences between evolutionary rates of different molecules and phenetic characters within the same evolving evolutionary line. The purpose of this chapter is to assess the extent and significance of these differences.

1.2   THE BIOLOGICAL UNIVERSALS THAT GOVERN EVOLUTION

Before discussing comparative evolution, one must first summarize the factors and principles that are essential to the process as a whole. First, what is biological or organic evolution? Among the numerous definitions that have been suggested, I reject both those that overemphasize the genetic component (evolution consists of alterations in gene frequencies within and between populations) and those that overemphasize population–environment interactions ('. . . the essence of evolution is the production of adaptive diversity,' Prosser, 1986, p. 21). An intermediate definition proposed by Futuyma (1986, p. 7) 'Biological evolution is change in the properties of populations of organisms that transcends the lifetime of a single individual', forms a reasonable basic parameter for making evolutionary comparisons between different kinds of organisms and processes.

Given this definition, five processes become essential, whichever organism and whatever characters we choose to compare:

1. Mutation in the broadest sense, including chromosomal changes;
2. Genetic recombination, both sexual and parasexual;
3. Selection, amplified to include differential survival not only of adult individuals, as Darwin proposed, but also of informational macro-molecules, organelles, cells, embryos, and under some conditions groups of individuals such as family groups or even species;
4. Chance events, such as genetic drift, and chance associations between particular populations and environments; and
5. Reproductive isolation that permits and sometimes furthers divergence between populations existing in the same habitat.

In this chapter, I shall present reasons for rejecting two extreme points of view: (a) that any one of these factors can sometimes be disregarded; and (b) that any one of them, even within a restricted category of organisms, is the only key to understanding its evolution.

At this point, I must call attention to a misconception that has become widespread: a supposed dichotomy between 'molecular' and 'phenetic' evolution, as maintained by Kimura (1983). One lesson that the molecular revolution in biology has taught evolutionists is that all hereditary change, and therefore all evolution in the sense of Futuyma (see above), has a molecular basis. For forging a new synthesis of evolutionary theory, a close coordination between the molecular and the organismal approach is essential.

A more instructive dichotomy is between *phenetic* and *cryptic* evolution, defined as follows. Phenetic evolution consists of all changes in populations that can be recognized by comparing structure, physiological properties and/or behavior of the component organism. Cryptic evolution refers to changes that can be recognized *only* by comparing linear sequences of information macromolecules: DNA, RNA or proteins. Both kinds include adaptive and neutral changes. The majority of phenetic changes are probably adaptive, but differences such as human fingerprints and some blood groups (e.g. M-N) are neutral. Cryptic differences are rarely if ever directly adaptive, but many of them, such as substitutions of base pairs at the third position of DNA–RNA triplet codons, may affect indirectly evolutionary competence, as is explained more fully below.

In addition to the five processes mentioned above, another sequence plays a dominant role almost universally in determining directions of evolution. It consists of (1) evolutionary competence, (2) organizational restraints, and (3) economy of adaptive shifts.

Evolutionary competence is here used as a more meaningful phrase for the condition that is sometimes called preadaptation or exaptation (Gould and Vrba 1982). The following three examples illustrate this property.

At the level of RNA–DNA information and the genetic code, six different triplets (CGU, CGC, CGA, CGG, AGA, AGG) code for the amino acid arginine. If a population is monomorphic at a particular position for one of them (e.g. CGU or CGC) a single mutation at that position can substitute a codon for any of six other residues (cysteine, serine, glycine, histidine, proline, leucine) but no more. If, however, random drift and neutral fixation have made the population polymorphic for all six of the codons for arginine, the evolutionary competence of that population is doubled with respect to mutations that affect an arginine residue. Twelve rather than six different residues could be substituted at a position by different single-step mutations.

Examples of increased evolutionary competence at the structural level in animals are the acquisition of a tightly fitting, somewhat flexible chitonous exoskeleton in primitive marine arthropods, and of a notochord in primitive chordates. The arthropod exoskeleton increased the competence of these animals to evolve limbs, that could easily be transformed into the homologous limbs of terrestrial arthropods. The chordate notochord served as the foundation about which evolved the internal skeleton of fishes. Mollusks, which never evolved structures leading to flexible skeletons, could evolve into the highly mobile predatory marine cephalopods, but could invade the land only in the form of proverbially slow-moving snails.

Among higher plants, evolutionary competence is well illustrated by the competence of some insect-pollinated flowers to become adapted for pollination by hummingbirds. Many such examples exist in western North America, involving flowers that have very different structural plans. Two kinds of change are necessary. Flower color must change from the blue, violet, lavender, yellow or white, which is most conspicuous to insect eyes, to the red that is most conspicuous to birds. The shape of the flower must be altered from one that fits the insect's body or proboscis to one that fits the bird's beak. The only flowers that have become modified for bird pollination are those in which colors and structures adapted to entomophily are altered with relative ease to fit ornithophily. Among the numerous white or yellow flowers found in various families and genera, none is closely similar to an ornithophilous flower. In addition, the entomophilous species that most closely resemble ornithophilous species have either erect rather than spreading petals, or petals bearing spurs that fit the insect proboscis, or short, vase- or urn-shaped corollas or calyx tubes. Dish-shaped or bowl-shaped flowers appear never to have evolved adaptations for pollination by birds.

Restraints on the direction of evolution are imposed by the adaptive properties of existing phenotypes. The oft-repeated statement: 'evolution results from random mutations followed by natural selection based upon the immediate environment of the population' is such a gross oversimplification that it is a caricature of the actual situation. The most recent discussions of this subject emphasize the importance of developmental restrictions (Maynard Smith *et al.*, 1985). On the other hand, Charles-worth, Lande and Slatkin (1982) maintain that these constraints have been exaggerated. They call attention to numerous breeds of domestic animals and varieties of cultivated plants, that have been created by artificial selection, and are developmentally normal, but cannot survive in nature without human aid. Darwin referred to such an example in the 'Origin of Species,' stating that the fantail breed of pigeons transcends the morpholog-ical limits of its family with respect to the number of tail feathers. Pigeons having this excess number are perfectly healthy and normally reproductive in the dovecote, but are relatively inefficient fliers. All the highly productive breeds of cattle, sheep and other food animals, as well as non-shattering varieties of cereal grains are likewise at a total loss as competitors with their wild relatives. The same restriction is valid for many of the laboratory mutants of *Drosophila* and other organisms familiar to geneticists. Selective con-straints may be far more widespread than developmental constraints.

Although the discussion reviewed by Maynard Smith *et al.* mentions chiefly developmental constraints, some of the examples cited could be interpreted as the results of developmental patterns acquired by the action of natural selection on ancestral forms, and therefore as resulting from selective constraints. The lack of branching in most species of palms, and some species of dicotyledons, such as arboreal lobeliads in the mountains of Central Africa and arboreal species of *Sonchus* on the Canary Islands, may well be due to the origin of these tree-like species from branched shrubby ancestors (Carlquist, 1974). If so, they could be based upon selective as well as developmental constraints.

In view of this situation, I believe that the term *organizational constraints* is a more neutral, less interpretative and therefore more desirable term than either developmental or selective constraints. Organizational restraints can be defined as follows: constraints upon the direction of evolutionary change imposed by the complexity of particular structures in the evolving popula-tion. Organizational constraints are links between evolutionary competence and genetic economy of adaptive shifts, as explained below.

The principle, economy of adaptive shifts, is a logical consequence both of evolutionary competence and of the fact that when a new adaptive niche is opened up, several different evolutionary lines may be competent to become adapted to it. The populations that can fill the open niche most quickly are most likely to capture the niche and exclude other competitors. The principle was well stated by the botanist, W. F. Ganong, in 1902 (Stebbins, 1950): adaptive change follows the path of least resistance, based upon structural and physiological properties. I have modified Ganong's term by substituting the term genetic resistance, i.e. the fewest mutational changes necessary to give rise to the new adaptation.

The examples of adaptive shifts in flowers from entomophily to ornithophily are good illustrations of this principle. In the genus *Delphinium*, most species have blue or purple flowers which have a relatively short spur on one of the petals and the other petals and sepals are spread in an oblique or horizontal fashion. The ornithophilous *D. cardinale* and *D. nudicaule* have narrower and straighter spurs, while their perianth parts are also narrower, erect, and overlapping to form a tube adapted to the bird's beak. The largely entomophilous species of *Astragalus* lack spurred petals, but in the ornithophilous *A. coccineus* the petals are narrow and elongated, and the calyx tube is longer than in related species. Genera belonging to several different families, including: *Ipomopsis*, *Salvia*, *Mimulus*, *Penstemon*, *Castilleja*, *Beloperone* and *Cirsium*, possess corolla tubes adapted to entomophily that in a few species have evolved red color and shapes adapted to ornithophily. In another genus, *Zauschneria* (*Epilobium* acc. to Raven) the calyx tube has become similarly modified.

A good example of Ganong's principle among animals is the defense mechanism of the skunk (*Mephitis*). All members of the family Mustelidae are equipped with anal glands that exude a strong smelling liquid, used either for marking territory or possibly as a minor source of defense for these relatively small, agile carnivores (MacDonald, 1984). Skunks appear to have evolved increased size, slower movement, and the capacity to spray approaching predators with the noxious liquid.

A number of characteristics contributed to the competence of larger tailless primates to evolve into tool-dependent, highly social humans. These are diurnal activity, spectroscopic color vision, prehensile forefeet, social aggregates, relatively large size and a brain large enough to permit a well-developed memory and associations between sense impressions. No other group of mammals even approaches the great apes with respect to this particular form of evolutionary competence.

1.3 THE MOST SIGNIFICANT KINDS OF COMPARISONS

Four different kinds of comparisons reveal different patterns of emphasis with respect to evolutionary processes: anagenetic, cladogenetic, mosaic and dynamic. Anagenetic comparisons attempt to answer the question: How does emphasis upon the five basic processes differ between the least-advanced and the most-advanced organisms on the evolutionary scale? Cladogenetic comparisons examine this question with respect to different branches of the evolutionary tree, at both similar and different levels of advancement. Mosaic comparisons show that rates of evolution vary greatly between different characters within the same evolutionary line. Dynamic comparisons recognize that the overall evolution of the same or related lines may vary greatly from one period in time or stage of advancement to another.

Anagenetic comparisons

Anagenetic comparisons show that the evolutionary importance of genetic recombination relative to genic or chromosomal mutation increases as organisms become larger and more complex, and the time span of each generation is correspondingly increased. The larger and more complex the organism, the more it relies upon stored variation, produced by past mutations, as the source upon which selection acts.

Prokaryotes include the smallest, structurally simplest and shortest-lived organisms. Their populations are much larger than those of most eukaryotes. The volume of an individual *E. coli* cell is of the order of 2×10^{-16} ml; that of an amoeba or *Chlamydomonas* cell about 10^{-13} ml; a *Drosophila* fly about 2×10^{-3} ml, and a mouse about 50 ml. A flask (or an animal intestine) having a volume of 100 ml can easily accommodate a population of 10^{10} individuals of *E. coli* or 10^7 protozoa, but such a container can hold only 10^2–10^3 *Drosophila* flies, and not even a single pair of mice. Due to limitations of space, populations of microorganisms can be millions of times larger than those of most multicellular animals or plants. This means that mutations at individual loci, that occur at rates of 10^{-6}–10^{-5} per individual per generation, are likely to occur in one or more individuals in each generation or two in populations of microorganisms, but only once every 10^3–10^4 generations in the fly. Since adaptive evolution is geared to the tempo of environmental change, particular new mutations, such as resistance to disease or greater ability to parasitize a host, are likely to appear in a few hours to a day in bacteria but only once in several weeks or months in a fly. Particular new mutations can be expected only once in several years in a mouse or any other animal or plant in which populations consist of hundreds or thousands of individuals, and generation times are measured in weeks, months or years.

The relative importance of stored mutational differences and of recombination becomes even greater when one considers different degrees of complexity of an individual organism. Unfortunately, reliable quantitative estimates of differences in complexity are difficult or impossible to obtain, but the following can serve as useful guidelines. The greater complexity of cell structure in eukaryotes as compared to prokaryotes would suggest that pathways from genes to phenotypic characters could be at least ten times as complex in *Amoeba* as compared to *E. coli*. Since developmental processes bring about the differentiation of 150–200 different kinds of cells in an insect such as *Drosophila*, and regulatory mechanisms control a large number of different interactions between cells, one might conclude that the genetic developmental system that would be altered by the most adaptively significant mutations is hundreds of times more complex in *Drosophila* than *Amoeba*, and that the difference in this respect between mice or any other mammal and a unicellular eukaryote is even greater.

These differences are so great that, while single mutations might be expected often to bring about adaptive adjustments in bacteria, and occasionally in unicellular eukaryotes, they would much more rarely have significant effects by themselves in multicellular animals or plants. In these organisms, a single mutation is likely to contribute to a new adaptation only when associated with modifiers, that must be already present in the genotype, or must arise while the mutation is still being stored in some individuals of the population.

The above conclusions about bacteria are well supported by experiments on populations in chemostats. Chao and Cox (1983) compared the success of a mutagenic strain 'Treffers' with that of a strain having normal mutation rates, using large populations that were maintained for 120 generations. They found that at all relative frequencies of the two strains, these frequencies persisted without change for 50–60 generations. Later, dramatic changes took place that were frequency-dependent. When introduced at low initial frequencies, the Treffers strain eventually declined and disappeared, leaving only the normal strain. At relatively high frequencies, the balance between the two strains was upset after about 70 generations by the increase in Treffers as compared to the normal strain. The experiment was repeatable, and similar results were obtained after reciprocal shifts in the marker genes that were used to identify the strains. Under the experimental conditions, gene transfer between strains was either lacking or insignificant. These results support a generally held opinion: most genes have adverse effects, but given a new environment, mutations that adjust organisms to the new conditions may occasionally arise. They are more likely to appear if the overall mutation rate is elevated.

Other experiments (Mortlock, 1982, 1984; Hall, 1982, 1983) have shown that individual mutations, including short duplications of nucleotides as well as point mutations, can contribute significantly to adaptive shifts. Important shifts such as the ability to digest a newly unused sugar can be acquired via two or three such mutations. These observations and experiments show that prokaryote populations are so large and generation times so short that absence or rarity of genetic recombination is not a serious hindrance to adaptive radiation via mutation-induced adaptive shifts.

Another fact that emerges from anagenetic comparisons is that while all kinds of characteristics have participated in adaptive radiation throughout the course of evolution, the three principal grades or levels of complexity of organisms, prokaryotes, unicellular eukaryotes and multicellular organisms differ with respect to the characteristics that play the largest roles in both adaptive radiation and increasing complexity. At the prokaryote grade, adaptive diversity evolves chiefly via differentiation of enzyme proteins. The way in which this diversity arises is just beginning to become clear.

Although point mutations play a large role, that of duplications and inversions of base pair sequences within genes are apparently more important. The experimental results reported by Rigby, Burleigh and Hartley (1974), Hall (1982, 1983), Hartley (1979, 1984), Crawford (1982), Mortlock (1982, 1984), and Belfaisa *et al.* (1986) all point in this direction. Nevertheless, intragenic structural change must be accompanied by point mutations, which may be more significant in determining changes in enzyme function. Sequence comparisons reveal homologies between enzymes having very different functions, in both bacteria and multicellular organisms (Stebbins, 1982).

The greatest progress in understanding prokaryote evolution will come from comparative sequencing of homologous proteins having different but related functions. Enzymes that include as prosthetic groups various metalloporphyrins, such as cytochromes, ferredoxin, chlorophyll, carotenes, phytochromes and rhodopsin are of particular importance in this connection. If phylogenetic connections between these proteins become firmly established, their existence will provide strong evidence in favor of the hypothesis, suggested by Zuckerkandl (Stebbins, 1982), Doolittle (1979) and others, that no new genes or protein molecules have evolved since the appearance of early forms of life. With respect to eukaryotes, the generalization – all new genes evolve from preexisting genes – may be as true as that recognized a century ago – all cells evolve from preexisting cells.

Although enzyme differentiation continued throughout eukaryote evolution during the evolution of unicellular forms it declined in importance relative to the evolution of intracellular structure. The most dramatic change at this level was the entrance of prokaryotes to form the chloroplasts and mitochondria of the nucleated cell. Of equal importance was the evolution of the mitotic apparatus, particularly tubulin and microtubules.

Cladistic comparisons

The second kind of valuable comparison is between groups that form the ends of clades or branches of the evolutionary tree. They become significant to the extent that the organisms compared differ with respect to the following characteristics.

Developmental pattern
Two characteristics are important: degree of complexity and location plus timing of differentiation. The importance of complexity is based upon the following correlation: the more complex is the adult organism, the greater is the complexity of the pathway from gene to character. A basic tenet of developmental genetics is that in multicellular organisms the differentiation of cells and tissues depends upon regulatory mechanisms that activate or

inhibit the action of particular genes at particular stages and locations. They can exert their effects at levels of either transcription, translation or post-translational processes. The larger the number of cells differentiated during development, the more complex are these gene-controlled regulatory mechanisms. Hence the more complex is the organism and its development, the larger is the number of genes that cooperate in producing a phenotypic trait such as size, fecundity or metabolic activity.

The importance of location and timing of cellular, tissue and organ differentiation is best exemplified by the comparison between higher animals and plants, that is presented below. It is based on the fact that under certain circumstances, mutations can be eliminated or spread by selective processes at the level of molecular or cellular replication as well as that of the organism.

Size and longevity
The size and longevity of organisms greatly affect the dynamics of the populations to which they belong. Large organisms occupy more space and use more resources than small ones. Their interbreeding populations are therefore smaller, often by several degrees of magnitude. They also take longer to develop, so that length of generations is much longer than in small organisms. Other things being equal, larger organisms also have smaller gene pools than small organisms, and given the same rate of change in gene frequency via differential selection, will require more time to evolve new adaptive syndromes.

Fecundity
This trait can vary from one to three offspring per pair per generation, as in some birds and mammals, up to thousands or millions, as in pelecypod mollusks or orchids among plants. Since in stable biotic communities, levels of population size remain constant, differential fecundity greatly affects the amount of loss that a population can tolerate, either by accidental death or adverse selection.

Vagility
Since adaptive shifts often accompany migration into new habitats, the ability of a population to send successful migrants into them is of major importance for evolutionary change. Vagility is not necessarily correlated with mobility. Since the pioneer synthesis of Ridley (1930), many plant geographers have shown via both direct and indirect evidence that passive dispersal of seeds and spores is equally or more effective over long distances than is active migration of animals. Moreover, successful migration is a two-step process. Establishment in the new environment is as important as transportation or migration to it. A well-known fact of biogeography is that

some species of sedentary higher plants, such as cattails (*Typha*) and reed grass (*Phragmites*) have far wider natural distributions, independent of human transport, than do any species of terrestrial animals.

In addition to the four primary characteristics just reviewed, the following secondary characteristics, based upon differences with respect to the first four, exert strong effects upon rates and directions of evolution.

Dominance in biotic communities

The relation between dominant vs. subordinate positions in ecosystems greatly affects the way in which environmental change affects evolution. Among autotrophic plants, the dominant species, whether they are forest trees, shrubs of semi-xeric scrub associations, or grassland herbs, have large populations, are often highly fecund, and greatly affect the lives of other plants and many animals. With exceptions, such as susceptibility of grasses to grazing pressure and the ravages of disease, the survival and spread of autotrophic dominants depends largely upon stability vs. changes in abiotic climatic or edaphic factors rather than associated organisms. For heterotrophic organisms, both animals and fungi, adaptive evolution depends primarily upon biotic factors. Dominant animals are predators that occupy the head of the food chain. Their populations are necessarily smaller than those of dominant plants in any one area, and they can acquire large gene pools only via extensive migration over one or a few generations. Their success depends on that of their prey which, in turn, depends for success upon that of associated plants.

Frequency of uniparental reproduction

In all organisms, the relative importance of selection as compared to chance depends greatly upon the size and breeding structure of populations. Organisms that have relatively simple developmental patterns are less likely than more complex organisms to acquire a load of recessive lethal or semi-lethal mutations. This is because simpler patterns can more easily tolerate new mutations, that in more complex organisms are likely to upset a finely tuned harmony of epistatic gene interactions. Consequently, reversion from obligate outcrossing to facultative or almost obligate self-fertilization is far more common in plants than in animals, and among animals is relatively common in simpler phyla such as nematodes and Platyhelminthes as compared to arthropods and vertebrates.

Behavioral patterns

Among higher animals, natural selection is greatly affected by the evolution of complex behavioral patterns. The aphorism of Wilson (1975), behavior is the pacemaker of evolution, is surely true for terrestrial vertebrates and arthropods, and to a great extent sets them apart from other organisms. Its consequences have been explored extensively by evolutionists interested in the rise of humanity, but as further facts are obtained, even more intensive study of this problem will be necessary.

Plant vs. animal evolution
The most important example of the application of the above differences is a comparison of evolution in insects, birds, mammals and flowering plants. With respect to developmental pattern, animals are at least 50 times as complex as plants. This may be a gross underestimate rather than an overestimate. By careful study of reviews of anatomy and histology in angiosperms and vertebrates, I have concluded that the maximum number of different kinds of differentiated cells possessed by an adult plant is about 30, while in mammals this number is about 250. In addition, the number of interactions between different kinds of cells and tissues is far greater in animals than in plants.

A second and even more important difference between animals and plants is between embryos and embryonic meristems. In essence, cell differentiation and tissue differentiation in animals is stage specific and occurs only once during the lifetime of an individual. In plants it is positionally specific, and the same sequence of differentiation processes is repeated many times during the lifetime of a single individual. A developing animal is successively an embryo, a fetus, a juvenile and an adult. Each of these stages includes sequences of processes taking place simultaneously, each of them is unique to that individual, and is not repeated until the next sexual generation. The only unique event in plant embryogeny is the differentiation of seed leaves or cotyledons, that are discarded from the adult plant. Embryonic cells that are capable of differentiation do not disappear, as they do at fetal and adult stages of animals. They persist and proliferate. Newly formed branches are tipped by embryonic meristems that are capable of differentiating new leaves and floral organs. A pine tree, creosote bush or buffalo grass clone that is hundreds or thousands of years old bears hundreds of separate meristems, each of which can produce its own particular seeds or pollen grains. Because of somatic mutation, genetic differences between gametes derived from neighboring branches of the same tree, shrub or herbaceous clone may occasionally arise.

The consequences of these differences are as follows:

Segregation of germ and somatoplasm
Weismann's argument against the Lamarckian concept of the inheritance of acquired adaptations does not hold for plants. Somatic mutations are well known to occur in plant meristems and are the basis of occasional 'bud sports' in cultivated trees. Self-perpetuated genetic mosaics that give rise to variegated leaves are well known, and may sometimes give rise to seedlings that are homogeneous for a mutated gene. The origin of some well-known

horticultural varieties, such as nectarine from peach, is believed to have occurred in this fashion.

Even though separate germ lines do not exist in them, extensive experimentation has shown that adaptive characters acquired via the direct influence of the environment are not inherited in plants any more than in animals. This is because the informational molecules, DNA and RNA, are strongly buffered against the effects of the environment, such as frost hardening, physiological drought resistance, and the effects of photoperiod on flowering. The amino acid sequences of the proteins involved in the regulatory mechanisms that are responsible for such adjustments are probably also buffered against change. At any rate, the genetic–developmental mechanisms of organisms are well-enough known to make a specific reverse influence from superficial environmental effects on the phenotype to a corresponding alteration of the genetic information virtually inconceivable.

Accumulation of deleterious mutations

The geneticist H. J. Muller (1964) pointed out an important result of the segregation of the germ line in animals, that makes periodic meiosis and sexual reproduction necessary to cleanse the germ line of deleterious mutations. Since cells of the germ line do not participate in phenotypic adaptation of the individual, they can persist, generation after generation, until the germ line has become so loaded with deleterious mutations that most of the eggs and sperm produced by meiosis are lethal. The only way in which this damage can be avoided is continuous sexuality, by means of which in every generation meiosis cleans the germ line of damaging mutations by selective elimination of the gametes, if any, that bear them. In perennial plants that possess multiple meristems from which meiocytes arise, this effect is strongly diluted by competition between different meristems produced by the same individual. Those meristems that give rise to a significantly higher proportion of defective gametes do not participate in reproduction, and the number of gametes produced by reproductive organs differentiated from undamaged meristems is more than enough to ensure adequate contribution to the gene pool of the population. This means that the phenomenon termed 'Muller's ratchet' does not operate in plants to limit the evolutionary life of asexually reproducing apomictic genotypes. Factual evidence that this principle does not operate is the existence of a relic species, *Houttuynia cordata*, belonging to a phylogenetically isolated monotypic genus and, so far as is known, completely apomictic (Stebbins, 1950). Also significant in this connection is that in annual angiosperms, in which the number of competing reproductive meristems may be very few, apomixis is unknown, although genetic constancy and near homogeneity of populations brought about by constant self-fertilization is very common (Stebbins, 1958).

An important effect of the greater complexity of development in animals is the comparative infrequency of chromosomal aberrations. Trisomic individuals, in which one chromosome belonging to the gametic set is present in triplicate rather than in duplicate, are well known in several plant species (Stebbins, 1950; Khush, 1973), and in *Clarkia unguiculata* their effects on the phenotype are so slight that they are hardly distinguishable (Vasek, 1963). Among animals, trisomics are inviable in *Drosophila* except for attached-X females and flies trisomic for very small chromosomes like the fourth of *D. melanogaster*. In humans, all trisomics are lethal or nearly so in the embryo and fetus, except for those involving the small chromosome responsible for Down's syndrome, the deleterious effects of which are well known. Apparently, the complexities of gene interaction during development are so much greater in animals than in plants that the imbalance of gene dosage brought about by trisomy cannot be tolerated in them.

The same effect is probably responsible at least in part for the relative scarcity of polyploids in higher animals as compared to plants. The explanation of this scarcity offered long ago by H. J. Muller, and repeated in many textbooks, is that polyploidy interferes with the sex ratio, and so cannot be expected in animals that nearly always have separate sexes. This explanation has limited validity, since in several plant genera, such as *Salix*, *Rumex* and *Antennaria*, that have separate sexes, polyploid species or cytotypes within species are common, widespread, and have normal 1 : 1 sex ratios. In *Drosophila*, triploid individuals have occasionally been found, but tetraploids, have never been observed, in spite of numerous efforts to locate them. My suspicion is that tetraploid genotypes of *Drosophila* are lethal as embryos, larvae or pupae, because of genetic imbalance. In humans and probably other mammals, genetic imbalance may well be responsible for the absence of polyploids. Triploid embryos have occasionally been found but are invariably lethal. The problem of polyploidy in mammals has been well summarized by Beatty (1957).

An even more dramatic difference between higher animals and plants is in their response systems. The growth substances and membranes of plants permit adaptive responses such as tropisms, the opening and closing of flowers in response to light, temperature and moisture, and in only a few species greater activity such as the capture of insects by carnivorous plants. There is no 'central office' that directs the plant's activities. On the other hand, the most persistent trend of progressive complexity in all but the most primitive of animal phyla lies in the increasing cephalization of the nervous system. Two characteristics culminate this trend: courtship patterns and socialization. In many animals, the modification of natural selection known as sexual selection brings about the most striking structural differences that distinguish their species. Distinctive shapes of horns, antlers and other appendages, striking colors of the body and its coverings, such as bristles, scales, hairs or plumage serve both to help animals recognize their own species, and for human naturalists to sort them into appropriate categories. Distinctive courtship patterns reinforce both the distinctness of species and the ease with which naturalists can recognize them. The various ways in which differences between higher plant and animals affect the species problem are summarized below.

To the evolutionary biologist who is interested in both plants and animals, the differences of opinion between zoologists and botanists as to the nature of species are a real hindrance to a unified synthetic theory. Some zoologists believe that the inability of botanists to apply their biological species concepts to plants, such as oaks, is due to ignorance or adherence to outworn dogmas, while many botanists find so many difficulties in applying to plants the concept that was first based upon animals that they discard altogether any biological concept of species. The only way to reconcile these differences is to recognize the very real differences in the nature of species boundaries in animals and plants, and either to establish two kinds of concepts, or a modified biological concept that may not be perfect for either group but will enable all naturalists to speak the same language. The basis of the differences is now well understood, and can be summarized as follows:

1. Differences between animal species in courtship patterns raise barriers to gene exchange that are absent in plants.
2. The greater complexity of development and gene interaction in higher animals means that independently evolved developmental patterns are much less likely to give rise to harmonious combinations when put together by hybridization.
3. The limited life span of animals greatly reduces the fitness of hybrids that have lowered fertility, whereas in vegetatively reproducing plants highly sterile hybrids can not only persist for hundreds of years, but often are so aggressive that by vegetative reproduction they can crowd out their less-aggressive parents. Even if such hybrids have seed fertility of 1% or less, they maintain acceptable Darwinian fitness.
4. The greater ability of plants to tolerate the altered gene dosages that result from chromosomal changes opens up to them the reticulate course of evolution provided by hybridization and polyploidy. The ability of polyploids to simulate their diploid ancestors with respect to diagnostic characters of morphology is a serious barrier to classification of reproductively isolated genotypes. The complexities raised by these differences will be explored in another essay.

1.4 SUMMARY AND CONCLUSIONS

Discoveries made during the past twenty years demand extensive revision of previous evolutionary syntheses. The five pillars of the mid-20th century synthesis: mutation, genetic recombination, selection, chance and reproductive isolation, still stand, but taken by themselves are insufficient to account for the intricate interactions between genes, in internal environment of the organism and external environment, both physical and biotic, that affect fitness of organisms. The necessary amplification must be pluralistic rather than monolithic, because of the great differences in lifestyle found in such organisms as bacteria, protists, higher animals and plants. Integration is achieved not only by the five classic pillars, but also by a sequence of events found in all evolutionary lines that restricts and in part determines the direction of evolution. This consists of organizational, developmental and selective constraints on the acceptance of new genetic variants and the resulting particular genetic competences, that make some directions of change more probable than others.

Evolutionary competence differs greatly from one kind of organism to another. It can be understood in its entirety only by comparing rates and directions of evolution between different kinds of organisms. Four kinds of comparisons are valuable:

1. Between different organisms having greatly different degrees of complexity;
2. Between those at the ends of different major branches of the evolutionary tree;
3. Between rates of evolution affecting different structures, molecules and different parts of domains of molecules within the same organism; and
4. Between different periods of evolution within the same evolutionary line.

What new research is needed the most in order to provide a satisfactory new synthesis? First, more knowledge is needed about protein homologies among primitive prokaryotes, particularly the anaerobic forms that are most similar to the supposed primordial cells. At this level an answer may be found to the question: How often have totally new enzymes evolved? Second, the bits of evidence now available that suggest a much greater amount of gene interaction in the synthesis of supramolecular structures that are responsible for pattern and form, need to be corroborated and amplified. In particular, they must also be integrated with factors that determine such properties as polarity and membrane permeability, and may be at least in part biophysical rather than biochemical in nature. Finally, our present rudimentary knowledge about the action of regulatory mechanisms, that activate and deactivate gene systems during development, must be greatly amplified.

In other words, the most important discipline for evolutionary genetics in the future is developmental genetics in its broadest sense. This discipline needs to be supplemented by additional ecological knowledge that will deepen our understanding of the relationships between organism and environment that govern natural selection. The path towards a complete synthesis will be long and arduous, but recent knowledge has opened the way to a vast new field of perspectives for research that will move evolutionists toward their goal.

1.5 REFERENCES

Beatty, R. A. (1957) *Parthenogenesis and Polyploidy in Mammalian Development.* Cambridge University Press, Cambridge.

Belfaisa, J., Paesot, C., Martel, A., Bouthier de la Tour, C., Margarita, D., Cohen, G. N. and Saint-Girons, I. (1986) Evolution in biosynthetic pathways: the enzymes catalyzing consecutive steps originate from a common ancestor and possess a similar regulatory region. *Proc. Natl. Acad. Sci. USA*, 83, 867–91.

Carlquist, S. (1974) *Island Biology.* Columbia University Press, New York.

Chao, L. and Cox, E. C. (1983) Competition between high and low mutating strains of *Escherichia coli. Evolution*, 37, 125–34.

Charlesworth, B., Lande, R. and Slatkin, M. (1982) A neodarwinian commentary on macroevolution. *Evolution*, 36, 474–98.

Crawford, I. P. (1982) Nucleotide sequences and bacteriological evolution. In *Perspectives in Evolution*, (ed. R. Milkman), Sinauer Assoc., Sunderland, MA.

Doolittle, R. F. (1979) Protein evolution. in *The Proteins* (eds H. Neurath and R. L. Hill), 3rd edn. Academic Press, New York, pp. 1–118.

Futuyma, D. J. (1986) *Evolutionary Biology*, 2nd edn, Sinauer Assoc., Sunderland, MA.

Gould, S. J. and Vrba, E. (1982) Exaptation – a missing term in the science of form. *Paleobiology*, 8, 4–15.

Hall, B. G. (1982) Evolution in a petri dish: The evolved galactosidase system as a model for studying acquisitive evolution in the laboratory. *Evol. Biol.*, 15, 85–150.

Hall, B. G. (1983) Evolution of new functions in laboratory organisms, in *Evolution of Genes and Proteins* (eds M. Nei and R. K. Koehn), Sinauer Assoc., Sunderland, MA, pp. 234–357.

Hartley, B. S. (1979) Evolution of enzyme structure. *Proc. Roy. Soc. London Ser. B*, 205, 443–52.

Hartley, B. S. (1984) The structure and control of the pentitol operons. in *Microorganisms as Model Systems for Studying Evolution* (ed. R. P. Mortlock), Plenum Press, New York, pp. 55-1-7.

Khush, G. S. (1973) *Cytogenetics of Aneuploids.* Academic Press, New York and London.

Kimura, M. (1983) *The Neutral Theory of Molecular Evolution.* Cambridge University Press, Cambridge.

MacDonald, D. (ed.) (1984) *Encyclopedia of Mammals*, Facts on File Publ., New York.

Maynard Smith, J., Burian, R., Kauffman, S., Alberch, P., Campbell, J., Lande, R., Raup, D. and Wolpert, L. (1985) Developmental constraints and evolution. *Quart. Rev. Biol.*, 60, 265–87.

Mortlock, R. P. (1982) Regulatory mutations and the development of new metabolic pathways by bacteria. *Evol. Biol.*, **14**, 205–68.

Mortlock, R. P. (1984) The utilization of pentitols in studies of the evolution of enzyme pathways. in *Microorganisms as Model Systems for Studying Evolution* (ed. R. P. Mortlock), Monogr. Evolutionary Biol., Plenum, New York, pp. 1–21.

Muller, H. J. (1964) The relation of recombination to mutational advance. *Mutat. Res.*, **1**, 2–9.

Prosser, C. L. (1986) *Adaptational Biology: Molecules to Organisms*. Wiley, New York.

Ridley, H. N. (1930) *The Dispersal of Plants Throughout the World*. Reeves, Ashford, Kent, UK.

Rigby, P. W., Burleigh, B. D. and Hartley, B. S. (1974) Gene duplication in experimental enzyme evolution. *Nature*, **251**, 200–4.

Stebbins, G. L. (1950) *Variation and Evolution in Plants*. Columbia University Press, New York.

Stebbins, G. L. (1958) Longevity, habitat and the release of genetic variability in the higher plants. *Cold Spring Harbor Symp. Quant. Biol.*, **23**, 365–78.

Stebbins, G. L. (1982) *Darwin to DNA: Molecules to Humanity*. W. H. Freeman, San Francisco and New York.

Vasek, F. (1963) Phenotypic variation in trisomics of *Clarkia unguiculata*. *Am. J. Bot.*, **50**, 308–14.

Wilson, E. O. (1975) *Sociobiology: The New Synthesis*. Harvard University Press, Cambridge, MA.

PART 2

Molecular evolution and species phylogeny

CHAPTER 2

Evolution and variation in plant chloroplast and mitochondrial genomes

C. WILLIAM BIRKY, Jr.

2.1 INTRODUCTION

G. Ledyard Stebbins has had a pervasive influence on plant evolutionary biology, in part because of his broad interests. These include the evolution of genome structure, especially as it is seen by the methods of cytogenetics, phylogenetics, and asexual reproduction. Ledyard has not worked on the genomes of mitochondria or chloroplasts, but it is appropriate that these organelle genomes be included in a symposium honoring him. We know a great deal about the evolution of the structure of the chloroplast and mitochondrial DNA molecules, i.e. about organelle cytogenetics at the molecular level. Mitochondrial cytogenetics has proved to be much like nuclear cytogenetics in many respects. Chloroplast cytogenetics is a powerful tool for making phylogenetic inferences, as is nuclear cytogenetics. And of course organelles are essentially asexual genetic systems wherever they are found, even in sexually reproducing plants.

In keeping with Stebbins' interests, this review is limited to multicellular plants; algae are not included. But in contrast to most of his work, 'evolution' means molecular evolution, changes in the structure and base sequence of the DNA molecule. 'Variation' means diversity in DNA structure and sequence within a species. In most cases we do not know what effects, if any, the molecular variation has on the morphology, physiology, or adaptation of the organism. We do know that mutations in organelles usually affect fitness rather than morphology, because organelle genomes are specialized for the control of photosynthesis and respiration as opposed to morphogenesis. Of course, molecular studies of evolution are important in their own right. For example, the molecular clock of base-pair substitution may be erratic, but where the fossil record is poor or non-existent, it is

23

the only game in town. Restriction fragment length polymorphisms make superb genetic markers for studying population structure and dispersal precisely because they often have little effect on morphology or fitness.

I will begin with cytogenetics, i.e. the evolution of organelle genome structure. Then I will review the data on rates of base-pair substitution and on genetic diversity in plant organelles. Finally, I will discuss some of the problems of population and evolutionary genetic theory that are raised by the differences between organelle and nuclear genomes. Comparisons will be made between mitochondrial, chloroplast and nuclear genomes at every step. It is particularly fascinating that three different genomes, embedded in and serving the same organism and interacting with each other in numerous ways, can evolve very differently.

2.2 TEMPOS AND MODES OF ORGANELLE EVOLUTION

The use of plurals in the title is deliberate, because mitochondria and chloroplasts appear to be evolving at very different rates and in very different ways. There are several excellent reviews of organelle evolution in plants and algae, especially thorough in the coverage of the evolution of genome structure and function (Palmer, 1985a,b, 1987; Bohnert and Michalowski, 1988; Sederoff, 1987). The reader is referred to these for details and complete documentation. I will give references mainly to more recent papers or those which need special emphasis.

Evolutionary changes in function and structure of the chloroplast genome

The plant cpDNA molecule is a genome frozen in time. It has probably had over 400 million years in which to change but did so very little compared with the nucleus or the mitochondrion. In this time the size of the haploid nuclear genome (the C-value) has come to vary in size by several hundred-fold, with large variations within families, genera, and even within single species (reviewed by Tanksley, Chapter 3). Large and variable proportions of this DNA (sometimes the majority of it) belong to families of repeated sequences, some of which have huge numbers of members (Mitra and Bhatia, 1986). Moreover the nuclear genome has been divided among a few to many chromosomes; the numbers of copies of genes coding for rRNA has varied immensely; and protein coding genes have been expanded into gene families of widely varying sizes (Flavell, 1980; Sorenson, 1984).

A useful baseline for discussions of the cpDNA molecule has been provided by Ohyama *et al.* (1986) and Shinozaki *et al.* (1986), who in a remarkable *tour de force* have sequenced the entire cpDNA molecules of the liverwort *Marchantia polymorpha* and of tobacco, respectively. A diagram of the major features of the *Marchantia* cpDNA molecule is given in

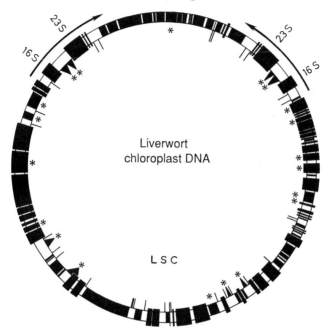

Fig. 2.1 Structure of the *Marchantia polymorpha* chloroplast genome (modified from Ohyama *et al.*, 1986). Identified genes are shown as solid regions projecting from the circle: rRNA genes are labelled, tRNA genes are narrow bars (or triangles if they contain introns), and the rest code for proteins. Long open reading frames are solid regions that do not project from the circle. Asterisks indicate genes with introns. The arrows show the long inverted repeats.

Fig. 2.1. It consists of 121 024 bp. One segment of about 10 kbp (kilobase pairs) is present as an inverted repeat separated by small (19.8 kbp) and large (81 kbp) single copy regions. The molecule carries genes for a complete set of four rRNA molecules and 32 tRNAs; 54 genes for known proteins; and at least 28 more long open reading frames that may code for additional proteins. Most of the proteins are involved in transcription, translation, or photosynthesis. Tobacco cpDNA is remarkably similar in all respects. It is larger (155 844 bp), with inverted repeats of 25.3 kbp and single-copy regions of 18.5 and 86.7 kbp; it has 52 identified genes and 38 unidentified open reading frames. However, the set of identified genes is almost identical. Among the differences, tobacco cpDNA has 30 tRNAs and codes for one fewer protein of the large ribosomal subunit, but one more of the large subunit; *Marchantia*, but not tobacco, cpDNA codes for three ferredoxins. Not only do tobacco and *Marchantia* have almost identical sets of genes, they are in nearly the same order on the map. The principal difference is an inversion of about 30 kbp in the large single-copy region.

About half the size difference is due to the inverted repeat. In tobacco, the repeat has engulfed two ribosomal proteins and a tRNA that were adjacent to it on the large single-copy region of *Marchantia*, and includes a number of unidentified long open reading frames whose homologies in *Marchantia* are unknown. The genes of these two species also differ somewhat in their complements of introns.

Tobacco and *Marchantia* have had most of the evolutionary history of plants in which to diverge, and they illustrate most of the extent, and the kinds, of variation to be found among plant chloroplast genomes. The size of the cpDNA molecule is very uniform. Palmer (1985b) has reviewed data from 230 species of angiosperms, one gymnosperm, three ferns, and two bryophytes. The total genome size ranges from 120 to 217 kbp; all but three plants fall between 120 and 160 kbp. Moreover, two-thirds of the variation in size is accounted for by changes in the extent of the inverted repeat, which shrinks or expands to incorporate genes from the single-copy regions. Some plant chloroplast genes contain introns, some of which are present in some species but not others. Although the size of the large inverted repeat varies considerably, it is present in almost all plant chloroplast genomes and always contains the rRNA genes. It was lost only once, in the common ancestor of three tribes of legumes (Palmer *et al.*, 1987). Other repeated sequences are found in the cpDNA of some species, but these are generally few in number and short, a few hundred bp or less (e.g. Palmer *et al.*, 1987).

Other plants have the same kinds of genes found in tobacco and *Marchantia*. Moreover, Palmer and his associates found that the arrangement of homologous sequences is essentially identical in 24 out of 30 families of angiosperms, a gymnosperm, and a fern. Some other species differ from this primitive groundplan by one or two large inversions and sometimes by additional small rearrangements. More extensive inversions and deletions are found in a few species, including the geranium and several legumes. Originally it was thought that the loss of one copy of the large inverted repeat in some legumes was correlated with extensive rearrangements, but this correlation has recently been disproved (Palmer *et al.*, 1987).

Evolutionary changes in structure and function of the mitochondrial genome

In stark contrast to the chloroplast genome, the mitochondrial genome of plants is extremely variable in organization and size, although perhaps not in function. Because of the large size of the mtDNA molecule and the complicating effects of intramolecular crossing-over, the analysis of this molecule has lagged behind that of cpDNA. However, restriction maps have now been made of mtDNA from several species and a number of genes have

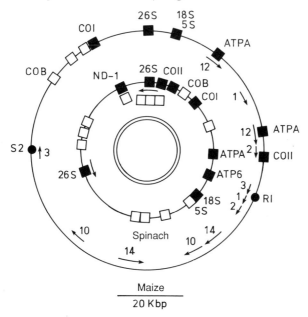

Fig. 2.2 Comparison of the mitochondrial genomes of maize and spinach [based on data from Dawson *et al.* (1986) and Stern and Palmer (1986)]. The outer two circles are the mtDNA molecules of maize and spinach, respectively. Genes are labelled solid boxes. Open boxes are regions of homology to cpDNA. Arrows indicate repeats; those in the maize genome are labelled with numbers indicating their sizes in kbp. Solid circles on the maize mtDNA molecule are the integrated plasmids R1 and S2. The molecules are drawn to scale, but the boxes indicate only the positions of the genes, not their sizes. For size comparison, the two innermost circles show the cpDNA molecules of spinach and maize.

been mapped on the mitochondrial genomes of two species, maize and spinach. These molecules are shown in Fig. 2.2. Both are circular, and each has been shown to carry genes for ribosomal RNAs, cytochrome *b*, and subunits of ATPase and cytochrome oxidase.

These molecules are also alike in remarkable features that they share with other plant mitochondrial genomes. First, they contain large direct repeats that can undergo intramolecular crossing-over to produce smaller circular molecules carrying partial genomes; this is illustrated in Fig. 2.3 for spinach. Maize is more complicated, with five different pairs of direct repeats; at least four of which recombine at varying frequencies within molecules (Lonsdale *et al.*, 1984). Only one plant species has been found to lack such repeats (Palmer and Herbon, 1987). In addition, there is recombination beween repeats in different molecules and also between the inverted 14 kbp repeats. The mechanism of recombination is unknown, but some recombinogenic

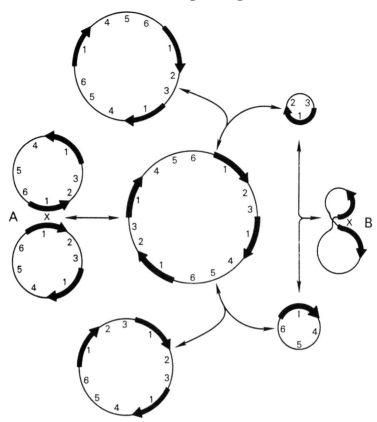

Fig. 2.3 Some of the genomes that could be produced by recombination in an organism in which the unit genome has one direct repeat. Genes (or sets of genes) are labelled 1 in the repeat and 2–6 in the remainder of the genome. Intramolecular crossing-over is shown in B. Intermolecular crossing-over is shown in A as involving the repeats, but it might also occur between other regions of homology.

repeats in *Oenothera* are only 10 bp long (Manna and Brennicke, 1986), which suggests a site-specific mechanism. Recombination also occurs between molecules at multiple sites (Vedel *et al.*, 1986). Second, the mtDNA molecules of both species have extensive sequences that apparently originated in the chloroplast. Maize mtDNA contains sequences homologous to at least six regions of the cpDNA molecule, while spinach mtDNA has at least 13 (Fig. 2.2). Some contain known chloroplast genes. It is not known if these are expressed; differences between organelles in the code and codon usage might make this difficult if not impossible.

The resemblances between these species, and among mtDNA molecules in general, end there. First, maize mtDNA is 74% larger (Fig. 2.2). In the mtDNA of other plants, we find a size range of 11-fold (218–2500 kbp) among angiosperms in general and 7-fold within one family, the Cucurbitaceae. The size variation of mitochondrial genomes is much greater than that of plant chloroplast genomes but is only about one-fiftieth the range of plant nuclear genomes. It is generally believed that the size difference is not accompanied by large differences in the number of different genes (Forde and Leaver, 1980; Hack and Leaver, 1984; Stern and Newton, 1985). As is evident from the figure, the maize and tobacco mitochondrial genomes differ in the order of genes as well as in size. There is evidence that genome rearrangements may be extensive even between closely related species of maize (Sederoff *et al.*, 1981). Different plant species have some of the direct repeats involved in intramolecular crossing-over in common but also have some unique repeats. The shared repeats may be active in recombination in some species but not others (Stern and Palmer, 1984). The repeats often contain genes, but the genes differ among species (e.g. Dawson *et al.*, 1986). Thus different species have different gene dosage relationships; however, it is not known if both copies of a repeated gene are functional. The size variation of mitochondrial genomes is not all due to repeated sequences, since the largest and smallest cucurbit mitochondrial genomes have about the same proportion (5–10%) of repeated sequences (Ward, Anderson and Bendich, 1981). In fact, although the repeated sequences have a remarkable effect on genome structure, they are modest in extent compared to nuclear genomes which often include 50% or more of repeated sequences.

If we assume that the mitochondria of different plant species have similar numbers and kinds of genes, how can we account for the rest of the DNA, which varies greatly in amount among different species? This is the C-value paradox, recognized long ago for nuclear genomes. Some of the paradoxical DNA in the nucleus is highly repetitive, but there is none of this in the mitochondria. Some nuclear DNA consists of untranscribed 'spacer' sequences, most of unknown function lying between genes. This may occur in the mitochondria of some but not all plants; muskmelon and watermelon transcribe 30% and 60% of their mitochondrial genomes, respectively (Ward and Bendich, personal communication), whereas maize appears to transcribe most of its mtDNA molecule (Carlson *et al.*, 1986; Bedinger and Walbot, 1986). Many nuclear genes contain introns and produce very long transcripts, large portions of which are discarded. Plant mitochondria also have introns which may be present in some species but not others (Fox and Leaver, 1981; Hiesel and Brennicke, 1983; Stern *et al.*, 1986). Moreover, transcripts are much larger than the coding sequence, and appear to be extensively spliced (Isaac *et al.*, 1985a; Braun and Levings, 1985; Dewey *et al.*, 1985). Of course, some cpDNA genes also have introns and long transcripts. We suspect, but do not know with certainty, that the amount of DNA that is transcribed but not translated is larger and more variable in plant mitochondria than in chloroplasts or the mitochondria of other organisms. Also, we suspect that much of that DNA is not necessary or even desirable for the organism, given the variation between similar species.

Another potentially important feature of plant mitochondria is that they often contain plasmids, while none have been found in chloroplasts (reviewed by Sederoff and Levings, 1985). The number and kind of plasmids is highly variable. In some races of *Zea mays*, there are two plasmids, R1 (= S1) and R2, that behave as episomes, being normally inserted in the mitochondrial genome at defined sites (Fig. 2.2) but capable of excision (correlated with cytoplasmic male sterility) and independent replication or reintegration (at different sites).

Compared to other organelle genomes, the plant mitochondrial genome is much more nucleus-like, although compared to nuclear genomes, plant mitochondrial genomes are small and conservative in function, size, and organization. But they are much larger than most chloroplast genomes, and larger than mitochondrial genomes in any other group of organism. They are much more variable than chloroplast genomes in organization and size (unless one includes the unique and poorly studied cpDNA of *Acetabularia*). Compared to mitochondria in other groups, it is notable that in less than 150 My angiosperms achieved more than 11-fold variation in size and substantial variation in organization. Meanwhile, there has been no change in vertebrate mtDNA size or gene order since the vertebrate phyla radiated about 400 My ago. Echinoderms and vertebrates show relatively modest changes of gene order and no significant changes in genome size in the 500 My since these groups diverged. One must turn to the ancient and highly diverse protists and fungi to find similar variability in the size and organization of the mitochondrial genome (reviewed by Sederoff, 1984; Brown, 1985; Birley and Croft, 1986).

Contrasting modes of structural evolution in chloroplast, mitochondrial and nuclear genomes

Why have the structure and function of chloroplast genomes evolved so slowly, while plant and nuclear genomes have changed so much? We know of three general mechanisms for structural changes in genomes: the activity of transposable elements; recombination; and transfer of sequences between genomes. I will consider each in turn.

Transposition
This is the movement of sequences from one site to another on the same or a different molecule. Transposable elements not only move themselves, but also cause structural rearrangements including deletions, inversions, co-

integration of two circles into one, and resolution of one circle into two. When transposition is duplicative, it not only changes the order of genes on the molecule but also causes chromosome rearrangements. As described above, some of the maize plasmids are probably transposable. There is also evidence for transposition of several different kinds in fungal mitochondria (e.g. Zinn and Butow, 1985; Michel and Lang, 1985). In contrast, there is as yet no good evidence for transposable elements in chloroplasts.

Recombination

Recombination, or more specifically, crossing-over, between repeated sequences is believed to be a major cause of chromosome rearrangements in nuclear genomes. As we have seen, recombination of plant mitochondrial genomes occurs with high frequency, both within and between molecules. Figure 2.3 shows that recombination within and between molecules containing a single set of direct repeats can produce all possible combinations of duplications and deletions of the sequences between the repeats. If there are more pairs of repeats, as in maize, or more copies of the same repeat, the number of kinds of genomic changes increases greatly. Recombination with plasmids may be another source of variability. The free S1 and S2 plasmids of male-sterile maize recombine with the mitochondrial genome at several different homologous sites (Braun *et al.*, 1986) and linearize the genome (Schardl *et al.*, 1984) as well as causing other kinds of rearrangements (Isaac, Jones and Leaver, 1985). Modified genomes arising in a population of molecules within a cell can become established in a population of plants just like point mutations do.

Recombination may be much more limited in plant chloroplasts. Crossing-over occurs with high frequency between the large inverted repeats, but this high-frequency event may have a unique enzymatic mechanism (as it does in the yeast 2 μm plasmid) and may require specialized sequences. It may serve to regulate copy number as in the 2 μm plasmid (Volkert and Broach, 1986). Intramolecular crossing-over between repeats does not necessarily signal the existence of a general mechanism for crossing-over in cpDNA molecules. The best evidence for homologous recombination is the fact that the large inverted repeats are always identical within an individual plant; this requires a copy correction mechanism, most likely gene conversion between repeats within a molecule, but possibly gene conversion or crossing-over between molecules. On the other hand, there is genetic evidence that recombination between different cpDNA molecules is extremely rare in angiosperms. Several investigators have looked for recombination between chloroplast mutations in crosses where the chloroplasts are inherited biparentally or in plants regenerated from fused somatic cells (R. A. E. Tilney-Bassett, personal communication; Chiu and Sears, 1985 and references therein); only a single recombinant has been found, recovered by very stringent selection (Medgyesy, Fejes and Maliga, 1985). However, it is not known whether the failure to detect recombination between molecules is due to the absence of an effective enzymatic mechanism or to the failure of the plastids to fuse and exchange molecules. The search for cpDNA recombination in gymnosperms has just begun; preliminary results suggest that it may occur (D. B. Neale, personal communication).

The hypothesis that recombination events are responsible for changes in chloroplast genome structure is testable by DNA sequencing, because it predicts that deletions and inversions will have homologous regions at their endpoints. This prediction has already been verified in at least one case, but with a surprising and revealing twist. Howe (1985) sequenced the endpoints of a 20 kb region that is inverted in wheat cpDNA relative to spinach and most other plants. The endpoints contain 70 bp sequences that show 85% homology to each other, thus looking like repeats that have diverged somewhat since their formation. More remarkably, each repeat contains a sequence very similar to the 15 bp lambda *att* core sequence at which the phage lambda undergoes site-specific integration into the *E. coli* chromosome. Lambda will integrate into these cpDNA sequences *in vitro*. Thus there are repeats where recombination could have occurred to produce the inversion, but the mechanism could have been either homologous crossing-over or a site-specific process. Bowman and Dyer (1986) found additional repeats in the wheat chloroplast genome, reversal of which may have been associated with rearrangements.

In principle, much of the mitochondrial C-value paradox can be explained by a high frequency of intra- and intermolecular recombination between repeated sequences, so as to generate additional repeats or delete segments. If so, why do not hybridization experiments detect a much larger proportion of repeated sequences, especially in the larger genomes? Perhaps it is because repeated sequences accumulate mutations and lose their homology over time. On the other hand, even rare gene conversion events would tend to make repeated sequences evolve in concert, counteracting their tendency to diverge. Also gene conversion can create new repeats by taking sequences that have some degree of homology and making them more similar. A possible example has been described by Dron *et al.* (1985).

Transfer

Transfer of sequences is defined here as transposition between different genomes (reviewed by Kemble, Gabay-Laughnan and Laughnan, 1985; Lonsdale, 1985; Timmis and Scott, 1985). Plant nuclei contain sequences homologous to mtDNA and cpDNA, and plant mitochondria are veritable dumping grounds for chloroplast sequences. In contrast, no 'foreign' sequences have been found in chloroplasts. Chloroplast sequences might be transcribed in the mitochondria, but their products are not likely to be functional in that environment. Consequently the transferred sequences are probably evolving rapidly and very ancient importations would no longer be recognizable because they would have no detectable sequence homology to the chloroplast genome. Also it is conceivable that mitochondria import nuclear DNA sequences; no one has screened mtDNA for sequences homologous to specific nuclear genes or non-genic sequences. The *kalilo* factor that causes senescence in *Neurospora* is a nuclear plasmid that is transferred to the mitochondria and inserted in the mtDNA (Bertrand *et al.*, 1987). Transfer of sequences is clearly an attractive possibility for explaining why the chloroplast genome is relatively small and uniform while the mitochondrial and nuclear genomes are large and variable, since only the latter are known to import sequences. The mechanism of transfer is unknown; however, it is interesting that sequences homologous to the maize mitochondrial plasmid S-1 are found in the nucleus (Kemble, Gabay-Laughnan and Laughnan, 1985), and that S-1 itself contains a fragment of the chloroplast *psb*A gene (Sederoff *et al.*, 1986).

It is possible that changes in chloroplast genome structure are rare because transposition, recombination, and importation are rare in the chloroplast. But there are other possibilities (e.g. Palmer, 1985a). Perhaps genome rearrangements occur frequently but are rarely fixed in the population because they are detrimental. Much of the chloroplast genome is transcribed (cf. Shinozaki *et al.*, 1986), so that rearrangements will often break up a transcription unit, with potentially disastrous consequences for gene expression. The nuclear and mitochondrial genomes may have more places where rearrangement breakpoints or transposon insertions can be tolerated. This hypothesis does not explain why alterations of cpDNA molecules have been permitted to occur frequently in a few lineages but not in most. Several striking examples have been documented in the legumes by Palmer *et al.* (1987). The data suggest that inversions are limited to certain lineages, within which they occurred in bursts over evolutionarily short time periods. In the Vicieae, the cpDNA of broad bean incurred two inversions relative to the primitive cpDNA map, whereas that of the pea has undergone about a dozen inversions. Michalowski, Breunig and Bohnert (1987), give an alternative interpretation of the differences between broad bean and other plants, which differs in detail but still postulated a small number of events.) The dozen inversions that differentiate the pea from other plants occurred after *Pisum* diverged from *Vicia* but before the recent divergence of the interfertile *Pisum* species and the domestication of the pea. In the Trifolieae, no detectable changes occurred in alfalfa while the cpDNA of subclover became so scrambled that it has not been possible to determine the kinds and numbers of events. Why have the chloroplast genomes in a few plant lineages become effectively destabilized? The case of subclover contains a possible clue. That cpDNA molecule is unusual in containing a family of at least five dispersed repeats of several hundred bp. One can imagine that these represent a transposable element that 'got loose' in the ancestral subclover genome and caused wholesale rearrangements, either in the course of transposition or by providing numerous repeats which could undergo recombination.

I conclude that most or all of the C-value paradox and the extensive variation in genome structure in plant mitochondria can be explained by a combination of three mechanisms: transposition of sequences within and between molecules, transfer of sequences from other genomes, and extensive duplication/deletion/inversion due to recombination. These events may occur much more frequently in plant mitochondria than in plant chloro-plasts, where there is no evidence for transposition or transfer, and at least some kinds of recombination events are rare. Plant chloroplasts do not have plasmids, which could provide increased opportunities for any or all of these processes. The plant mitochondrial genome not only has more opportunity for structural change but is probably more tolerant of such changes; if it has already acquired some extensive non-functional spacer sequences between genes it should be better able to acquire addition sequences and rearrange-ments without disrupting gene expression. Of course there must be some upper limit on genome size, perhaps set by the time needed to complete replication. Plant mitochondria must be able to replicate very large genomes quickly; they may achieve this with multiple replication origins. If plant mtDNA molecules did not have at least one origin for each possible subgenomic molecule, all the sequences present in the same subgenomic molecule(s) as the replication origin would be present in more copies than the rest of the genome; restriction analyses show that this does not happen.

In this respect as in most others, mitochondrial genomes may be useful model systems for the nucleus, since they share most of the complexity and variability but on a smaller scale – sufficiently small, in fact, that entire genomes could be sequenced and understood in their most intimate details.

Sequence evolution in mitochondria and chloroplasts

Increasing numbers of molecular biologists are discovering that plant chloroplast genes are interesting, experimentally tractable, and may even be economically important. As a result, chloroplast genes are being sequenced at a rapid rate. The plants and genes are usually chosen for their economic importance or ease of manipulation rather than for their evolutionary interest, but the sequences are nevertheless a gold mine of evolutionary information that has not yet been fully exploited. Mitochondrial genes have attracted less interest until recently, and only a few genes have been sequenced in more than one species. I have done a partial analysis of many

MISSOURI BOTANICAL GARDEN (MO)

Family:

Collector:

Determination:

Country:

Acronym:
(return chit to)

Determined by: Al Gentry, 199___   Acronym: MO

To be completed by MO staff.

☐ Recorded on Computer   ☐ Typed on Duplicates   ☐ Entered in Field Book   ☐ Recorded on Mounted MO Sheet (or if none)

#Sppl/genus Meeting
vs.
#Sppl/genus Africa

↳ is diff =
total # of
denominator

of the available mitochondrial and chloroplast sequences. Aligned sequences were compared to determine the proportion of base pair differences, omitting any segments that are deleted in either genome relative to the other. The frequency of base pair substitutions (d) was corrected for the possibility of two successive substitutions at the same base pair in the simplest way: corrected frequency of base pair substitutions = $K = -(3/4)\ln(1 - 4d/3)$. Next I calculated the rate of base pair substitution: $k = K/2T$, where T is the time of divergence of the species, determined from the fossil record. The correction for multiple hits is overly simplistic, but the resulting error is small compared to the uncertainty in the estimates of T from the fossil record. An analogous procedure can be used to calculate the rate of amino acid substitutions for protein-coding genes.

Table 2.1 summarizes the results of this exercise, comparing the rates for plant organelle genomes to rates for mammalian mitochondrial and nuclear genes taken from the literature. Rates are expressed as base pair (or amino acid) substitutions per base pair (amino acid) per 10^9 years. For protein-coding genes, a distinction is made between rates of replacement substitutions (K_r), which change amino acids, and synonymous substitutions (K_s), which do not. The following important generalizations can be made. First, synonymous substitution rates are nearly always higher than replacement rates, and pseudogenes rates are highest of all. This is expected because substitutions that do not change amino acids are less likely to be selectively disadvantageous than those that do change amino acids. Changes in pseudogenes should be selectively neutral because pseudogenes are unexpressed duplicates of functional genes. The second generalization is now well known: in mammals, mitochondrial genes evolved substantially faster than nuclear genes. Third, chloroplast and mitochondrial genes in plants evolve much more slowly than mammalian mitochondrial genes. In this limited sample, plant chloroplast genes also appear to evolve somewhat more slowly than mammalian nuclear genes. For ribosomal RNA genes, the evolutionary rates appear to be similar in the nuclei, chloroplasts, and mitochondria of plants. Sequence data are available for a smaller number of plant mitochondrial genes; the calculated rates overlap with those of the chloroplasts. However, when one directly compares plant mitochondria and chloroplasts and looks only at monocotyledon-dicotyledon divergences so that all divergence times are the same, the chloroplast genes seem to be evolving more quickly than those in the mitochondria. When these data were compiled and presented there were no comparable data for plant nuclear genes that encode proteins. Most nuclear genes are members of multigene families which show substantial divergence within the family, so that one cannot distinguish divergence between species from divergence between family members.

Another approach that has provided some useful information about

Table 2.1 Rates of molecular evolution in different genomes. Rates of amino acid substitution (k_a) and base pair substitutions (k_0, pseudogenes; k_r, replacement substitutions; k_s, synonymous substitutions) are given in base pair (amino acid) substitutions per base pair (amino acid) per 10^9 years.

Mammals		*Nuclei*	*Mitochondria*
pseudogenes	(k_0)	4.1–6.7*	—
genes: synonymous substitutions (k_s)		1.4–11.8*	34.5–94.5‡
genes: replacement substitutions (k_r)		0.004–2.8*	1.2–20‡
small RNA genes		—	2.3–3.0‡§
tRNA genes†		0.01	1.7

Plants‖	*Nuclei*	*Chloroplasts*	*Mitochondria*
genes			
synon. sub's (k_s)	—	0.021–0.84	0.041–0.33
replac. sub's (k_r)	—	0.002–0.49	0.035–0.25
aa sub's (k_a)	—	0.039–1.39	0.089–0.65
large rRNA	0.16–0.39	0.38	0.20
small rRNA	0.16	0.088	0.29–0.82

*Li, Luo and Wu (1985).

†Brown *et al.* (1982).

‡Brown and Simpson (1982); Brown (1983); Miyata *et al.* (1982).

§Hixson and Brown (1986).

‖Plant data consist of 26 pairwise comparisons of sequences for 10 chloroplast protein genes, two of chloroplast rRNAs, 9 comparisons for two mitochondrial protein genes, and six comparisons for mitochondrial rRNAs. Values are ranges for this data set. The proportion of substitutions, d, was corrected for multiple hits with the formulae $K = -(3/4)\ln(1 - 4d/3)$ for base pairs or $K = -(19/20)\ln(1 - 20d/19)$ for amino acids. The substitution rate $k = K/2T$ where T is the time of divergence of the species being compared. For comparing monocots and dicots, T was assumed to be 110 My (Hughes, 1977); for barley and maize, $T = 50 - 65$ My following Stebbins as cited in Zurawski, Bottomley and Whitfield (1984). For dicot taxa B and C, T was estimated to lie midway between the time of appearance of the earliest fossil pollens (Muller, 1981) for their common ancestral taxon A and the age of the earliest fossil pollen of B or C, whichever is smaller. Cited numbers are from sequences in Brennicke, Möller and Blanz (1985), Chao, Sederoff and Levings (1984), Eckenrode, Arnold and Meagher (1985), Edwards and Kossel (1981), Fox and Leaver (1981), Grabau (1985, 1986), Hiesel and Brennicke (1983), Kao, Moon and Wu (1984), Link and Langridge (1984), Manna and Brennicke (1985), Messing *et al.* (1984), Moon, Kao and Wu (1985), Spencer, Schnare and Gray (1984), Schwarz and Kossel (1980), Shinozaki, *et al.* (1983), Sugita and Sugiura (1983), Takaiwa and Sugiura (1982), Tohdoh and Sugiura (1982), Zurawski and Clegg (1984), Zurawski, Bottomley and Whitfield (1982, 1984).

organelle evolutionary rates is to measure total sequence divergence by restriction analysis. This method has the advantage of looking at the entire genome rather than a very small and probably biased sample. Restriction analyses of the cpDNA of some Solanaceae (Palmer and Zamir, 1982) and Leguminosae (Palmer *et al.*, 1983) can be used to estimate minimum divergence rates of $0.065-0.79 \times 10^{-9}$, assuming that these groups are 54 My and 94 My old, respectively (based on the earliest fossil pollen records; Muller, 1981). These values are substantially lower than the usual estimate of about 10^{-9} for single-copy nuclear DNA in animals. Sytsma and Schaal (1985) compared restriction fragment patterns for nuclear rDNA and total cpDNA among different populations of the *Lisianthius skinneri* species complex. They found that estimated sequence divergences between pairs of populations were about five times greater for nuclear rDNA (a rather conservative nuclear gene) than for total cpDNA. McClean and Hanson (1986) found that mtDNA appears to be evolving more rapidly than cpDNA among *Lycopersicum* species, but Palmer and Herbon (1987) reached the opposite conclusion for *Brassica*. Together, these data suggest that plant cpDNA and mtDNA evolves a bit more slowly than plant nuclear and mitochondrial DNA, and substantially slower than animal nuclear or mitochondrial DNA.

The rate estimates for individual genes are subject to large statistical sampling errors because of the limited sizes of the coding sequences and the small numbers of substitutions. Moreover, the divergence times might well be in error by a factor of 2. Nevertheless, some interesting if tentative conclusions can be drawn. First, there may be substantial differences in evolutionary rates between genes. Second, there may be substantial differences in rates between different evolutionary lineages. An example is shown in Table 2.2, where the corrected frequency of amino acid substitution (K_a)

Table 2.2 Evolutionary divergences of some chloroplast genes in different lineages. Corrected frequencies of amino acid substitutions (K_a) and synonymous base pair substitution (K_s) are given for genes coding for the large subunit of rubisco (*rbc*L) and for the beta and epsilon subunits of ATPase (*atp*B and *atp*E).

	K_a			K_s		
	*rbc*L	*atp*B	*atp*E	*rbc*L	*atp*B	*atp*E
dicot–dicot						
spinach–tobacco	0.083	0.056	0.129	0.065	0.068	0.074
monocot–dicot						
maize–spinach	0.095	0.091	0.295	0.123	0.112	0.090
barley–spinach	0.078	0.104	0.305	0.103	0.108	0.101
monocot–monocot						
maize–barley	0.056	0.035	0.022	0.038	0.035	0.035

Sequence data from Zurawski *et al.* (1981, 1982, 1984), Krebbers *et al.* (1982), Zurawski and Clegg (1984), Shinozaki *et al.* (1983), Shinozaki and Sugiura (1982), and McIntosh, Poulsen and Bogorad (1980, as modified in Poulsen 1982) were analyzed as described in the text and in Table 2.1.

is greater for *atp*E than for *atp*B and *rbc*L, in comparisons between monocots and dicots or between two dicots. But the three genes have diverged to approximately the same extent between two monocots. The most interesting interpretation of these data is that the rate of amino acid substitution is (or was) higher in dicots than in monocots. Of course the higher rate may not be characteristic of all dicots; it could be limited to the evolutionary lineage of spinach. Note that the K values in this table are divergences, not evolutionary rates as in Table 2.1. Consequently the differences between genes cannot be artefacts due to errors in determining divergence times. Table 2.2 also shows that these differences in evolutionary divergences disappear when one looks at synonymous substitutions (K_s). This suggests that the differences may be largely or entirely due to selection, as opposed to differences in mutation rates between genes. Rodermal and Bogorad (1987) and Wolfe *et al.* (1988) also found evolutionary rate differences between monocots and dicots, with the dicots evolving more rapidly in some genuses and lineages but more slowly in others.

There are not enough good fossil data, and these analyses of sequence divergence have not been thorough enough, to decide if chloroplast gene sequences will provide a useful molecular clock. The data suggest that amino acid substitutions should be used with caution, if at all, for evolutionary studies; silent substitutions are almost certainly better. More sophisticated analyses by Ritland and Clegg (1987) and others suggest that chloroplast gene sequences can be used with confidence for phylogenetic analyses even if they do not behave as perfect clocks. Of course the results may not agree in detail with those based on morphology (e.g. Sytsma and Gottlieb, 1986); molecular and morphological evolution are not always strongly coupled.

2.3 INTRASPECIFIC DIVERSITY OF ORGANELLE GENOMES

One of the major tasks of population genetics is to determine the amount, kind, and distribution of genetic diversity within individual species or populations of organisms. For nuclear genes, diversity is being measured at all levels, from fitness and morphology to electromorphs and DNA base sequences. For organelles, virtually all the studies have been at the DNA level, principally using restriction analysis, and have focused on animal mitochondrial genomes (reviewed by Avise and Lansman 1983; Wilson *et al.*, 1985; Birley and Croft, 1986). A useful measure of molecular diversity is nucleotide diversity, pi, which is the mean frequency of base pair differences between individuals (Nei and Li, 1979). Some nucleotide diversity values (summarized by Nei, 1983) for nuclear DNA are 2×10^{-3} for human beta globin and 6×10^{-3} for the ADH region in *Drosophila*. For comparison, the nucleotide diversity of primate mtDNA is $4 - 13 \times 10^{-3}$

and of *Drosophila* mtDNA, $4 - 7 \times 10^{-3}$. Thus mtDNA may be somewhat more polymorphic than nuclear DNA in animals. Animal populations also appear to show more local differentiation for mitochondrial than for nuclear genes. However, measures of population subdivision such as G_{ST} have not been calculated for both genomes in the same population, so it is not clear whether mitochondrial alleles are more localized, or the localization is more readily seen because of the higher diversity.

There have been relatively few quantitative studies of chloroplast and mitochondrial genetic diversity in plants. Variation in restriction fragments has been found in several species of wheat (Terachi, Ogihara and Tsunewaki, 1984; Bowman, Bonnard and Dyer, 1983), *Lycopersicon peruvianum* (Palmer and Zamir, 1982), several *Pisum* species (Palmer, Jorgensen and Thompson, 1985), *Nicotiana debneyi* (Scowcroft, 1979), maize (Timothy *et al.*, 1979), *Daucus* (Matthews, Wilson and DeBonte, 1984), Douglas fir (Neale, Wheeler and Allard, 1986), and lodgepole and jackpine (Wagner *et al.*, 1987), but not in pearl millet (Clegg *et al.*, 1983). Most of these studies had rather small sample sizes and did not quantitatively analyze the data. In the first large-scale survey of chloroplast diversity in a wild plant, Banks (Banks and Birky, 1985) examined the cpDNA of *Lupinus texensis* from 21 populations throughout the range of the species in Texas. Restriction analysis sampled 804 bp in 100 plants: 88 plants were identical. Among the other 12 plants, three different restriction sites were gained or lost and a small deletion was detected. Nucleotide diversity was $0.26 \times 10^{-3} \pm 0.93 \times 10^{-3}$ if the deletion is counted. These data suggest that cpDNA diversity in plants is one or two orders of magnitude lower than mtDNA or nuclear DNA diversity in animals. There are no data for nuclear DNA diversity in plants from which nucleotide diversity can be calculated. However, *L. texensis* is one of the most variable of all plants with respect to electrophoretic variability of nuclear enzymes, and is much more variable in this respect than are most animals (Babbel and Selander, 1974; Hamrick *et al.*, 1979; Selander, 1976). Therefore, chloroplast DNA is probably much less variable than nuclear DNA in this species. Holwerda, Jana and Crosby (1986) surveyed the diversity of both cpDNA and mtDNA in wild barley (*Hordeum spontaneum*) and domestic barley (*H. vulgare*). DNA samples were obtained from 12–13 locations throughout the Middle East, usually from only one plant per location. Restriction analysis sampled 2280 bp out of approximately 132 kbp in the chloroplast genome; nucleotide diversities were 0.56×10^{-3} and 0.59×10^{-3} for *H. spontaneum* and *H. vulgare* respectively. Restriction of the mitochondrial genome sampled 5160 bp out of approximately 420 kbp; nucleotide diversities were 0.60×10^{-3} and 0.98×10^{-3} for *H. spontaneum* and *H. vulgare*. An earlier and smaller-scale study by Clegg, Brown and Whitfield (1984) differed in finding less polymorphism in domestic than in wild barley. Most barley variants found by Holwerda, Jana and Crosby appeared to be changes in fragment sizes due to insertions or deletions, as opposed to the gain or loss of single restriction sites by point mutations; consequently the calculated nucleotide diversities are best viewed as maximum estimates. A survey of *Hordeum spontaneum* by D. B. Neale, M. A. Saghai-Maroof, R. W. Allard, and Q. Zhang (personal communication) involved 215 plants from 25 populations in Israel and 30 plants from five populations in Iran. Restriction analysis of the cpDNA sampled about 2000 base pairs: three base pair substitutions and one size variant were detected. The data were collected in such a way that only a minimum estimate of the nucleotide diversity can be calculated; it is 0.18×10^{-3}.

These studies agree in showing low diversity in plant cpDNA, much of which is due to small insertions or deletions. The barley data suggest that mitochondrial polymorphism is similarly low in plants. Note, however, that some of the observed variation could be due to variations in the frequencies of different subgenomic circles produced by recombination, which would cause variation in the intensity of the restriction fragments that include recombination junctions (Borck and Walbot, 1982) and might even make some of them undetectable by the methods of Holwerda, Jana and Crosby. It is tempting to conclude that plant organelle genomes show less intra-specific variability than nuclear genomes or the animal mitochondrial genome. But the conclusion is dangerous, because we have quantitative data for only two wild species and those data are not directly comparable to most of the data for nuclear genes. It is sometimes argued that the chloroplast genome ought to show low levels of intraspecific diversity because it shows low levels of interspecific diversity. But variation within and between species is subject to different (albeit overlapping) pressures. The danger of confounding intra- and interspecific diversity is highlighted by the mtDNA data, which show low diversity within barley but high diversity among species.

2.4 MECHANICS AND QUANTITATIVE THEORY FOR SEQUENCE DIVERSITY AND EVOLUTION IN ORGANELLES

Population and evolutionary genetic theory for organelle genes is not well developed, but it has reached the stage where it can begin to tell us what parameters are most important, what kinds of data are needed and how they can be obtained, and what kinds of explanations for evolutionary phenomena should be taken seriously. In this section I will outline some of the important results, especially as they pertain to plants, and point to some of the large dark areas where more theory is urgently needed. The focus is on quantitative theory relating rates of molecular evolution and amounts of sequence diversity to population size, mutation rates, migration rates, and the peculiarities of organelle gene inheritance. The theory applies to any kind of mutation, including changes in genome structure, but will be discussed in the context of base pair substitutions where it is most often applied.

Organelle gene inheritance

Much of population and evolutionary genetic theory for nuclear genes is not applicable to organelle genes because they do not obey Mendel's laws (for review and references, see Birky, 1983). Their behavior differs from that of nuclear genes in several important respects. The first is uniparental inheritance. Chloroplast and mitochondrial genes are transmitted preferentially, or exclusively, by the female gamete during sexual reproduction in angiosperms. In the few gymnosperms studied, chloroplast genes are transmitted preferentially via the pollen, but mitochondrial genes show maternal transmission (David B. Neale, personal communication). The second unique feature of organelle genes is vegetative segregation: heteroplasmic cells, which contain two different alleles of an organelle gene due to mutation or to biparental transmission, become homoplasmic or produce homoplasmic daughters. Vegetative segregation is very rapid in plants, usually being completed within one sexual generation of an annual plant, and the segregation of one pair of alleles is always complete before another mutation occurs in the same gene in the same lineage. The result of vegetative segregation and uniparental inheritance is that most plants are homoplasmic most of the time. Consequently, when population geneticists count the number of copies of a gene in a population, it is simply equal to the number of individuals who can transmit organelle genes. This is in contrast to nuclear genes, where heterozygosity is common and each individual is counted as having two copies of each gene.

Molecular evolution: rates of sequence divergence

Evolution at the molecular level takes place when a mutation occurs and the mutant allele is fixed in the population by selection, random drift, or some combination of these. We can separate the mutation and fixation steps with a simple equation: $E = MF$, where E is the rate of base pair substitution, M is the total mutation rate, and F is the probability that the mutation will be fixed. Note that $M = Nu$, where N is the number of copies of the gene in the population and u is the mutation rate per gene. For organelle genes, N is simply the number of individuals in the population. For nuclear genes, $M = 2Nu$. As is well known, the case of neutral alleles is simple and elegant: $F = 1/N$ (or $1/2N$ for nuclear genes) so that $E = u$. For alleles under selection, on average F is less than $1/N$ so $E < u$. Consequently, if we could identify a set of base pair substitutions that we knew were absolutely neutral, we could determine the mutation rate by measuring the rate of substitution. Pseudogenes are almost certainly evolving without selection, and are thus much prized as subjects for evolutionary studies. Unfortunately there are no known pseudogenes in plant chloroplasts. Fig. 2.1 shows that most of the cpDNA molecule is occupied by known genes or long open reading frames, so I am not sanguine about finding pseudogenes. Nevertheless, they would be so valuable that it would be worthwhile searching for them by hybridization at low stringency, or for the completely sequenced genomes, by computer searches. On the other hand, excellent candidates for chloroplast pseudogenes are to be found in the mitochondria and the nucleus, in the form of transferred sequences. Of course the mutation rate observed in these pseudogenes would be that of mitochondrial DNA, not chloroplast DNA.

Most plant cells have two copies of each nuclear gene, but of the order of a thousand copies of each organelle DNA molecule. It is often suggested that this should increase the mutation rate, but this is not so. Paradoxically, it is more likely to decrease the mutation rate. The reason is that for a mutation to be expressed and transmitted efficiently to progeny cells, it must be fixed in a cell, i.e. its frequency must increase from $1/n$, where n is the number of organelle DNA molecules per cell, to 1 or nearly so. From the viewpoint of organismal population genetics, the mutation rate is the rate of appearance of cells that have already fixed a mutation. To think about this process, we can consider the cell as a small population of genes and use the same formula that is applied to populations of organisms, written as $u = vnf$ where u is the rate of appearance of cells fixed for a mutation, v is the mutation rate per gene within a cell, and f is the probability of fixation. If the mutant DNA molecule has no selective advantage or disadvantage in reproductive rate within the cell relative to the wild-type allele, its fate will be determined by intracellular random drift (random changes in gene frequencies due to random selection of organelle DNA molecules for replication and for partitioning to daughter cells). Then $f = 1/n$ and $u = v$, i.e. the mutation rate per cell is equal to the mutation rate per gene. Now consider mutations that are subject to intracellular selection. The majority of these will be detrimental; for them, $f < 1/n$. There will be fewer mutations that have a replicative advantage within the cell, so that $f > 1/n$. Consequently, the mean fixation probability for all mutations will be less than $1/n$ and u will be less than v, not greater. For more discussion of how cells fix organelle mutations, see Backer and Birky (1985).

We can say nothing more concrete because we lack good estimates of the parameters. There are measurements of n, but these have usually been average values for whole plants or leaves, whereas the important place to know n is in the germ line: the fertilized egg, embryo cell, meristem cells, etc. The simplest way to measure v is by measuring the pseudogene substitution rate; it could also be done by measuring the rate of appearance of mutant cells under strong selection, e.g. with antibiotics in cell cultures, so that f = 1 and u = vn. Selection within cells, and between cells in a plant, has been detected but not quantified; neither the experimental techniques nor the theory for studying intracellular selection are well developed.

The fixation of a partial mitochondrial genome formed by recombination is a related but slightly different problem. Consider a cell containing complete mitochondrial genomes and two different partial genomes, maintained in steady-state frequencies by balanced inter- and intramolecular recombination (possibly modified by differential replication). How frequently will a partial genome become fixed in a cell? Random partitioning of genomes during cell division will occasionally do this by chance, with a probability of $p(1/n_e)$ per cell division, where p is the frequency of the partial genome in the cell and n_e is the effective number of organelle genomes per cell; n_e is always less than n (Birky, Fuerst and Maruyama, 1983). In a simple case with one complete genome and two different partial genomes present in equal frequencies, because n is on the order of 10^3, the probability of fixing one partial genome or the other is $> 6.7 \times 10^{-4}$. The total 'mutation rate' for the deletions, i.e. the frequency of gametes that will contain only one or the other partial genome, will be at least this great. Thus the mutation rate for changes in genome structure due to recombination is potentially very high in plant mitochondria. A possible case where a minor molecular class becomes predominant in a cell culture line has been described by Morgens, Graban and Gesteland (1984). Note that n_e is very small, on the order of 10, for plant chloroplasts; it has not been measured for plant mitochondria.

What can we say about the relative roles of mutation and selection in determining the differences in evolutionary rates between the various genomes in Table 2.1? From the pseudogene rate for animal nuclei, we can estimate $u = 11 \times 10^{-9}$. From the synonymous rates, u is greater than $35–95 \times 10^{-9}$ for animal mitochondria and greater than about $0.3–0.8 \times 10^{-9}$ for plant chloroplasts and mitochondria. Clearly, the more rapid evolution of animal mitochondria relative to the other genomes is largely, if not entirely, due to a higher mutation rate. We might conclude that plant organelles have a lower mutation rate than animal nuclei, but only if we are willing to make the dangerous assumption that selective pressures are similar.

There are two factors that would affect organelle and nuclear genomes differently; neither has been carefully studied. First, uniparental inheritance and vegetative segregation will tend to reduce the effectiveness of selection on organelle genes, relative to nuclear genes. This is because alleles are effectively neutral when the product of the effective population size (number of genes) and the selection coefficient is much less than one. But in many populations the effective number of genes will be greater for nuclei than for organelles. Consequently a greater proportion of organelle mutations will be neutral and this will tend to increase the overall rate of base substitution. The second factor is linkage. Recall that uniparental inheritance and vegetative segregation prevent recombination between genomes from different individuals, which is the kind of recombination of most importance for evolution. Brown *et al.* (1983) and Cann, Brown and Wilson (1984) suggested that the complete linkage of organelle genomes might influence the rate of molecular evolution. Bruce Walsh and I used computer simulations and probability theory to evaluate this suggestion. We found that tight linkage increases the fixation probability of detrimental mutations, decreases the fixation probability of advantageous mutations, and has no effect on neutral mutations. The net effect will be a small increase in evolutionary rate.

Sequence diversity within species

There have been a number of theoretical investigations of sequence diversity in organelle genes (see Birky, Fuerst and Maruyama, 1983; Takahata and Palumbi, 1985, and references therein). These have dealt almost exclusively with animals, but some features are easily extended to plants. In doing so, we will consider only neutral alleles. This is reasonable because the data on organelle diversity are all from restriction analyses and it is likely that many RFLPs are nearly if not quite neutral. It is also necessary because there is no general selection theory for organelles. The problem has been neglected because of the mathematical and conceptual difficulties of dealing with selection that can occur at three or four levels: between molecules or organelles within cells; between cells within individuals; and between individuals. We will also assume that one is studying a long gene or other segment of DNA so that the number of alleles is effectively infinite. To illustrate the effects of uniparental inheritance and vegetative segregation, consider the simplest case: strictly maternal inheritance and a sex ratio of 1. Genetic diversity can be measured by the effective number of alleles, which is the number of alleles that would be maintained if all had the same frequency. The amount of diversity goes to an equilibrium determined by mutation and random drift. We will begin with dioecious plants. The effective number of nuclear alleles at equilibrium is equal to $1 + 2(2N)u$, i.e. by 1 plus two times the number of genes in the population times the mutation rate u; the number of genes is twice the population size N. But the effective number of organelle alleles is $1 + 2N_f u$, because the number of genes is simply equal to the number of females. Because the number of genes is smaller, random drift is stronger, eliminating alleles more rapidly; consequently genetic diversity would be lower for organelle genes. This prediction will not be true if there are eight or more breeding female plants for every male plant, or if the mutation rate is eight or more times higher for organelle than for nuclear genes.

Uniparental inheritance makes no difference for monoecious plants, where every individual can transmit genes as a female or as a male, and the number of genes is always N. The effective number of alleles for organelle genes is now $1 + 2Nu$. For nuclear genes it is also $1 + 2NU$ in selfing plants (because selfing eliminates heterozygosity), and increases with increasing amounts of outcrossing to $1 + 4Nu$ for random mating or obligatory outcrossing plants.

Unfortunately the situation is not this simple, because the genetic diversity of a population is determined by many other factors. The complexity of the real world (or at least a more realistic theoretical world), is illustrated by the following:

1. Dispersal is the only other factor that has been investigated theoretically (Takahata and Palumbi, 1985; Birky, Fuerst and Maruyama, unpublished). These investigations have dealt with animals, but the basic principles are easily extended to plants. For simplicity, assume that organelle genes are transmitted strictly maternally. Dispersing seeds, or whole plants, are counted as carrying two copies of each nuclear gene to a new colony in dioecious or monoecious outcrossing species. But each plant or seed only carries one copy of each organelle gene. And in dioecious species, only female seeds or plants are counted as carrying organelle genes because only they can transmit them to progeny. Dispersing pollen carry one copy of each nuclear gene and, if inheritance is maternal, no organelle genes. Thus dispersal will generally be less, and population will exhibit more subdivision, for organelle genes than for nuclear genes in the same species. The difference will be even greater in species that disperse primarily via pollen. This is intuitively obvious and well known. It is not as widely appreciated that the reduced migration of organelle genes will also affect the genetic diversity in the population. If one deliberately samples one individual from each colony, measuring only between-colony diversity, or if one samples randomly throughout the species range, measuring total diversity, that diversity will actually be increased by the lesser dispersal of organelle genes.
2. The effects of linkage on genetic diversity are well known: diversity is lowered by the hitchhiking effect. This is seen in asexually reproducing microorganisms as a striking phenomenon called periodic selection. When a favorable mutation occurs and is fixed by selection, it carries with it one allele of every other gene, reducing genetic diversity to zero. New mutations gradually build the diversity back up until the next favorable mutation occurs and sweeps one genotype to fixation. Plant chloroplasts may be expected to show this behavior, since they do not

recombine even if they are inherited biparentally; although mitochondria may be able to recombine, they will also show periodic selection so long as they are inherited uniparentally.

3. The history of a population is always of potential importance. For example, if a population has recently passed through a bottleneck of reduced size, its genetic diversity may not have recovered to the level expected for the present number of individuals. As Wilson *et al.* (1985) have noted, bottlenecks will have more severe effects on organelle genes than on nuclear genes (whenever organisms have smaller effective numbers of organelle genes than of nuclear genes).

Clearly the next theoretical problem in population genetics is to begin evaluating the relative strengths of the reduced number of genes and of hitchhiking, which reduce diversity, and reduced migration, which will increase diversity. This will require simultaneous experimental studies to provide better estimates of population size, mutation rates, and dispersal.

2.5 ACKNOWLEDGEMENTS

I am grateful to Steve Kuhl, Dawne and Karl Kipp, and Lori Humphrey for assisting with the sequence analyses, and to Hans Bohnert, Michael Clegg, Masahiro Sugiura and David Neale for sharing unpublished manuscripts.

2.6 REFERENCES

Avise, J. C. and Lansman, R. A. (1983) Polymorphism of mitochondrial DNA in populations of higher animals. in *Evolution of Genes and Proteins* (eds M. Nei and R. Koehn), Sinauer Assoc., Sunderland, MA, pp. 147–64.

Babbel, G. R. and Selander, R. K. (1974) Genetic variability in edaphically restricted and widespread plant species. *Evolution*, **28**, 619–30.

Backer, J. S. and Birky, C. W. Jr. (1985) The origin of mutant cells: mechanisms by which *Saccharomyces cerevisiae* produces cells homoplasmic for new mitochondrial mutations. *Curr. Genet.*, **9**, 627–40.

Banks, J. A. and Birky, C. W. Jr. (1985) Chloroplast DNA diversity is low in a wild plant, *Lupinus texensis. Proc. Natl. Acad. Sci. USA*, **82**, 6950–4.

Bedinger, P., and Walbot, V. (1986) DNA synthesis in purified maize mitochondria. *Curr. Genet.*, **10**, 631–7.

Bendich, A. J. (1985) Plant mitochondrial DNA; Unusual variation on a common theme. in *Genetic Flux in Plants* (eds B. Hohn and E. S. Dennis), Springer-Verlag, Vienna, pp. 111–38.

Bertrand, H., Griffiths, A. J. F., Court, D. A. and Cheng, C. K. (1987) An extramitochondrial plasmid is the etiological precursor of kal DNA insertion sequences in the mitochondrial chromosome of senescent cells of *Neurospora intermedia. Cell*, **47**, 829–37.

Birky, C. W. Jr. (1983) Relaxed cellular controls and organelle heredity. *Science*, **222**, 468–75.

Birky, C. W. Jr., Fuerst, P. and Maruyama, T. (1983) An approach to population and evolutionary genetic theory for genes in mitochondria and chloroplasts, and some results. *Genetics*, **103**, 513–27.

Birley, A. J. and Croft, J. H. (1986) Mitochondrial DNAs and phylogenetic relationships. in *DNA Systematics* (Vol. I. *Evolution*) (ed. S. K. Dutta), CRC Press, Boca Raton, Fl. pp. 107–37.

Bohnert, H. J. and Michalowski, C. (1988) Organization of plastid genomes. *Am. J. Bot.*, (in press).

Borck, K. S., Walbot, V. (1982) Comparison of the restriction endonuclease digestion patterns of mitochondrial DNA from normal and male sterile cytoplasms of *Zea Mays* L. *Genetics*, **102**, 100–26.

Bowman, C. M., Bonnard, G. and Dyer, T. A. (1983) Chloroplast DNA variation between species of *Triticum* and *Aegilops*: location of the variation on the chloroplast genome and its relevance to the inheritance and classification of the cytoplasm. *Theor. Appl. Genet.*, **65**, 247–62.

Bowman, C. M. and Dyer, T. A. (1986) The location and possible evolutionary significance of small dispersed repeats in wheat ctDNA. *Curr. Genet.*, **10**, 931–41.

Braun, C. J. and Levings, C. S. III (1985) Nucleotide sequence of the F_1-ATPase alpha subunit gene from maize mitochondria. *Plant Physiol.*, **79**, 571–7.

Braun, C. J., Sisco, P. H., Sederoff, R. R. and Levings, C. S. III (1986) Characterization of inverted repeats from plasmid-like DNAs and the maize mitochondrial genome. *Curr. Genet.*, **8**, 625–30.

Brennicke, A., Möller, S. and Blanz, P. A. (1985) The 18S and 5S ribosomal RNA genes in *Oenothera* mitochondria: Sequence rearrangements in the 18S and 5S rRNA genes of higher plants. *Mol. Gen. Genet.*, **198**, 404–10.

Brown, G. G., Bell, G., Desrosiers, L. and Prussick, R. (1983) Variation in animal mitochondrial genes and gene products. in *Endocytobiology II: Intacellular Space as Oligogenetic Ecosystem* (eds H. E. A. Schenk and W. Schwemmler), de Gruyter, New York, pp. 247–61.

Brown, G. G. and Simpson, M. V. (1982) Novel features of animal mtDNA evolution as shown by sequences of two rat cytochrome oxidase subunit II genes. *Proc. Natl. Acad. Sci. USA*, **79**, 3246–50.

Brown, W. M. (1983) Evolution of mitochondrial DNA. in *Evolution of Genes and Proteins* (eds M. Nei and R. K. Koehn), Sinauer Assoc., Sunderland, MA., pp. 62–88.

Brown, W. M. (1985) The mitochondrial genome of animals. in *Molecular Evolutionary Genetics* (ed. R. J. MacIntyre), Plenum Press, New York, pp. 95–130.

Brown, W. M., Prager, E. M., Wang, A. and Wilson, A. C. (1982) Mitochondrial DNA sequences of primates: Tempo and mode of evolution. *J. Mol. Evol.*, **18**, 225–39.

Cann, R. L., Brown, W. M. and Wilson, A. C. (1984) Polymorphic sites and the mechanism of evolution in human mitochondrial DNA. *Genetics*, **106**, 479–99.

Carlson, J. E., Brown, G. L. and Kemble, R. J. (1986) *In organello* mitochondrial DNA and RNA synthesis in fertile and cytoplasmic male sterile *Zea mays* L. *Curr. Genet.*, **11**, 151–60.

Chao, S., Sederoff, R. and Levings, C. S. III (1984) Nucleotide sequence and evolution of the 18S ribosomal RNA gene in maize mitochondria. *Nuc. Acids Res.*, **12**, 6629–44.

Chiu, W.-L. and Sears, B. B. (1985) Recombination between chloroplast DNAs does not occur in sexual crosses of *Oenothera*. *Mol. Gen. Genet.*, **198**, 525–8.

Clegg, M. T., Brown, A. H. D. and Whitfeld, P. R. (1984) Chloroplast DNA diversity in wild and cultivated barley. *Genet. Res.*, **43**, 339–43.

Clegg, M. T., Rawson, J. R. Y. and Thomas, K. (1983) Chloroplast DNA evolution in pearl millet and related species. *Genetics*, **106**, 449–61.

Dawson, A. J., Hodge, T. P., Isaac, P. G., Leaver, C. J., Lonsdale, D. M. (1986) Location of the genes for cytochrome oxidase subunits I and II, apocytochrome *b*, alpha-subunit of the F_1-ATPase and the ribosomal RNA genes on the mitochondrial genome of maize (*Zea mays* L.). *Curr. Genet.*, **10**, 561–4.

Dewey, R. E., Schuster, A. M., Levings, C. S. III, Timothy, D. H. (1985) Nucleotide sequence of F_0-ATPase proteolipid (subunit 9) gene of maize mitochondria. *Proc. Natl. Acad. Sci. USA*, **82**, 1015–19.

Dron, M., Hartmann, C., Rode, A. and Sevignac, M. (1985) Gene conversion as a mechanism for divergence of a chloroplast tRNA gene inserted in the mitochondrial genome of *Brassica oleracea*. *Nuc. Acids Res.*, **13**, 8603–10.

Eckenrode, V. K., Arnold, J., Meagher, R. B. (1985) Comparison of the nucleotide sequence of soybean 18S rRNA with the sequences of other small-subunit rRNAs. *J. Mol. Evol.*, **21**, 259–69.

Edwards, K. and Kossel, H. (1981) The rRNA operon from *Zea mays* chloroplasts: Nucleotide sequence of 23S rDNA and its homology with *E. coli* rDNA. *Nuc. Acids Res.*, **9**, 2853–69.

Flavell, R. (1980) The molecular characterization and organization of plant chromosomal DNA sequences. *Ann. Rev. Plant Physiol.*, **31**, 569–96.

Forde, B. G. and Leaver, C. J. (1980) Nuclear and cytoplasmic genes controlling synthesis of variant mitochondrial polypeptides in male-sterile maize. *Proc. Natl. Acad. Sci. USA*, **77**, 418–22.

Forde, B. G., Oliver, R. J. C. and Leaver, C. J. (1978) Variation in mitochondrial translation products associated with male-sterile cytoplasms in maize. *Proc. Natl. Acad. Sci. USA*, **75**, 3841–5.

Fox, T. D. and Leaver, C. J. (1981) The *Zea mays* mitochondrial gene coding cytochrome oxidase subunit II has an intervening sequence and does not contain TGA codons. *Cell*, **26**, 315–23.

Grabau, E. A. (1985) Nucleotide sequence of the soybean mitochondrial 18S rRNA gene: evidence for a slow rate of divergence in the plant mitochondrial genome. *Plant Mol. Biol.*, **5**, 119–24.

Grabau, E. A. (1986) Cytochrome oxidase subunit II gene is adjacent to an initiator methionine tRNA gene in soybean mitochondrial DNA. *Curr. Genet.*, **11**, 287–95.

Hack, E. and Leaver, C. J. (1984) Synthesis of dicyclohexylcarbodiimide-binding proteolipid by cucumber (*Cucumis sativus* L.) mitochondria. *Curr. Genet.*, **8**, 537–42.

Hamrick, J. L., Linhart, Y. B., Mitton, J. B. (1979) Relationships between life history characteristics and electrophoretically detectable genetic variation in plants. *Ann. Rev. Ecol. Syst.*, **10**, 173–200.

Hiesel, R. and Brennicke, A. (1983) Cytochrome oxidase subunit II gene in mitochondria of *Oenothera* has no intron. *EMBO J.*, **2**, 2173–8.

Hixson, J. E. and Brown, W. M. (1986) A comparison of the small ribosomal RNA genes from the mitochondrial DNA of the great apes and humans: Sequence, structure, evolution, and phylogenetic implications. *Mol. Biol. Evol.*, **3**, 1–18.

Holwerda, B. C., Jana, S. and Crosby, W. L. (1986) Chloroplast and mitochondrial DNA variation in *Hordeum vulgare* and *Hordeum spontaneum*. *Genetics*, **114**, 1271–91.

Howe, C. J. (1985) The endpoints of an inversion in wheat chloroplast DNA are associated with short repeated sequences containing homology to *att*-lambda. *Curr. Genet.*, **10**, 139–45.

Hughes, N. F. (1977) Palaeo-succession of earliest angiosperm evolution. *Bot. Rev.*, **43**, 105–27.

Isaac, P. G., Brennicke, A., Dunbar, S. M. and Leaver, C. J. (1985) The mitochondrial genome of fertile maize (*Zea mays* L.) contains two copies of the gene encoding the alpha-subunit of the F_1-ATPase. *Curr. Genet.*, **10**, 321–8.

Isaac, P. G., Jones, V. P. and Leaver, C. J. (1985) The maize cytochrome *c* oxidase subunit I gene: sequence, expression and rearrangement in cytoplasmic male sterile plants. *EMBO J.*, **4**, 1617–23.

Kao, T.-H., Moon, E. and Wu, R. (1984) Cytochrome oxidase subunit II gene of rice has an insertion sequence within the intron. *Nuc. Acids Res.*, **12**, 7305–15.

Kemble, R. J., Gabay-Laughnan, S. and Laughnan, J. R. (1985) Movement of genetic information between plant organelles: mitochondria-nuclei. in *Genetic Flux in Plants* (eds B. Hohn and E. S. Dennis), Springer-Verlag, New York, pp. 79–87.

Krebbers, E. T., Larrinua, I. M., McIntosh, L. and Bogorad, L. (1982) The maize chloroplast genes for the β and ϵ subunits of the photosynthetic coupling factor CF_1 are fused. *Nuc. Acids Res.*, **10**, 4985–5002.

Levings, C. S. III and Pring, D. R. (1977) Diversity of mitochondrial genomes among normal cytoplasms of maize. *J. Hered.*, **68**, 350–4.

Li, W.-H., Luo, C.-C. and Wu, C.-I. (1985) Evolution of DNA sequences. in *Molecular Evolutionary Genetics* (ed. R. J. MacIntyre), Plenum Press, New York, pp. 1–94.

Link, G. and Langridge, U. (1984) Structure of the chloroplast gene for the precursor of the M_r 32 000 photosystem II protein from mustard (*Sinapis alba* L.) *Nuc. Acids Res.*, **12**, 945–58.

Lonsdale, D. M. (1985) Movement of genetic material between the chloroplast and mitochondrion in higher plants. in *Genetic Flux in Plants* (eds B. Hohn and E. S. Dennis), Springer-Verlag, New York, pp. 52–60.

Lonsdale, D. M., Hodge, T. P. and Fauron, C. M.-R. (1984) The physical map and organisation of the mitochondrial genome from the fertile cytoplasm of maize. *Nuc. Acids Res.*, **12**, 9249–61.

Manna, E. and Brennicke, A. (1985) Primary and secondary structure of 26S ribosomal RNA of *Oenothera* mitochondria. *Curr. Genet.*, **9**, 1505–16.

Manna, E. and Brennicke, A. (1986) Site-specific circularisation at an intragenic sequence in *Oenothera* mitochondria. *Mol. Gen. Genet.*, **203**, 377–81.

Matthews, B. F., Wilson, K. G. and DeBonte, L. R. (1984) Variation in culture, isoenzyme patterns and plastid DNA in the genus *Daucus*. *In Vitro*, **20**, 38–44.

McIntosh, L., Poulsen, C. and Bogorad, L. (1980) Chloroplast gene sequence for the large subunit of ribulose bisphosphatecarboxylase of maize. *Nature*, **288**, 556–60.

McLean, P. E. and Hanson, M. R. (1986) Mitochondrial DNA sequence divergence among *Lycopersicon* and related *Solanum* species. *Genetics*, **112**, 649–67.

Medgyesy, P., Fejes, E. and Maliga, P. (1985) Interspecific chloroplast recombination in a *Nicotiana* somatic hybrid. *Proc. Natl. Acad. Sci. USA*, **82**, 6960–4.

Messing, J., Carlson, J., Hagen, G., Rubenstein, I. and Olesen, A. (1984) Cloning and sequencing of the ribosomal RNA genes in maize: the 17S region. *DNA*, **3**, 31–40.

Michalowski, C., Breunig, K. D. and Bohnert, H. J. (1987) Points of rearrangements between plastid chromosomes: Location of protein coding regions on broad bean chloroplast DNA. *Curr. Genet.*, **11**, 265–74.

Michel, F. and Lang, B. F. (1985) Mitochondrial class II introns encode proteins related to the reverse transcriptases of retroviruses. *Nature*, **316**, 641–3.

Mitra, R. and Bhatia, C. R. (1986) Repeated DNA sequences and polyploidy in cereal crops. in *DNA Systematics*, (Vol. II: *Plants*) (ed. S. K. Dutta), Boca Raton, FL, pp. 21–43.

Miyata, T., Hayashida, H., Kikuno, R., Hasegawa, M., Kobayashi, M. and Koike, K. (1982) Molecular clock of silent substitution: At least six-fold preponderance of silent changes in mitochondrial genes over those in nuclear genes. *J. Mol. Evol.*, **19**, 28–35.

Moon, E., Kao, T.-H. and Wu, R. (1985) The cytochrome oxidase subunit II gene has no intron and generates two mRNA transcripts with different 5'-termini. *Nuc. Acids Res.*, **13**, 3195–212.

Morgens, P. H., Grabau, E. A. and Gesteland, R. F. (1984) A novel soybean mitochondrial transcript resulting from a DNA rearrangement involving the 5S rRNA gene. *Nuc. Acids Res.*, **12**, 5665–84.

Muller, J. (1981) Fossil pollen records of extant angiosperms. *Bot. Rev.*, **47**, 1–141.

Neale, D. B., Wheeler, N. C. and Allard, R. W. (1986) Paternal inheritance of chloroplast DNA in Douglas-fir. *Can. J. For. Res.*, **16**, 1152–4.

Nei, M. (1983) Genetic polymorphism and the role of mutation in evolution. in *Evolution of Genes and Proteins* (eds M. Nei and R. K. Koehn), Sinauer Associates, Sunderland, MA, pp. 165–90.

Nei, M. and Li, W.-H. (1979) Mathematical model for studying genetic variation in terms of restriction endonucleases. *Proc. Natl. Acad. Sci. USA*, **76**, 5269–73.

Ohyama, K., Fukuzawa, H., Kohchi, T., Shirai, H., Sano, T., Sano, S., Umesono, K., Shiki, Y., Takeuchi, M., Chang, Z., Aota, S.-I., Inokuchi, H. and Ozeki, H. (1986) Chloroplast gene organization deduced from complete sequence of liverwort *Marchantia polymorpha* chloroplast DNA. *Nature*, **322**, 572–4.

Palmer, J. D. (1985a) Evolution of chloroplast and mitochondrial DNA in plants and algae. in *Molecular Evolutionary Genetics* (ed. R. J. MacIntyre), Plenum Press, New York, pp. 131–240.

Palmer, J. D. (1985b) Comparative organization of chloroplast genomes. *Ann. Rev. Genet.*, **19**, 325–54.

Palmer, J. D. (1987) Chloroplast DNA evolution and biosystematic uses of chloroplast DNA variation. *Am. Nat.* **130**, (supplement) S6–S29.

Palmer, J. D. and Herbon, L. A. (1987) Unicircular structure of the *Brassica hirta* mitochondrial genome. *Curr. Genet.*, **11**, 565–70.

Palmer, J. D., Jorgensen, R. A. and Thompson, W. F. (1985) Chloroplast DNA variation and evolution in *Pisum*: Patterns of change and phylogenetic analysis. *Genetics*, **109**, 195–213.

Palmer, J. D., Osorio, B., Aldrich, J. and Thompson, W. F. (1987) Chloroplast DNA evolution among legumes: Loss of a large inverted repeat occurred prior to other sequence rearrangements. *Curr. Genet.*, **11**, 275–86.

Palmer, J. D., Singh, G. P., Pillay, D. T. N. (1983) Structure and sequence evolution of three legume chloroplast DNAs. *Molec. Gen. Genet.*, **190**, 13–19.

Palmer, J. D. and Zamir, D. (1982) Chloroplast DNA evolution and phylogenetic relationships in *Lycoperiscon*. *Proc. Natl. Acad. Sci. USA*, **79**, 5006–10.

Poulsen, C. R. (1982) Comments on the structure and function of the large subunit of the enzyme ribulose bisphosphate carboxylase-oxygenase. *Carlsberg Res. Comm.*, **46**, 259–73.

Quigley, F. and Weil, J. H. (1985) Organization and sequence of five tRNA genes and of an unidentified reading frame in the wheat chloroplast genome: evidence for gene rearrangements during the evolution of chloroplast genomes. *Curr. Genet.*, **9**, 495–503.

Ritland, K. and Clegg, M. T. (1987) Evolutionary analyses of plant DNA sequences. *Am. Nat.*, **130**, (supplement):S74–S100.

Rodermel, S. R. and Bogorad, L. (1987) Molecular evolution and nucleotide sequences of the maize plastid genes for the α subunit of CF_1 (*atpA*) and the proteolipid subunit of CF_0 (*atpH*). *Genetics*, **116**, 127–39.

Schardl, C. L., Lonsdale, D. M., Pring, D. R. and Rose, K. R. (1984) Linearization of maize mitochondrial chromosomes by recombination with linear episomes. *Nature*, **310**, 292–6.

Schwarz, Z. and Kossel, H. (1980) The primary structure of 16S rDNA from *Zea mays* chloroplast is homologous to *E. coli* 16S rRNA. *Nature*, **283**, 739–42.

Scowcroft, W. R. (1979) Nucleotide polymorphism in chloroplast DNA of *Nicotiana debneyi*. *Theor. Appl. Genet.*, **55**, 133–7.

Sederoff, R. R. (1984) Structural variation in mitochondrial DNA. *Adv. Genet.*, **22**, 1–108.

Sederoff, R. R. (1987) Molecular mechanisms of mitochondrial-genome evolution in higher plants. *Am. Nat.*, **130**, (supplement):S30–S45.

Sederoff, R. R. and Levings, C. S. III (1985) Supernumerary DNAs in plant mitochondria. in *Genetic Flux in Plants* (eds B. Hohn and E. S. Dennis), Springer-Verlag, New York, pp. 92–109.

Sederoff, R. R., Levings, C. S. III, Timothy, D. H. and Hu, W. W. L. (1981) Evolution of DNA sequence organization in mitochondrial genomes of *Zea*. *Proc. Natl. Acad. Sci. USA*, **10**, 5953–7.

Sederoff, R. R., Ronald, P., Bedinger, P., Rivin, C., Walbot, V., Bland, M. and Levings, C. S. III (1986) Maize mitochondrial plasmid S-1 sequences share homology with chloroplast gene *psb*A. *Genetics*, **113**, 469–82.

Selander, R. K. (1976) Genetic variation in natural populations. in *Molecular Evolution* (ed. F. Ayala) Sinauer Assoc., Sunderland, MA, pp. 21–45.

Shinozaki, K., Deno, H., Kato, A. and Sugiura, M. (1983) Overlap and cotranscription of the genes for the beta and epsilon subunits of tobacco chloroplast ATPase. *Gene*, **24**, 147–55.

Shinozaki, K., Deno, H., Wakasugi, T. and Sugiura, M. (1986) Tobacco chloroplast gene coding for subunit I of proton-translocating ATPase: comparison with the wheat subunit I and *E. coli* subunit b. *Curr. Genet.*, **10**, 421–3.

Shinozaki, K., Ohme, M., Tanaka, M., Wakasugi, T., Hayashida, N., Matsubayashi, T., Zaita, N., Chunwongse, J., Obokata, J., Yamaguchi-Shinozaki, K., Ohto, C., Torazawa, K., Meng, B. Y., Sugita, M., Deno, H., Kamogashira, T., Yamada, K., Kusuda, J., Takaiwa, F., Kato, A., Tohdoh, N., Shimada, H. and Sugiura, M. (1986) The complete nucleotide sequence of tobacco chloroplast genome: its gene organization and expression. *EMBO Journal*, **5**, 2043–50.

Shinozaki, K. and Sugiura, M. (1982) The nucleotide sequence of the tobacco chloroplast gene for the large subunit of ribulose-1,5-bisphosphate carboxylase/ oxygenase. *Gene*, **20**, 91–102.

Sorenson, J. C. (1984) The structure and expression of nuclear genes in higher plants. *Adv. Genet.*, **22**, 109–44.

Spencer, D. F., Schnare, M. N. and Gray, M. W. (1984) Pronounced structural similarities between the small subunit ribosomal RNA genes of wheat mitochondria and *Escherichia coli*. *Proc. Natl. Acad. Sci. USA*, **81**, 493–7.

Stern, D. B. and Lonsdale, D. M. (1982) Mitochondrial and chloroplast genomes of maize have a 12-kilobase DNA sequence in common. *Nature*, **299**, 698–702.

Stern, D. B. and Newton, K. J. (1985) Mitochondrial gene expression in *Cucurbitaceae*: conserved and variable features. *Curr. Genet.*, **9**, 395–405.

Stern, D. B. and Palmer, J. D. (1984) Extensive and widespread homologies between mitochondrial DNA and chloroplast DNA in plants. *Proc. Natl. Acad. Sci. USA*, **81**, 1946–50.

Stern, D. B. and Palmer, J. D. (1986) Tripartite mitochondrial genomes of spinach: physical structure, mitochondrial gene mapping, and locations of transposed chloroplast DNA sequences. *Nuc. Acids Res.*, **14**, 5651–66.

Stern, D. B., Bang, A. G. and Thompson, W. F. (1986) The watermelon mitochondrial URF-1 gene: evidence for a complex structure. *Curr. Genet.*, **10**, 857–69.

Sugita, M. and Sugiura, M. (1983) A putative gene of tobacco chloroplast coding for ribosomal protein similar to *E. coli* ribosomal protein S19. *Nuc. Acids Res.*, **11**, 1911–18.

Sytsma, K. J. and Schaal, B. A. (1985) Phylogenetics of the *Lisianthius skinneri* (Gentianaceae) species complex in Panama utilizing DNA restriction fragment analysis. *Evolution*, **39**, 594–608.

Sytsma, K. J. and Gottlieb, L. D. (1986) Chloroplast DNA evolution and phylogenetic relationships in *Clarkia* sect. *Peripetasma* (Onagraceae). *Evolution*, **40**, 1248–61.

Takahata, N. and Palumbi, S. R. (1985) Extranuclear differentiation and gene flow in the finite island model. *Genetics*, **109**, 441–57.

Takaiwa, F. and Sugiura, M. (1982) The complete nucleotide sequence of a 23S rRNA gene from tobacco chloroplasts. *Eur. J. Biochem.*, **124**, 13–19.

Terachi, T., Ogihara, Y. and Tsunewaki, K. (1984) The molecular basis of genetic diversity among cytoplasms of *Triticum* and *Aegilops*. III. Chloroplast genomes of the M and modified M genome-carrying species. *Genetics*, **108**, 681–95.

Timmis, J. N. and Scott, N. S. (1985) Movement of genetic information between the chloroplast and nucleus. in *Genetic Flux in Plants* (eds B. Hohn and E. S. Dennis), Springer-Verlag, New York, pp. 61–78.

Timothy, D. H., Levings, C. W. III, Pring, D. R., Conde, M. F. and Kermicle, J. L. (1979) Organelle DNA variation and systematic relationships in the genus *Zea*:Teosinte. *Proc. Natl. Acad. Sci. USA*, **76**, 4220–4.

Tohdoh, N. and Sugiura, M. (1982) The complete nucleotide sequence of a 16S ribosomal RNA gene from tobacco chloroplasts. *Gene*, **17**, 213–18.

Vedel, F., Chétrit, P., Mathieu, C., Pelletier, G. and Primard, C. (1986) Several different mitochondrial DNA regions are involved in intergenomic recombination in *Brassica napus* cybrid plants. *Curr. Genet.*, **11**, 17–24.

Volkert, F. C. and Broach, J. R. (1986) Site-specific recombination promotes plasmid amplification in yeast. *Cell*, **46**, 541–50.

Wagner, D. B., Furnier, G. R., Saghai-Maroof, M. A., Williams, S. M., Dancik, B. P., Allard, R. W. (1987) Chloroplast DNA polymorphisms in lodgepole and jack pines and their hybrids. *Proc. Natl. Acad. Sci. USA*, **84**, 2097–100.

Ward, B. L., Anderson, R. S. and Bendich, A. J. (1981) The mitochondrial genome is large and variable in a family of plants (Cucurbitaceae). *Cell*, **25**, 793–803.

Wilson, A. C., Cann, R. L., Carr, S. M., George, M., Gyllensten, U. B., Helm-Bychowski, K. M., Higuchi, R. G., Palumbi, S. R., Prager, E. M., Sage, R. D. and Stoneking, M. (1985) Mitochondrial DNA and two perspectives on evolutionary genetics. *Biol. J. Linnean Soc.*, **26**, 375–400.

Wolfe, K. H., Li, W.-H. and Sharp, P. M. (1988) Rates of nucleotide substitution vary greatly among plant mitochondrial, chloroplast and nuclear DNAs. *Proc. Natl. Acad. Sci. USA* (in press).

Zinn, A. R. and Butow, R. A. (1985) Nonreciprocal exchange between alleles of the yeast mitochondrial 21S rRNA gene: kinetics and the involvement of a double-strand break. *Cell*, **40**, 887–95.

Zurawski, G., Bohnert, H. J., Whitfield, P. R. and Bottomley, W. (1982) Nucleotide sequence of the gene for the M_r 32 000 thylakoid membrane protein from *Spinacia oleracea* and *Nicotiana debneyi* predicts a totally conserved primary translation product of M_r 38 950. *Proc. Natl. Acad. Sci. USA*, **79**, 7699–703.

Zurawski, G., Bottomley, W., Whitfeld, P. R. (1982) Structures of the genes for the β and ε subunits of spinach chloroplast ATPase indicate a dicistronic mRNA and an overlapping translation stop/start signal. *Proc. Natl. Acad. Sci. USA*, **79**, 6260–4.

Zurawski, G., Bottomley, W. and Whitfeld, P. R. (1984) Junctions of the large single copy region and the inverted repeats in *Spinacia oleracea* and *Nicotiana debneyi* chloroplast DNA: sequence of the genes for tRNA^{His} and the ribosomal proteins S19 and L2. *Nuc. Acids Res.*, **12**, 6547–8.

Zurawski, G. and Clegg, M. T. (1984) The barley chloroplast DNA *atpBE*, *trnM2*, and *trnVI* loci. *Nuc. Acids Res.*, **12**, 2549–59.

Zurawski, G., Perrot, B., Bottomley, W. and Whitfeld, P. R. (1981) The structure of the gene for the large subunit of ribulose 1,5-bisphosphate carboxylase from spinach chloroplast DNA. *Nuc. Acids Res.*, **9d**, 3251–70.

CHAPTER 3

Organization and evolution of sequences in the plant nuclear genome

STEVEN D. TANKSLEY and ERAN PICHERSKY

3.1 DNA CONTENT IN PLANTS

Higher plants vary more than 100-fold in DNA content per nucleus. The diploid *Arabidopsis thaliana* L. has a DNA content of 0.14 pg while some polyploid species exceed 200 pg. In general, compared with other organisms, plants have a wider range of DNA content and, on the average, they have more DNA per nucleus (Fig. 3.1).

Part of the great variation in DNA content in plants is due to polyploidization. It is estimated that perhaps half the species of higher plants are polyploid and some may have very high ploidy levels such as *Galium grande* which is 29-ploid (2n > 220).

The large variation in DNA content among plants does not appear to reflect differences in complexity. Diploid species in the genus *Vicia* vary 7-fold in DNA content (Bennett and Smith, 1976). That species of close phylogenetic relationship vary so dramatically in DNA content suggests that the nuclear genome can change size in relatively short periods of time. This raises the interesting question whether change in DNA content is bidirectional or unidirectional. In other words, have species with high DNA content evolved from ancestors having lower DNA content and *vice versa*. The prevalence of polyploidy makes it clear that higher values can evolve from lower ones. Gene duplication also increases the DNA content and is well documented in animals and plants (McIntyre, 1976; Gottlieb, 1982).

Whether species with high DNA content give rise to those with less DNA is less obvious. There is some evidence from animal studies that duplicate gene activity can be lost after polyploidization (Ferris and Whitt, 1977), but similar data in plants are few. Hexaploid wheat is the best documented case of polyploidization and its consequences in plants. However, wheat is a

Evolution of the nuclear genome

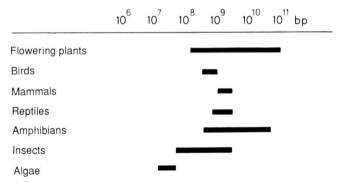

Fig. 3.1   Haploid DNA content of various groups of organisms. Plants vary more in DNA content than any other group of organisms (adapted from Lewin, 1985).

recent polyploid (ca. 10 000 yrs old) and evidence from isozyme and DNA level studies suggests that few if any structural genes have been deleted (Hart, 1983; M. Gale personal communication).

Indirect evidence regarding the question of evolutionary transition from more to less DNA in plants can be gathered from systematic studies. Eighteen diploid species of the genus *Lathyrus* vary threefold in nuclear DNA content. When these species are arranged from primitive to advanced, on the basis of traditional taxonomic characters, it appears that DNA loss accompanied species specialization (Rees and Hazarika, 1967). Similar comparative studies on diploid members of the tribe Chichorieae in the sunflower family suggest that both increases and decreases of DNA content have occurred (Price, 1976).

Direct experimental evidence for change from higher to lower DNA content is also available. Heritable changes can be induced in the DNA content of flax plants grown under different fertility regimes (Cullis, 1985). Stable genotrophs were synthesized that had either higher or lower DNA contents compared to the original variety, suggesting that evolution of DNA content can proceed either up or down.

Evolutionary significance of variation in DNA content

The fact that DNA content can change rapidly raises questions about the adaptive significance of such changes. We now know from studies to be discussed below that it is the repeated fraction of the genome that is largely responsible for rapid changes in DNA content in plants. It is difficult to propose a sequence-dependent, adaptive function for changes in the repeated portion of the plant genome; however, sequence-independent functions of repeated DNA might also be adaptive. Van'T Hoff and Sparrow (1963) showed that the duration of the mitotic cycle among plant species is positively correlated with the nuclear DNA content. Presumably the more DNA, the more time required for replication. DNA content may also be correlated with cell size, generation time, duration of meiosis and size of tissues and organs (Bennett, 1972). Bennett refers to these as nucleotypic effects and suggests that modulations in DNA content may, through these effects, have important adaptive significance. Cited as support for this hypothesis are data showing a correlation between DNA content in cultivated plants and the latitude at which they are grown (Bennett, 1976).

While correlations such as these are thought-provoking, they do not prove that most changes in DNA content are guided by selection or have an adaptive function. One might imagine that selection places a limit on the amount of DNA in the nucleus and, under certain conditions, intense selection for a reduction in generation time might favor a reduction in total DNA content.

The only reported direct tests of the effect on genome size on phenotype are by Hutchinson, Rees and Seal (1979) and Bachmann, Chambers and Price (1985). The work by Hutchinson and coworkers involved crossing two *Lolium* species with different DNA contents (C = 4.16 pg versus 6.23 pg) but with the same chromosome number and analyzing DNA content of the derived backcross progeny. As the species used in the crosses also differed for a number of morphological characters it was possible to look for correlations between DNA content in the offspring and each of these characters. No significant correlations were found between DNA content and any of the characters examined. Bachmann, Chambers and Price (1985) reported similar negative results in intraspecific crosses between strains of *Microseris douglasii* which differ in DNA content.

Although these experiments do not rule out the possibility that the differing DNA contents of the plants in question are adaptive, they do suggest that DNA content (sequence-independent effects) plays, at best, a minor role in determining phenotype as compared with the segregation of alleles at 'classical' genetic loci (sequence-dependent effects).

3.2 SEQUENCE ORGANIZATION IN THE NUCLEUS

Comparing DNA contents among species can provide information about the rate and direction of changes in nuclear DNA content, but it tells us nothing about the nature of the changes that accompany such evolutionary events. If we are to understand the nature of genomic changes and the mechanisms that give rise to those changes, we must first gain an understanding of how sequences are organized in the plant nucleus.

In the early 1960s it was discovered that the complementary strands of DNA could be separated and that, under the appropriate conditions in an aqueous solution, they would reassociate. It was soon discovered that the nucleotide sequence organization of an organism could be studied by such 'liquid hybridization' studies.

Liquid hybridization studies of genomic DNA have taught us several things about eukaryotic genome organization:

1. The nuclear genome contains sequences that are repeated many times.
2. Species with larger genomes normally have more repeated DNA and usually have a higher proportion of repeated DNA to single copy DNA.
3. Some regions of the chromosome are composed of tandem repeats often located at telomeres, centromeres or heterochromatic knobs.
4. A substantial proportion of the sequence of repeated DNA is interspersed with sequences of unique DNA.

Liquid hybridization studies in plants have yielded no great surprises relative to similar studies in animals. However, in general, plant genomes tend to have a larger proportion of repeated DNA. This is especially true of species with larger genomes such as wheat. Plant genomes usually possess a short interspersion pattern characteristic of eukaryotes. There are two striking examples among higher plants which violate the trend of the short interspersion: these are mung bean and *Arabidopsis* – both species of small genome size and both with long interspersion patterns.

Interspersed repeats

Members of interspersed repeated DNA families are evolving in concert as indicated by the fact that the members of a repeat family within a species are more similar in sequence than they are to homologous members of the same family in related species (Doolittle, 1985; Flavell *et al.*, 1980). If the repeats were not co-evolving, then members of a repeat family should be equally divergent within as they are between species. Unfortunately, few interspersed repeat families have been studied in detail in plants and most knowledge about interspersed repeated DNA comes from studies of animals.

In tomato, clones from two different interspersed repeat families have been characterized and sequenced (Zabel *et al.*, 1985). The repeat size of each is 452 and 281 bp respectively and both are characterized by unique sequences without obvious internal duplication. Short direct and indirect repeats are also found in both. Each family accounts for approximately 1.5% of the tomato genome. *In situ* hybridization of the 452 bp repeat to pachytene chromosomes indicated that members of this family can be found on all chromosomes.

Members of five related interspersed repeat families obtained at random have also been cloned in maize (Rivin, Cullis and Walbot, 1986). Repeat sizes ranged from 410 to 910 bp. The copy number of each of these clones was determined in ten different strains of maize. For all but one of these families, the copy number varied 2–3-fold. No correlation was observed for copy number changes among families. In other words, the copy number of the various repeat families seemed to vary independently in the different inbred strains. The results from this study are especially interesting since they suggest that copy number of interspersed repeat families varies even among individuals of the same species. How such variation is generated is still unknown.

Attempts have been made to implicate interspersed, repeated DNA in basic features of gene regulation. Thus far, demonstrating such a connection has proved elusive and there are reasons to doubt that much of the repeated DNA could be functioning in such a role (Murray, Peters and Thompson, 1981). The most difficult thing to explain in this regard is how closely related species of similar complexity can differ so dramatically in both genome size and organization if these two features are tightly coupled with regulation in any sequence-dependent manner.

Convincing evidence for the lack of a significant regulatory role for the majority of interspersed repeated DNA comes from studies of the wheat genome. Hexaploid wheat is composed of three different genomes, the A, B and D genomes. The majority of the highly repeated DNA, however, is found interspersed on chromosomes of the B genome (Gerlach and Peacock, 1980). Despite this, the B genome has an informational content very similar to the A and D genomes, proof of which is provided by the ability of homologous chromosomes from each of the genomes to compensate for each other in nullisomic–tetrasomic stocks (Sears, 1966).

With respect to interspersed repeated DNA in plants, we can abstract the following:

1. Interspersed repeats exist to some degree in all plants that have been thus far examined.
2. In general, they do not play a major role in gene regulation.
3. Unlinked members of a given family are co-evolving. The reasons for their existence and their mode of dispersal and co-evolution remain a mystery.

Tandem repeats

Highly repeated DNA has been isolated from a number of plant species and has, for the most part, been found to be composed of sequences repeated in tandem. In maize, the main satellite DNA isolated in CsC1 gradients corresponds to knob heterochromatin (Peacock *et al.*, 1981). Knobs are interstitial regions of heterochromatin easily visualized in prophase and prometaphase of both mitosis and meiosis. Knob heterochromatin consists of a 185 bp repeating unit which is found not only in maize, but also teosinte and *Tripsacum*, two close relatives. Considerable variation is found among repeat units both within and between species except for a 27 bp conserved region (Dennis and Peacock, 1984). It is also known that the knobs, which contain this repeat, are not essential to the genome, since some maize varieties possess no knobs. Species of the Old World genus of Maydeae, *Coix*, also possess knobs, but do not contain this 185 bp repeat.

The accumulated evidence indicates that this knob specific repeat is not essential either to maize or its relatives, leaving questions about its function and possible adaptive significance. The knobs containing this repeat are not entirely neutral with respect to plant phenotype since their presence can affect chromosomal recombination and lead to skewed segregation in progeny heterozygous for the presence of particular knobs (Rhoades, 1978). Whether such effects have any adaptive significance is unresolved.

Unexpressed single copy DNA

The proportion of single copy DNA in plant genomes (as determined by liquid hybridization ranges from approximately 25% in wheat and pea to nearly 100% in *Arabidopsis* (Thompson and Murray, 1981; Meyerowitz and Pruitt, 1985). Even based on liberal estimates, this amount of DNA is much greater than that which can be accounted for by the plant's genes. Direct experimental evidence derived from hybridizing saturating amounts of mRNA to nuclear DNA indicates that most of the single-copy DNA does not code for mRNA (Goldberg, Hoschek and Kamalay, 1978).

As with total nuclear DNA, the paradox exists that the total amount of single copy DNA also varies dramatically among plant species – even those of close taxonomic affinity (Thompson and Murray, 1981). For example, the two legumes, mung bean and pea, contain 0.4 pg and 1.4 pg of single copy DNA, respectively. There is no reason to believe that pea contains threefold more genes than mung bean. Murray *et al.* (1981) have estimated that less than 10% of the pea single copy DNA could possibly be composed of coding regions.

If genes do not account for most of the single copy DNA, then what does? Relative rates of sequence divergence between species indicate that non-coding single copy sequences are evolving at a faster rate than coding regions and therefore are under relaxed selection (Thompson and Murray, 1981; Britten, Cetta and Davidson, 1978). The difference in total amounts of single copy DNA between related species also indicates that the quantity of non-coding single copy DNA can change over relatively short periods of time. These features, rapid sequence divergence and modulations in quantity, are reminiscent of repeated DNA evolution. The similarities may not be coincidental. Murray *et al.* (1981), working with pea, demonstrated that most 'single copy' DNA is really composed of 'fossil repeats' – members of highly diverged and thus presumably ancient families of repeated sequences. It is only under non-stringent hybridization conditions that the repetitive nature of this fraction of DNA can be demonstrated. In other words, members of repeated DNA families, which diverge via mutation, but fail to undergo significant amplification and are not deleted, eventually become classified as single copy DNA.

3.3 MECHANISMS OF REPEATED DNA EVOLUTION

The change in copy number and sequence diversity of tandemly arranged repeated DNA is reasonably well explained by the unequal exchange model (Smith, 1973). This model involves recombination of tandemly repeated regions either between different members of a homologous chromosome pair or between two sister chromatids after DNA replication. When homologous regions containing tandem repeats pair out of register and a crossover occurs, unequal exchange will result in two chromatids, one with more copies of the repeat and one with less. Additional cycles of mispairing and crossing over would result in significant increases or decreases in the number of tandem repeats. This process for increasing and decreasing repeat number combined with divergence in repeats due to mutation (base substitution, additions, deletions) can account for most of the patterns of variation observed in tandemly repeated DNA.

While the unequal exchange hypothesis can accommodate most of the features of tandemly repeated DNA, it fails to provide a workable model for the distribution and concerted evolution of dispersed repeated elements. It is these elements that contribute to the interspersion patterns previously discussed. Since the unequal exchange hypothesis requires physical pairing of homologous chromosomes, it does not provide for concerted evolution of sequences in non-homologous regions of the same chromosome or different chromosomes.

In *Drosophila* movement and co-evolution of the majority of the interspersed repeated DNA can be explained by transposon activity. Nearly one-sixth of the *Drosophila* genome, and most of the interspersed repeated DNA, is composed of transcriptionally active repeats of average length 5 kbp (Young and Schwartz, 1981). Cloning and sequence analysis of several of these families (e.g. copia and P-elements) have revealed that many if not most are active transposons, and this can explain their variable locus positions (Doolittle, 1985).

Drosophila differs from most animals and plants in that most of the unique sequence DNA is found in very long stretches and the interspersed repeats (transposons) are much longer than those found in other organisms. In fact, the average interspersed repeat in other organisms (including plants) is approximately 500 bp, much too small to be autonomous transposons. While interspersed repeated DNA may not largely comprise autonomous transposons, the participation of transposons in creating or mobilizing this class of DNA cannot be ruled out. In maize there may be as many as several hundred copies of *Ds* elements. *Ds* elements are sequences derived from autonomous transposible elements (in this case *Ac*) which can no longer code for their own transposition but which can be induced to transpose when present in the same nucleus with an autonomous element. These *Ds* elements may be as short as a few hundred base pairs and would be classified as interspersed, moderately repetitive DNA.

Transposons have been identified in both monocots and dicots suggesting that they may be widespread in higher plants (for review see Nevers, Sheperd and Saedler, 1986). The role of transposons in creating mutations in expressed genes is undisputed and the potential evolutionary significance of this in producing variation in plants has been pointed out (Schwarz-Sommer *et al.*, 1985). However, the role of transposons in determining overall genome structure, and especially in explaining organization and behavior of interspersed repeated DNA, is still unknown and represents one of the most interesting questions in the study of plant genome organization and evolution.

Probably the most appealing model for the evolution of interspersed repeated DNA, which accommodates both dispersion and concerted evolution, is the one proposed by Flavell, O'Dell and Hutchinson (1981) and Thompson and Murray (1981). The hypothesis is based on comparative studies in both monocots (wheat and related species) and dicots (pea and mung bean). The basic tenet of the hypothesis is that much of the DNA in chromosomes 'turns over' through evolutionary time. Short sequences of DNA (a few hundred base pairs) are amplified and dispersed. The sites of integration of the translocated pieces are generally random and may fall in areas of unique sequence DNA as well as in the middle of other, unrelated repeats. In addition to amplification, a mechanism must also exist for deletion or the genome would continue to grow unabated. When two species diverge from a common ancestor, the amplification–translocation–deletion cycle, which proceeds in no directed manner, results in the amplification and spread of different sets of repeats in the independent lineages. Thus, new repeated DNA families are created, the members of which will be more closely related within a species than between species. The model accommodates many of the features of repeated DNA found in higher plants. Left unanswered are the mechanisms by which the repeats are amplified and deleted and how they transpose throughout the genome on such a grand scale.

3.4 CODING SEQUENCES

Coding sequences (i.e. genes) have paramount importance because they provide the cell with the myriad of molecular components necessary for growth differentiation and reproduction. Until recently, most studies of plant genes have not considered how the genes are organized in the genome. As a result, we have only a sketchy picture of gene organization and evolution in higher plants. Consequently, we, and other people, have been interested in how genes are organized in the nuclear genome and how this organization has changed through evolutionary time.

Most genes in diploid species are single copy

If a gene is present as a single copy in the nuclear genome, one expects allelic variants to segregate in monogenic ratios (i.e. $1:2:1$ in an F_2 or $1:1$ in a testcross). Classical genetic studies in a wide variety of higher plants have tested this hypothesis and the overwhelming evidence supports the hypothesis that most genes affecting morphology are single copy in diploid species. For example, in tomato, more than 200 single copy genes have been detected via segregation analysis.

If more than one gene codes for the same function, the changes of simultaneous mutations occurring in the several copies and thus giving rise to a detectable mutant phenotype, is remote. Thus, segregation studies would be unlikely to detect multicopy genes. To estimate the proportion of single copy genes to multicopy genes, it has been necessary to employ molecular techniques.

Isozymes are a direct end product of gene expression and by combining genetic studies with physical characterization of the individual isozymes, it is possible to determine the presence of duplicate genetic loci and to estimate the overall proportion of duplicate loci in the plant genome (Gottlieb, 1982). In tomato, an undisputed diploid species, genetic analysis of a large number of isozymes has indicated that about 12% of the genes are duplicated (Tanksley, 1987).

Estimates of the proportion of duplicated genes in tomato have also been made by genetic analysis of random, unique cDNA clones (Bernatzky and Tanksley, 1986a). The approach was to use random clones as probes against nuclear DNA cleaved with restriction endonucleases to determine how many different genomic restriction fragments hybridize to each unique cDNA probe. The segregation of hybridizing fragments was then monitored in progeny from a cross between two inbred lines which differed for restriction fragment length polymorphism (Fig. 3.2). Based on segregation data it was possible to determine whether the gene(s) corresponding to a particular cDNA are found in a single chromosomal locus or more than one

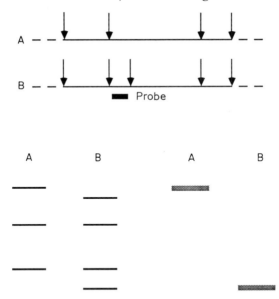

Fig. 3.2 Detection of restriction fragment length polymorphism (RFLP) using cDNA probe. DNAs from two inbred lines (A and B) are cleaved (arrows) with restriction endonuclease. Sequence differences between lines result in five cleavage sites in line B but only four in line A. Restriction fragments are separated according to size on an agarose gel (bottom left) and transferred to nylon membrane for probing. When nick-translated probe is added to filter it will hybridize to homologous sequences which can be detected by autoradiography (bottom right). In this case, the homologous sequence is found on different-sized restriction fragments in the two inbred lines. The restriction fragment length polymorphism can then be monitored in segregating progeny to confirm allelism and locus copy number.

locus. The results from this study indicate that the majority of the genes (53%) are single copy and the remainder are normally present in 2 to 5 copies (Fig. 3.3). Results from liquid hybridization studies have also been consistent and unequivocal in identifying expressed genes with the single copy DNA fraction (Murray *et al.* 1981).

Changes in organization and gene copy number

To determine the nature and extent of organizational changes in genes that have occurred in the evolution of higher plants, it is necessary to study homologous, or more importantly, orthologous* genes in different species.

---

*Orthologous genes are derived from a single gene found in the last common ancestor of species being compared.

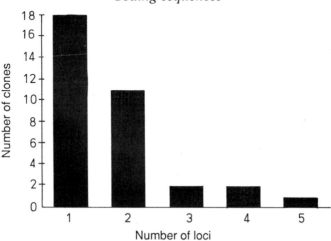

Fig. 3.3 Histogram depicting frequency of cDNA clones corresponding to various numbers of loci in tomato genome. Number of loci per clone was determined by Southern analysis in F2 population (from Bernatzky and Tanksley, 1986a).

With isozymes, it has been possible to accomplish this goal (for reviews see Gottlieb, 1982 and Tanksley and Orton, 1983).

Alcohol dehydrogenase (ADH)

ADH was one of the first plant isozymes subjected to both biochemical and genetic studies, and maize ADH is the best characterized plant isozyme (for review see Freeling and Bennett, 1985). Maize possesses two unlinked genes encoding ADH (*Adh1* and *Adh2*), both of which are inducible by anaerobiosis. Both maize *Adh* genes have been cloned and sequenced and demonstrate 82% and 87% sequence homology at the nucleotide and amino acid levels, respectively (Dennis *et al.*, 1984, 1985).

The genetics of ADH have been studied in wheat, pearl millet, sunflower, *Stephanomeria* and lupins (see Tanksley, 1983 for review). Despite the extensive research conducted on ADH in such a wide variety of species no clear pattern to the evolution of this gene has emerged. Most plants have two *Adh* genes. The duplicate genes may be tightly linked, as they are in wheat, barley and *Stephanomeria* or they may be on different chromosomes as they are in maize, tomato, sunflower and lupins. In some cases (e.g. maize, wheat, barley and millet) both genes are inducible by anaerobic stress. However, in tomato, sunflower, *Stephanomeria* and lupins only one gene is inducible. For most of the species studied, one of the *Adh* genes is expressed in pollen. For example, in tomato, the gene which is not inducible is expressed in pollen. However, for sunflower, the one that is inducible is the one found in pollen.

The fact that most species of plants have two *Adh* genes might suggest that two are required. However, *Arabidopsis* has only one gene (Chang and Meyerowitz, 1986) and studies with null mutations of maize *Adh* genes indicate that the plant does not normally require either gene product and even under anaerobic conditions, it is sufficient to have only one active gene (Freeling and Bennett, 1985).

We can only conclude from the ADH studies that, even though the amino acid sequence of this protein is fairly highly conserved among plants, regulation and patterns of chromosomal arrangement of ADH genes are not conserved among the different species suggesting that *Adh* genes may have experienced a number of independent modification events, including duplications, throughout higher plant evolution.

Gene duplication in Clarkia
The most comprehensive studies on the natural occurrence and consequences of gene duplication in plants have been those conducted by Gottlieb and coworkers on genes encoding glycolytic enzymes in species of the genus *Clarkia*. Duplications have been documented for cytosolic phosphoglucoisomerase (PGI) (Gottlieb, 1977) and both cytosolic and plastid forms of triose phosphate isomerase (TPI) (Pichersky and Gottlieb, 1983). The PGI and plastid TPI duplications appear to be confined to the genus *Clarkia* whereas the cytosolic TPI duplication is apparently widespread in the family Onagraceae. These *Clarkia* duplications have two especially interesting features: (1) the duplications of at least two of the genes (PGI and plastid TPI) are unlinked; (2) the duplications may have occurred after divergence (Gottlieb and Higgins, 1984). Both these features impact directly on our understanding of the nuclear plant genome and how it is evolving.

Tandem duplications are presumably generated by unequal crossing over; however, there are no well-documented mechanisms for the evolution of unlinked duplicate genes. Thus the discovery that the duplicated isozyme genes in *Clarkia* were unlinked was unexpected. While it cannot be absolutely ruled out that the genes duplicated first in tandem and then were separated by translocations/inversions, it seems unlikely that both duplications found in *Clarkia* could have been so efficiently separated. As an alternative, Gottlieb (1977) has proposed that these unlinked duplications were a direct result of chromosomal translocations which are known to be prevalent in the genus. The mechanism requires hybridization between individuals which differ by at least two reciprocal translocations involving the same chromosomes and was modelled after experiments conducted with translocation stocks in maize (Burnham, 1962). The hypothesis has three testable predictions:

1. The rate of translocation must be relatively high in or among crossable populations of the species in question.

2. Since the duplication is derived from the fixation of two different alleles, possibly from a wide (i.e. interspecific) cross, the duplicated genes probably will not be identical.
3. Because of the nature of the mechanism, the duplication will almost certainly include a small chromosome segment and not just a single gene.

Cytogenetic evidence in *Clarkia* fulfills the first prediction. Chromosomal rearrangements are common in *Clarkia* and many of the species differ by one or more translocations/inversions. Gottlieb and Higgins (1984) have demonstrated that duplicate PGI genes code for subunits of different apparent molecular weight – a finding consistent with prediction (2). Testing prediction (3) requires a linkage map of the plant genome in question in order to look for the presence of blocks of linked, duplicated genes. Being a wild species, no such map exists in *Clarkia*, thus the prediction has not been tested.

Unlinked gene duplications in tomato and related species
In tomato, we have evidence for four isozyme duplications, two are tandem (*Est-1,5,6,7* and *Prx-2,3*) and two are unlinked (*6-Pgdh-1,3* and *Adh-1,2*) (Tanksley, 1987). In addition, based on analysis with random cDNA clones, we have evidence for more than 20 other cases of unlinked gene duplication (Bernatzky and Tanksley, 1986b and unpublished data). A detailed map has been prepared of tomato chromosomes based on more than 100 loci detected by cDNA and single copy clones and isozymes (Fig. 3.4). It is therefore possible to test the prediction that duplicate genes will appear as linked sets representing the extra chromosomal piece generated by the translocation mechanism of duplication proposed by Gottlieb. The data collected thus far do not appear to fit this prediction. The only evidence we have found for duplication of an entire chromosomal piece is on chromosome 2, where a linkage group, consisting of approximately 20 map units, may have been duplicated. Included in the putative duplicated region is a ribulose, bisphosphate carboxylase small subunit (RBCS) locus and two loci detected by random cDNA clones (Fig. 3.4b). The other unlinked, duplicate loci appear in solitary arrangement and do not appear to be part of larger duplications (Fig. 3.4b).

Novel mechanisms for gene duplication
The preponderance of unlinked duplications in *Clarkia* and tomato indicates that a mechanism(s) exists, other than unequal crossing over, for producing gene duplications. Whether or not translocations lead to unlinked gene duplication is still uncertain. There is circumstantial evidence that this may have happened in *Clarkia*, but it does not seem likely that this mechanism could explain all or even most of the unlinked duplications in tomato. Given the early stage of investigations in this area, other hypotheses must be considered.

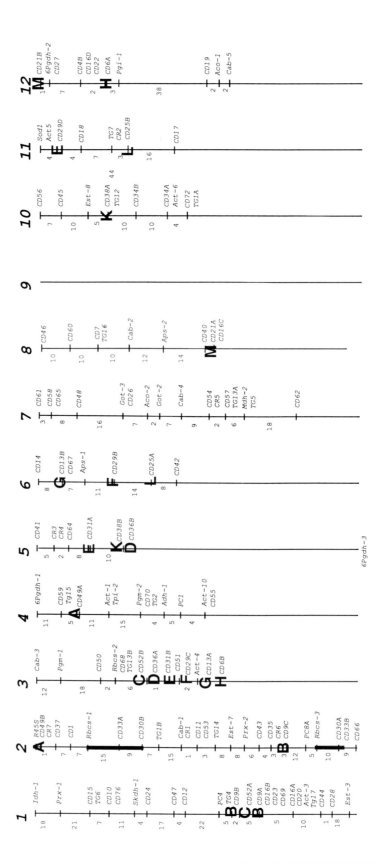

Studies of both plants and animals have shown that interspersed repeated DNA is mobile, both within chromosomes and between chromosomes. The size of the elements moving about the genome is apparently variable, ranging from a few hundred base pairs to several kilobases and movement is not necessarily sequence-dependent (Flavell *et al.*, 1980). It is possible that the mechanism that moves repeated DNA about the genome is not able to differentiate between genes and other non-coding sequences. If so, interspersed repeated DNA and gene duplications (especially unlinked duplications) may be manifestations of the same process.

Active transposons are mobile and do contain genes – the genes coding for their own transposition. Maize transposons have been linked with a variety of chromosomal rearrangements including creation of tandem or linked duplications of entire pieces of chromosomes (McClintock, 1950). A compound *Ds* element, isolated from the *shrunken* locus of maize, has been found to possess a 30 kbp piece of novel DNA – possibly transported from another region of the genome into the *sh* locus via the *Ds* element (Courage-Tebbe *et al.*, 1983). It has not yet been demonstrated, however, that plant transposons can 'carry' genes (other than their own) from one chromosomal locus to another, but the possiblity must be considered in a search for an explanation of the origin of unlinked gene duplications. Such a mechanism is not without precedent. Bacterial transposons, referred to as insertion sequences (IS), do have the ability to transport genes to new places in the genome (Kleckner, 1981).

3.5 GENE FAMILIES

From the above discussions it should be apparent that the majority of genes in higher plants are single copy. However, duplication is not uncommon and genes which are single copy in one species may be duplicated even in taxonomically closely related species. There also exist a different set of genes in plants that are generally found in multiple copies that encode identical or

---

Fig. 3.4 Molecular linkage map of tomato chromosomes. *CD*s = loci detected by Southern analysis with random cDNA clones. Letters after each locus indicate different loci possessing gene sequences homologous to the same cDNA. For example cDNA clone pCD49 was found to have homologous loci on chromosome 2 (*CD49B*) and chromosome 4 (*CD49A*). Other loci are RBCS, CAB, actin (*Act*). *TG* = loci detected by hybridization with single copy genomic sequences (not necessarily genes). Large letters on chromosomes (e.g. A,B,C etc.) show chromosomal sites of some of the known gene duplications in tomato detected by hybridization with cDNA clones. Note that most gene duplications are unlinked. Bold lines on chromosome 2 indicate a possible duplication of a chromosomal fragment containing an RBCS gene and two genes detected by analysis with random cDNAs (*CD33* and *CD30*).

almost identical gene products. Such groups of genes have been termed 'gene families'.

The term 'gene family' has been used to describe both a set of genes encoding identical or almost identical gene products and a set of genes encoding structurally and functionally different, but clearly related, gene products. Thus, the two maize ADH isozymes, and even different dehydrogenases, could fit the latter definition. In addition, some groups of genes fit both definitions (e.g. CAB genes, p. 76) in that they encode several different, but related, types of gene products, and each type is encoded by more than one gene.

The existence of gene families raises several questions regarding genome and gene volution. The first is why is there more than one copy when the products are functionally equivalent? The corollary question is how are the multiple copies maintained in the genome? In the following section, we describe gene families in higher plants and discuss research that attempts to answer questions concerning the genomic organization and evolution of gene families.

Genes encoding ribosomal RNAs

Eukaryotic cytoplasmic ribosomes each comprise a large and small subunit. The large subunit is composed of three discrete RNA molecules: 28S (4700 bp), 5.8S (160 bp) and 5S (120 bp). The small subunit contains a single RNA species referred to as 18S (1900 bp). The 28S, 5.8S and 18S rRNAs are transcribed as a single RNA transcript (often referred to as 45S) that is cleaved to form the three functional molecules. 5S rRNA is normally transcribed from sequences physically independent from the 45S genes. Normally, both types of rRNA genes are found as multiple copies in long tandem arrays with each repeating unit consisting of a transcribed region and a non-transcribed spacer. The 45S genes are usually associated with the cytologically identifiable nucleolar organizer. 45S and 5S genes are differentiated from most other genes in the nucleus in that they are transcribed by RNA polymerase I and RNA polymerase III, respectively, and not RNA polymerase II.

The 45S rRNA genes (often referred to as the ribosomal genes) have been studied in a number of plant species. Repeating units average approximately 10 kbp, about one-third of which is non-transcribed spacer. There are several features of interest with respect to plant ribosomal genes:

1. They are among the few plant genes present in relatively high copy number.
2. Despite the fact that most plants possess several thousand copies, they are normally found in only one or two chromosomal loci.

3. The actual number of gene copies is highly variable, not only between species, but even among different individuals in the same species.
4. The multiple copies of the gene within a given locus appear to evolve in concert. That is to say, there is more uniformity among the tandem repeats within a locus than between different loci (if they exist in the same nucleus) or between homologous repeats present in members of related populations or species (Coen, Thoday and Dover, 1982; Saghai-Maroof *et al.*, 1984).

The relatively large number of ribosomal genes per nucleus can be attributed to the need for large outputs of transcripts to form the cytosolic ribosomes. Undoubtedly, the overall output from the ribosomal genes is the largest for any type of gene found in the plant nucleus. It is not immediately obvious why the multiple copies of the gene should be arranged in tandem in only one or two chromosomal loci. This may be related to the constraints of trying to place many physically independent genes in contact with the nucleolar organizer (site of transcription). In this regard, the most convenient arrangement would be a single locus. The rarity of more than one or two ribosomal gene clusters may also be related to a need to maintain many *identical* copies of the gene. If the arrangement of these genes in tandem arrays is related to their ability to evolve in concert, then selection may be for many tandem copies at a single locus and against multiple copies scattered around the genome which might now readily evolve independently. The most plausible mechanism for explaining concerted evolution of ribosomal genes is unequal crossing over which is based on the existence of tandem copies (Smith, 1973).

Genes encoding abundant proteins

Although ribosomal genes are interesting, they are exceptions among plant nuclear genes. Unlike most genes, they are not transcribed by RNA polymerase II, they do not code for a protein product, they are present in very high copy numbers, and they are arranged in tandem, co-evolving units. In contrast, most protein-coding genes are present in but a few copies and as we shall see, when there are multiple copies, they are not always in tandem array.

Among the first and best-studied plant genes are those which code for prevalent proteins. Part of this bias stems from the fact that the more abundant proteins had been the subject of much prior biochemical and physiological research, so that when the techniques became available for cloning genes, it made sense to clone these genes first. Storage proteins and proteins of the photosynthetic apparatus are two cases in point.

Storage protein genes

The organization of seed storage protein genes has now been studied in a number of agronomically important plant species. In maize, the major seed storage zein accounts for more than 50% of the total protein in the mature seed. Based on molecular weight, the zein polypeptides can be separated into four distinct classes. The largest of these two groups corresponds to molecular weights of 19 000 and 22 000 (Wilson and Larkins, 1984). Genes for the 19 000 MW zeins are found in loci on chromosome 4 and 7 and genes for the 22 000 MW zeins are on chromosome 4 and 10. There is good evidence from Southern blotting and reconstruction experiments with cDNAs that at least 54 and 24 genes code for the 19 000 and 22 000 MW polypeptides, respectively (Wilson and Larkins, 1984).

Nearly half of barley seed protein is composed of the storage protein hordein. The hordeins of barley are encoded by two linked chromosomal loci *Hor-1* and *Hor-2*. The exact number of genes per locus is unclear, but the number is high (Shewry *et al.*, 1980; Forde *et al.*, 1981). The genes within each locus code for similar polypeptides, whereas there are significant sequence differences between the protein products encoded by the separate loci (Forde *et al.*, 1981).

Seed storage protein genes have been cloned from other species and with few exceptions they exist in multiple copies and in most cases are arranged in multiple loci (e.g. wheat, Payne *et al.* (1984); pea, Domoney, Ellis and Davies (1986)). The demand on seed storage protein genes must be exceptional. They are limited in expression to the window between pollination and seed maturation and the polypeptides they encode can comprise as much as 50% of the dry seed. Few other genes are required to produce such an output in the life cycle of the plant and it has been suggested that it is this demand that may be the cause of their multiplicity (Hall *et al.*, 1983).

Photosynthetic protein genes

Photosynthesis is probably the most utilized bioreaction in nature. It should not be surprising then to find that two of the major polypeptides involved in this process account for the majority of leaf protein. These components are the small subunit of ribulose bisphosphate carboxylase (RBCS) and the chlorophyll a/b binding polypeptides (CAB).

Because of their important roles in photosynthesis, the RBCS and CAB proteins have been investigated extensively by plant physiologists and protein biochemists. The nuclear genes encoding these chloroplast proteins were among the first plant genes to be cloned and characterized. Higher plants possess multiple copies of genes encoding both RBCS and CAB proteins. This explains why the initial attempts to work out the genetic systems of RBCS and CAB using protein variants as markers were not entirely successful. Our two laboratories have been involved in a collaborative effort to examine systematically the genomic organization of members of these two gene families and to describe the extent of structural divergence within each family.

Ribulose 1,5-bisphosphate carboxylase/oxygenase (small subunit (RBCS))
The primary CO_2-fixing enzyme in the photosynthetic cell, ribulose 1,5-bisphosphate carboxylase/oxygenase (RubisCO), is composed of eight small subunits and eight large subunits. The large subunit, which contains the catalytic site, is encoded by a gene on the chloroplast genome, *rbcL*. The small subunit, whose function is still undetermined, is encoded by multiple nuclear genes (RBCS genes).

We have mapped the tomato RBCS genes to three independent loci (Vallejos, Tanksley and Bernatzky, 1986, Fig. 3.4). Two of the loci, *Rbcs-1* and *Rbcs-2*, contain one copy of the gene whereas *Rbcs-3* contains three copies arranged in tandem (Pichersky *et al.*, 1987a; Sugita *et al.*, 1987). *Rbcs-1* and *Rbcs-3* reside 86 map units apart on chromosome 2 whereas *Rbcs-2* is on chromosome 3.

The eight RBCS genes of petunia have also been cloned, and they too fall into three groups (Dean *et al.*, 1985). One of the groups contains six copies of the gene, again arranged in tandem, and the other two groups each contain a single gene. Genetic mapping experiments indicate that at least two of these groups, including the one with six copies, map to independent loci (Tanksley and Pietraface, unpublished data). If the three groups of RBCS genes in petunia do indeed map to three independent loci, then petunia and tomato share the same number of RBCS loci. The actual number of genes, however, is larger in petunia (8 versus 5 in tomato). This is because one of the petunia loci contains six copies of the genes whereas its putative counterpart in tomato contains three. In *Pisum sativum* (family Leguminosae), the only plant species outside Solanaceae in which the genetic organization of the RBCS gene family has been worked out, all the RBCS genes (also five) are found in a single locus (Polans, Weeden and Thompson, 1985).

The molecular studies of RBCS genes in petunia and tomato tentatively suggest that the three loci were present in their last common ancestor. However, the copy number of the gene has not been conserved. Preliminary genetic studies of species in other solanaceous genera, *Capsicum* (pepper) and *Datura* (jimsonweed) support the hypothesis that three RBCS loci was the ancestral condition (Tanksley *et al.*, unpublished data).

The nucleotide sequences of all five tomato RBCS genes have been determined (Pichersky *et al.*, 1987a, Sugita *et al.*, 1987) and the values of nucleotide divergence among them are summarized in Fig. 3.5A. These results suggest that the three RBCS gene lineages are ancient since there is

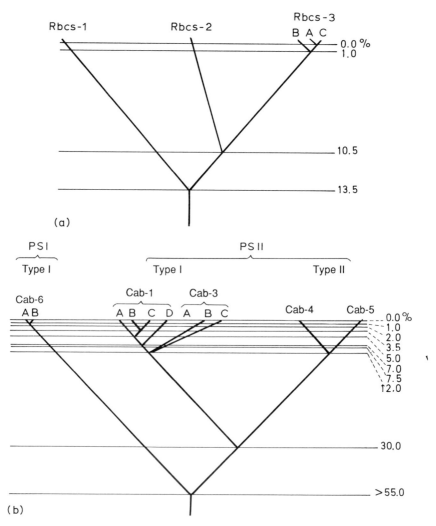

Fig. 3.5 Phylogenetic trees of the CAB and RBCS gene families based on comparisons of the coding sequences of tomato genes. A. The RBCS gene family in tomato consists of three lineages with a total of five genes. Although the nucleotide sequences of RBCS genes from different lineages are divergent, all five genes encode mature RBCS proteins which are almost identical. B. Branches of the tomato CAB gene family which have been identified and physically analyzed thus far. In complex loci, neighboring genes are labelled consecutively with letters. Numbers on the right indicate percentage divergence. Only coding sequences are compared. Deletions/insertions, which are few and usually very short (2–3 codons), are ignored in these comparisons.

The lengths of the branches of these phylogenetic trees are not proportional to the divergence levels and should not be compared with one another. The intention is to display, schematically, the results of nucleotide sequence comparisons, and the comparisons are not directly converting into evolutionary time scales.

considerable divergence at the nucleotide level. Most of the nucleotide substitutions are 'silent', i.e. they do not result in amino acid substitutions. As a consequence, the five tomato RBCS genes encode mature proteins that differ from one another by no more than four amino acids (3.5%) (the amino acid sequence in the transit peptide sequences are somewhat higher). This situation also occurs in petunia and other plants: for example, the mature RBCS polypeptides encoded by the petunia genes whose sequences have been reported to date vary by no more than three amino acids (Tumer *et al.*, 1986). The conservation of RBCS protein sequences within the genome of a given species is in strong contrast to the high rate of divergence between species. For example, mature RBCS proteins from tomato and petunia differ by as much as 20%.

Prior to our work on the RBCS genes in tomato and other solanaceous species, RBCS protein and DNA sequences were available only from distantly related species (e.g. pea, petunia and wheat). The observation was made from these data that, not only was the within-genome RBCS protein sequence variation small, but so was the within-genome RBCS DNA variation. One explanation for the relatively high intragenomic sequence homogeneity was that the genome of each species possesses a set of recently duplicated (and thus more closely related) RBCS genes. The demonstration of the ancient origin of the three RBCS gene lineages of tomato and other solanaceous species allows this explanation to be rejected. The alternative hypothesis is that selection operates to maintain homogeneity at the protein level within a given genome *regardless* of the specific sequence (Pichersky *et al.*, 1986a). Why this is so is not clear and awaits the elucidation of the function of the RBCS protein in the holoenzyme.

In contrast, the evolution of the three RBCS genes within the compound locus *Rbcs-3* seems to be dominated by the process of gene conversion. The nucleotide sequences of *Rbcs-3A* and *Rbcs-3C* are identical throughout the coding region and the two intervening sequences, but this homology begins and terminates abruptly at 10 bp upstream from the first ATG codon and directly downstream from the TAA stop codon, respectively. Beyond these borders, homology cannot be detected, except for regulatory elements (which comprise only a small portion of the non-coding flanking regions). In the case of *Rbcs-3B*, evidence for gene conversion is even stronger since the 5′ end of the area of putative gene conversion begins within the gene, about 30 nucleotides downstream from the first ATG. Thus, in the first 30 nucleotides, there are seven substitutions relative to *Rbcs-3A,C* and in the rest of the gene only nine such substitutions (and two short deletions/ insertions in the second intron).

The solanaceous RBCS genes also provide evidence for a recent introduction of an intron. Sequence comparisons indicate that the RBCS-1 lineage diverged from the RBCS-2 and RBCS-3 lineages first (Fig. 3.5A), and the latter two lineages then diverged from each other. Yet, the genes in the RBCS-1 and RBCS-3 lineages all have two introns (as do all other RBCS genes in non-solanaceous dicots reported to date). The single gene of the RBCS-2 lineage in tomato (Sugita *et al.*, 1987), petunia (Dean *et al.*, 1985) and tobacco (Mazur and Chui, 1985) contains an additional intron, which must have been added after the RBCS-1 lineage split from the other RBCS genes. In monocots, there are also examples of intron loss in these genes (Wimpee, 1984; Broglie *et al.*, 1983).

Chlorophyll a/b-binding protein genes (CAB)
The chlorophyll a/b-binding proteins are constituents of the photosynthetic apparatus. As their name implies, they bind chlorophyll a and b molecules and other pigments, and they function in capturing light energy and in the distribution of excitation energy between photosystem I (PS I) and photosystem II (PS II). Both photosystems contain CAB polypeptides, and comparative studies using immunological and other biochemical techniques have identified at least two types of CAB proteins in PS II and four types in PS I, all of which are structurally related but are encoded by different genes (Evans and Anderson, 1986). The functional differences among these types are not yet known, and the structural differences are only now being determined through the nucleotide sequencing of the corresponding genes. PS II Type I and Type II CAB genes have been cloned from tomato and several other species, both monocots and dicots. PS I type I CAB genes have been cloned only from tomato. The genes belonging to the other branches of the CAB gene family have not yet been identified.

Due to the combination of genetic mapping and physical cloning, our knowledge of the genomic organization and evolution of the CAB gene family in tomato exceeds that of this system in any other species. Our results are summarized in Fig. 3.5B. Tomato PS II Type I CAB genes are found in three independent loci, *Cab-1*, *Cab-2* and *Cab-3* (Pichersky *et al.*, 1985; Vallejos, Tanksley and Bernatzky, 1986). *Cab-1* includes four genes, all arranged in tandem. *Cab-3* locus contains at least three genes (and possibly one or two more). Of the three genes in the *Cab-3* locus, two are arranged in tandem and one in opposite orientation. A partially deleted, non-functional CAB gene has also been identified in the *Cab-3* locus. The *Cab-2* locus appears to contain a single gene. Thus, the total number of PS II Type I CAB genes in tomato is at least 8–10.

The mature CAB polypeptides encoded by genes in the *Cab-1* locus are divergent from those encoded by the *Cab-3* genes in a region of 13–15 amino acids at the N-termini, but otherwise the proteins are identical. The functional significance of this difference in the N-termini is not known; and CAB-3 polypeptides might represent different subtypes of PS II Type I CAB polypeptides (Pichersky *et al.*, 1985). PS II Type I CAB genes have been isolated from several other plant species, both monocots and dicots, but the sequences of more than one or a few genes from any given species have seldom been reported. In petunia, there are at least 16 such genes, whereas *Arabidopsis thaliana* has only three, all arranged in tandem in one locus. None of the PS II Type I CAB genes contain introns.

There are two PS II Type II CAB genes in tomato found in independent loci, *Cab-4* and *Cab-5*, and DNA clones have been isolated for each (Vallejos, Tanksley and Bernatzky, 1986; Pichersky *et al.*, 1987a). Although their nucleotide sequences have diverged considerably from each other (12%), they encode mature CAB proteins which differ by only four amino acids (1.7% divergence). The tomato PS II Type II CAB genes are 30% divergent from PS II Type I CAB genes, and the corresponding polypeptides are about 15% divergent. The genome of the monocot *Lemna gibba* has at least three PS II Type II CAB genes (but this species is a tetraploid); the sequence of only one gene has been reported and it is not known if these genes are linked (Karlin-Neumann *et al.*, 1985). PS II Type II CAB genes contain one intron.

The tomato genome possesses one locus on chromosome 5 containing two genes, designated *Cab-6A* and *Cab-6B*, encoding PS I Type I CAB polypeptides (Hoffman *et al.*, 1987; Pichersky *et al.*, 1987b). This branch of the CAB gene family has not yet been characterized in other species. The genes contain three introns although none of them in the same position as the single intron found in the PS II Type II CAB genes. The proteins encoded by *Cab-6A* and *Cab-6B* are identical and are substantially divergent from PS II CAB proteins. Overall homology with any PS II CAB protein is about 30% (ignoring two large deletions in the PS I Type I CAB relative to the PS II proteins); however, the segments, each about one quarter of the length of the protein, show 50% and 65% homology, respectively, to the corresponding segments in the PS II CAB proteins.

The CAB gene family is more complex than the RBCS gene family but the arrangement and evolution in the tomato genome of the two families follow similar patterns. PS II Type I CAB genes are distributed in three loci, two of which contain more than one gene. PS II Type II CAB genes consist of two single-copy loci, while the two PS I Type I CAB genes are found in a single locus. In all cases, multiple genes encode each type of CAB polypeptide. Their mode of evolution is also similar to that of the RBCS genes. When the genes are dispersed (e.g. *Cab-1* and *Cab-3* genes for Type I PS II CABs and *Cab-4* and *Cab-5* genes for Type II PS II CABs) the driving force that maintains the uniformity of coding information may be selection. Within loci (e.g. *Cab-1* genes, *Cab-3* genes, *Cab-6* genes), gene conversion seems to be a stronger force. One interesting observation, however, is that the presence of introns seems to be correlated with the frequency of gene conversion. For example, the level of divergence of the intronless genes in the *Cab-1* and *Cab-3* loci is greater than that of the *Cab-6* or the *Rbcs-3* genes which both have introns (Fig. 5 A,B). However, at present, this is only a correlation and any cause and effect relationship between introns and gene conversion remains to be demonsrated.

In conclusion, it seems that the organization of the RBCS and CAB gene families in tomato and other plants is brought about by the action of general mechanisms (duplication, transposition and sequence conversion) operating within the nucleus. Since these genes are represented by multiple copies in most plants, one must speculate that selection is involved in modulating copy number. One can imagine that this might occur in one of two ways. Either there is selection pressure to maintain several copies of the gene (i.e. one copy is not enough) or duplication of these genes is a fairly common phenomenon and selection pressure is only exerted when the copy number gets too high. In this last scenario, one would expect an equilibrium to be reached where the copy number varies up to some maximum. The current data are not sufficient to differentiate these two hypotheses.

In order to gain a better understanding of gene family organization and evolution and to uncover the mechanisms determining that organization, it is important to reconstruct the phylogenies of the RBCS and CAB families. Were *Cab-4* and *Cab-5* tandemly linked prior to dispersion, or did the duplication that gave rise to the two also create the two independent loci in one step? How often do gene conversion events occur in the complex CAB and RBCS loci? When were the three RBCS gene lineages in Solanaceae created? Did *Pisum* lose RBCS loci or have solanaceous species gained some or have both events occurred? When did each of the different branches of the CAB gene family, including the ones whose genes have not yet been identified and cloned, begin to diverge? To answer all these questions, it is necessary to follow individual genes, their genomic organization and molecular structure, and to identify orthologous genes in both closely and distantly related species.

In some cases, it may not be possible to reconstruct evolutionary events from sequence comparisons. For example, because of the concerted evolution of RBCS genes (and also CAB genes), it might be impossible to identify orthologous genes and loci from distantly related species. Even with 'silent' substitutions at saturation level, the RBCS genes in any given species do not diverge more than 15% (this is below the theoretical value because of the pronounced bias in the codon usage of these genes). Thus, tomato RBCS genes from all three RBCS lineages are more similar to each other than to any pea RBCS gene. It is not therefore possible to determine, from the pea–tomato comparison alone, which tomato RBCS locus (if any) is orthologous to the pea locus. Examination of intermediate species might solve this problem.

3.6 CONCLUDING REMARKS

Plant chromosomes have been observed for nearly a century and a vast collection of literature exists describing their form and cellular activities. It has been only recently, however, that we have gained a glimpse into how the nuclear genome is organized at the DNA sequence level. However, although some knowledge about the patterns of sequence organization in a few plants has been gained, almost nothing is known about the mechanisms and rules that govern genome structure. There is no doubt that gene duplication has played a major role. In plants, unlinked duplications seem to be common, yet there are few clues to the source of these duplications. The mode of interspersed sequence movement and its effect on adaptation is unknown, yet work on wheat by Flavell and coworkers has shown that these sequences are rapidly turning over and 're-colonizing' the genome such that recently diverged species have many species-specific interspersed repeats. The potential consequences of this process on reproductive isolation have been pointed out (Flavell, O'Dell and Hutchinson, 1981) but are as yet untested.

While it is apparent that plant linkage groups are being rearranged by translocations and inversions it is not known if these events themselves cause other events such as gene duplication. Perhaps chromosomal rearrangements and gene duplication are not causally related but are both manifestations of genomes in flux due to mobilization of repeats or transposons. Transposons are known to lead to chromosome breakage in maize and the possibility exists that transposons may be vectors for gene duplication.

We now have evidence that environmental conditions can trigger heritable genomic changes in plants (Cullis, 1985), and McClintock (1984) has proposed the genomic stress hypothesis that stress (tissue culture, irradiation, and possibly other environmental factors) triggers genomic reshuffling which may lead to mutations that become useful to overcome the stressful conditions. This hypothesis has important implications. First, the environment is implicated in inducing heritable genetic changes that may be adaptive in meeting the challenges of that environment. Second, the changes induced by the environment are inextricably tied to chromosomal rearrangements. Thus the mechanisms that are responsible for determining overall genomic architecture may then be important in creating the genetic variation over which selection has to choose. If this is true, then our ignorance about plant genome organization and evolution may be causally related to the gaps in our knowledge about how the diversity of plant forms have come into existence and the rules that dictate their evolutionary options.

3.7 REFERENCES

Bachmann, K., Chambers, K. L. and Price, H. J. (1985) Genome size and natural selection: observations and experiments in plants. in *The Evolution of Genome Size* (ed. T. Cavalier-Smith), Wiley, New York, pp. 267–76.

Bennett, M. D. (1972) Nuclear DNA content and minimum generation time in herbaceous plants. *Proc. R. Soc. Lond. B*, **191**, 109–35.

Bennett, M. D. (1976) DNA amount, latitude and crop plant distribution. in *Current Chromosome Research* (eds K. Jones and P. E. Brandham), North Holland, Amsterdam, pp. 151–8.

Bennett, M. D. and Smith, J. B. (1976) Nuclear DNA amounts in angiosperms. *Phil. Trans. R. Soc. London, Ser. B*, **274**, 227–73.

Bernatzky, R. and Tanksley, S. D. (1986a) Majority of random cDNA clones correspond to single loci in the tomato genome. *Mol. Gen. Genet.*, **203**, 8–14.

Bernatzky, R. and Tanksley, S. D. (1986b) Towards a saturated linkage map in tomato based on isozymes and cDNA clones. *Genetics*, **112**, 887–98.

Britten, R. J., Cetta, A. and Davidson, E. H. (1978) The single copy DNA sequence polymorphism of the sea urchin, *Strongylocentrotus purpuratus. Cell*, **15**, 1175–86.

Broglie, R., Coruzzi, G., Lamppa, G., Keith, B. and Chua, N.-H. (1983) Structural analysis of nuclear genes coding for the precursor to the small subunit of wheat ribulose bis-phosphate carboxylase. *Bio/Technology*, **1**, 55–61.

Burnham, C. R. (1962) *Discussions in Cytogenetics*, Burgess Publishing Co., Minneapolis.

Chang, C. and Meyerowitz, E. M. (1986) Molecular cloning and DNA sequence of the *Arabidopsis thaliana* alcohol dehydrogenase gene. *Proc. Natl. Acad. Sci. USA*, **83**, 1408–12.

Coen, E. S., Thoday, J. M. and Dover, G. (1982) Rate of turnover of structural variants in the rDNA gene family of *Drosophila melanogaster. Nature*, **295**, 564–8.

Courage-Tebbe, U., Doring, H.-P., Federoff, N. and Starlinger, P. (1983) The controlling element Ds at the Shrunken locus in *Zea mays*: structure of the unstable sh-m5933 allele and several revertants. *Cell*, **34**, 383–93.

Cullis, C. A. (1985) Experimentally induced changes in genome size. in *The Evolution of Genome Size* (ed. T. Cavalier-Smith), Wiley, New York, pp. 197–209.

Dean, C., van den Elzen, P., Tamaki, S., Dunsmiur, P. and Bedbrook, J. (1985) Linkage and homology analysis divides the eight genes for the small subunit of petunia ribulose bisphosphate carboxylase into three gene families. *Proc. Natl. Acad. Sci. USA*, **82**, 4964–8.

Dennis, E. S., Gerlach, W. L., Pryor, A. J., Bennetzen, J. L., Inglis, A., Llewellyn, D., Sachs, M. M., Ferl, R. J. and Peacock, W. J. (1984) Molecular analysis of the alcohol dehydrogenase (*Adh1*) gene of maize. *Nucl. Acids Res.*, **12**, 3983–4000.

Dennis, E. S. and Peacock, W. J. (1984) Knob heterochromatin homology in maize and its relatives. *J. Mol. Evol.*, **20**, 341–50.

Dennis, E. S., Sachs, M. M., Gerlach, W. L., Finnegan, E. J. and Peacock, W. J. (1985) Molecular analysis of the alcohol dehydrogenase 2 (Adh2) gene of maize. *Nucl. Acids Res.*, **13**, 727–43.

Domoney, C., Ellis, T. H. and Davies, D. R. (1986) Organization and mapping of legumin genes in *Pisum. Mol. Gen. Genet.*, **202**, 280–5.

Doolittle, W. F. (1985) The evolutionary significance of middle-repetitive DNAs. in *The Evolution of Genome Size* (ed. T. Cavalier-Smith), New York.

Evans, P. K. and Anderson, J. M. (1986) The chlorophyll A/B-proteins of PSI and PSII are immunologically related. *FEBS Lett.*, **199**, 227–33.

Ferris, S. D. and Whitt, G. S. (1977) Loss of duplicate gene expression after polyploidization. *Nature*, **265**, 258–60.

Flavell, R. B., O'Dell, M. and Hutchinson, J. (1981) Nucleotide sequence organization in plant chromosomes and evidence for sequence translocation during evolution. *Symp. Quant. Biol.*, **45**, 501–8.

Flavell, R., Rimpau, J., Smith, D. B., O'Dell, M. and Bedbrook, J. R. (1980) The evolution of plant genome structure. in *Genome Organization and Expression in Plants* (ed. C. J. Leaver) Plenum Press, New York.

Forde, B. G., Kreis, M., Bahramian, M. B., Matthews, J. A. and Miflin, B. J. (1981) Molecular cloning and analysis of cDNA sequences derived from poly A+ RNA from barley endosperm: identification of B hordein clones. *Nucl. Acids Res.*, **9**, 6689–707.

Freeling, M. and Bennett, D. C. (1985) Maize *Adhl. Ann. Rev. Genet.*, **19**, 297–323.

Gerlach, W. L. and Peacock, W. J. (1980) Chromosomal locations of highly repeated DNA sequences in wheat. *Heredity*, **44**, 269–76.

Goldberg, R. B., Hoschek, G. and Kamalay, J. C. (1978) Sequence complexity of nuclear and polysomal RNA in leaves of the tobacco plant. *Cell*, **14**, 123–31.

Gottlieb, L. D. (1977) Evidence for duplication and divergence of the structural gene for phosphoglucoisomerase in diploid species of *Clarkia. Genetics*, **86**, 289–307.

Gottlieb, L. D. (1982) Conservation and duplication of isozymes in plants. *Science*, **216**, 373.

Gottlieb, L. D. and Higgins, R. C. (1984) Evidence from subunit molecular weight suggests hybridization was the source of the phosphoglucose isomerase gene duplication in *Clarkia. Theor. Appl. Genet.*, **68**, 369–73.

Hall, T. C., Slighton, J. L., Ersland, D. R., Scharf, P., Barker, R. F., Murray, M. G., Brown, J. W. S. and Kemp, J. D. (1983) in *Advances in Gene technology: Molecular Genetics of Plants and Animals*. Academic Press, New York.

Hart, G. E. (1983) Hexaploid wheat (*Triticum aestivum* L. em Thell). in *Isozymes in Plant Genetics and Breeding* (eds S. D. Tanksley and T. J. Orton), Elsevier, Amsterdam.

Hoffman, N. E., Pichersky, E., Malik, V. S., Castresana, C., Ko, K., Darr, S. C., Arnsen, C. J. and Cashmore, A. R. (1987) A cDNA clone encoding a photosystem I protein with homology to photosystem II chlorophyll A/B binding polypeptides. *Proc. Natl. Acad. Sci. USA* (in press).

Hutchinson, J., Rees, H. and Seal, A. G. (1979) An assay of the activity of supplementary DNA in *Lolium. Heredity*, **43**, 411–21.

Karlin-Neumann, G. A., Kohorn, B. D., Thornber, J. P. and Tobin, E. M. (1985) A chlorophyll a/b protein encoded by a gene containing an intron with characteristics of a transposable element. *J. Mol. Appl. Genet.*, **3**, 45–61.

Kleckner, N. (1981) Transposable elements in prokaryotes. *Ann. Rev. Genet.*, **15**, 341–404.

Lewin, B. (1985) *Genes* 2nd edn, Wiley, New York.

MacIntyre, R. J. (1976) Evolution and ecological value of duplicate genes. *Ann. Rev. Ecol. Syst.*, **7**, 421–68.

Mazur, B. and Chui, C. F. (1985) Sequence of genomic DNA clone for the small subunit of ribulose bis-phosphate carboxylase-oxygenase from tobacco. *Nucl. Acids Res.*, **13**, 2373–86.

McClintock, B. (1950) Mutable loci in maize. *Carnegie Inst. Washington Year Book*, **49**, 157–67.

McClintock, B. (1984) The significance of responses of the genome to challenge. *Science*, **226**, 792–801.

Meyerowitz, E. M. and Pruitt, R. E. (1985) *Arabidopsis thaliana* and plant molecular genetics. *Science*, **229**, 1214–24.

Murray, M. G., Peters, D. L. and Thompson, W. F. (1981) Ancient repeated sequences in the pea and mung bean genomes and implications for genome evolution. *J. Mol. Evol.*, **17**, 31–42.

Nevers, P., Sheperd, N. S. and Saedler, H. (1986) Plant transposable elements. *Adv. Bot. Res.*, **12**, 103–203.

Payne, P. I., Holt, L. M., Jackson, E. A. and Law, C. N. (1984) Wheat storage proteins: their genetics and their potential for manipulation by plant breeding. *Phil. Trans. R. Soc. Lond. B*, **304**, 359–71.

Peacock, W. J., Dennis, E. S., Rhoades, M. M. and Pryor, A. J. (1981) Highly repeated DNA sequence limited to knob heterochromatin in maize. *Proc. Natl. Acad. Sci. USA*, **78**, 4490–4.

Pichersky, E., Bernatzky, R., Tanksley, S. D., Breidenbach, R. B., Kausch, A. P. and Cashmore, A. R. (1985) Molecular characterization and genetic mapping of two clusters of genes encoding chlorophyll a/b-binding proteins in *Lycopersicon esculentum* (tomato). *Gene*, **40**, 247–58.

Pichersky, E., Bernatzky, R., Tanksley, S. D. and Cashmore, A. R. (1986) Evidence for selection as a mechanism in the concerned evolution of *Lycopersicon esculentum* (tomato) genes encoding the small subunit of ribulose-1,5-bisphosphate carboxylase/oxygenase. *Proc. Natl. Acad. Sci. USA*, **83**, 3880–4.

Pichersky, E. and Gottlieb, L. D. (1983) Evidence for duplication of the structural genes coding plastid and cytosolic isozymes of triose phosphate isomerase in diploid species of *Clarkia*. *Genetics*, **105**, 421–36.

Pichersky, E., Malik, V., Bernatzky, R., Hoffman, N., Tanksley, S. D. and Cashmore, A. R. (1987a) Tomato *Cab-4* and *Cab-5* genes encode a second type of CAB polypeptide localized in photosystem II. *Plant Mol. Biol.*, **9**, 109–20.

Pichersky, E., Hoffman, N. E., Bernatzky, R., Piechulla, E., Tanksley, S. D. and Cashmore, A. R. (1987b) Molecular characterization and genetic mapping of DNA sequences encoding the TYPE I chlorophyll a/b binding polypeptide of photosystem I in *Lycopersicon esculentum* (tomato). *Plant Mol. Biol.*, **9**, 205–16.

Polans, N. O., Weeden, N. F. and Thompson, W. F. (1985) Inheritance, organization and mapping of rbcS and cab multigene families in pea. *Proc. Natl. Acad. Sci. USA*, **82**, 5083–7.

Price, J. H. (1976) Evolution of DNA content in higher plants. *Bot. Rev.*, **42**, 27–52.

Rees, H. and Hazarika, M. H. (1967) Chromosome evolution in *Lathyrus*. in *Chromosomes Today* (eds C. D. Darlington and K. R. Lewis), Vol. 2, Plenum Press, New York.

Rhoades, M. M. (1978) Genetic effects of heterochromatin in maize. in *Maize Breeding and Genetics* (ed. D. B. Walden), Wiley, New York. pp. 641–72.

Rivin, C. J., Cullis, C. A. and Walbot, V. (1986) Evaluating quantitative variation in the genome of *Zea mays*. *Genetics*, **113**, 1009–19.

Saghai-Maroof, M. A., Soliman, K. M., Jorgensen, R. A. and Allard, R. W. (1984) Ribosomal DNA spacer-length polymorphisms in barley: mendelial inheritance, chromosomal location and population dynamics. *Proc. Natl. Acad. Sci. USA*, **81**, 8014–18.

Schwarz-Sommer, Z., Gierl, A., Cuypers, H., Peterson, P. A. and Saedler, H. (1985) Plant transposable elements generate the DNA sequence diversity needed in evolution. *EMBO J.*, **4**, 591–7.

Sears, E. R. (1966) Nullisomic-tetrasomic combinations in hexaploid wheat. in *Chromosome Manipulations in Plant Genetics* (eds R. Riley and K. R. Lewis), *Heredity* (Suppl.), **20**, 29–45.

Shewry, P. R., Faulks, A. J., Pickering, R. A., Jones, I. T., Finch, R. A. and Miflin, B. J. (1980) The genetic analysis of barley storage proteins. *Heredity*, **44**, 383–9.

Smith, G. P. (1973) Unequal crossover and the evolution of multigene families. *Cold Spring Harbor Symp. Quant. Biol.*, **38**, 507–14.

Sugita, M., Manzora, T., Pichersky, E., Cashmore, A. R. and Gruissem, W. (1987) Genomic organization, sequence analysis and expression of all the live genes encoding the small unit of ribulose bis-phosphate carboxylase-oxygenase from tomato. *Mol. Gen. Gen.* (in press).

Tanksley, S. D. (1983) in *Isozymes in Plant Genetics and Breeding. Part A.* (eds S. D. Tanksley and T. J. Orton), Elsevier, Amsterdam. pp. 111–40.

Tanksley, S. D. (1987) Nuclear genome organization in tomato and related diploid species. *Am. Naturalist*, **130**, 546–61.

Tanksley, S. D. and Orton, T. J. (eds) (1983) *Isozymes in Plant Genetics and Breeding*. Elsevier, Amsterdam.

Thompson, W. F. and Murray, M. G. (1981) The nuclear genome: structure and function. in *The Biochemistry of Plants*. Vol. 6. Academic Press, New York.

Tumer, N. E., Clark, W. G., Tabor, G. J., Hironaka, C. M., Fraley, R. T. and Shah, D. M. (1986) The genes encoding the small subunit of ribulose-1,5-bisphosphate carboxylase are expressed differentially in petunia leaves. *Nucl. Acids Res.*, **14**, 3325–42.

Vallejos, C. E., Tanksley, S. D. and Bernatzky, R. (1986) Localization in the tomato genome of DNA restriction fragments containing sequences homologous to the rRNA (45S), the major chlorophyll a/b binding polypeptide and the ribulose bisphosphate carboxylase genes. *Genetics*, **112**, 93–105.

Van't Hoff, J. and Sparrow, A. H. (1963) A relationship between DNA content, nuclear volume and minimum mitotic cycle time. *Proc. Natl. Acad. Sci. USA*, **49**, 897–902.

Wilson, D. R. and Larkins, B. A. (1984) Zein gene organization in maize and related grasses. *J. Mol. Evol.*, **20**, 330–40.

Wimpee, C. W. (1984) Organization and expression of light-regulated genes in *Lemna gibba*. PhD. Thesis, University of California, Los Angeles, CA.

Young, M. W. and Schwartz, H. E. (1981) Nomadic gene families in *Drosophila*. *Cold Spring Harbor Symp. Quant. Biol.*, **45**, 629–40.

Zabel, P., Meyer, D., van de Stolpe, O. van der Zaal, V., Ramanna, M. S., Koornneef, M., Krens, F. and Hille, J. (1985) Towards the construction of artificial chromosomes for tomato. in *Molecular Form and Function of the Plant Genome* (eds L. van Vloten-Doting, G. Groot and T. Hall), Plenum, New York.

CHAPTER 4

Onagraceae as a model of plant evolution

PETER H. RAVEN

4.1 INTRODUCTION

Onagraceae are a well-defined family of flowering plants, consisting of seven tribes, 16 genera, and approximately 652 species of worldwide distribution (Table 4.1). With very few exceptions, the family has been studied in biosystematic detail, breeding systems and pollinators have been investigated, chromosome numbers are mostly known, the vegetative and floral anatomy has been studied in detail, and flavonoids have also been investigated in the great majority of Onagraceae. Because of this, Onagraceae are clearly the best-known plant family of their size.

Given the rich stock of information that is available about Onagraceae, they appear to be an ideal group for the application of modern genetic and biochemical techniques as a way to elucidate their phylogeny. These methods are powerful in answering phylogenetic questions; for example, in *Clarkia*, a much-studied genus, species that possess the same cytosolic PGI duplication are monophyletic regardless of differences in their morphology and chromosome numbers (Gottlieb and Weeden, 1979). The four species in sect. *Peripetasma* that have lost one or both 6PGD duplications are monophyletic and represent two pairs of sister species, respectively (Odryzkoski and Gottlieb, 1984), a result verified by restriction enzyme analysis of their chloroplast DNAs (Sytsma and Gottlieb, 1986a,b). Similarly, the demonstrated relationship of the former monotypic genus *Heterogaura* to a particular section, and perhaps even a particular species, of *Clarkia*, attests to the power of the method (Sytsma and Gottlieb, 1986a). Such results, however, have meaning only in a context in which the relationships of the plant groups being considered are well understood; therefore, data sets such as those available for Onagraceae are particularly valuable. With the current availability of cladistic methods of analysis that exceed those employed in the past in their rigor and ability to compare different subsets of informa-

Table 4.1 Tribes and genera of Onagraceae, with estimated numbers of species and geographical distribution of each.

Tribe	Genus	Estimated number of species	Geographical distribution
I. Jussiaeeae	1. *Ludwigia*	82	Pantropical, best represented in South America; 3 sects. with 21 spp. in temperate North America, 2 monotypic sects. in temperate Asia; 7 spp. endemic to Africa, 1 to tropical Asia, and 2 endemics common; total of 25 spp. in Old World, 12 endemic
II. Fuchsieae	2. *Fuchsia*	102	Andean South America, with 12 spp. endemic in Mexico and Central America, 2 on Hispaniola 7 in coastal Brazil, 1 in Tahiti, and 3 in New Zealand
III. Circaeeae	3. *Circaea*	7	North temperate forests and bordering alpine areas, 5 spp. endemic to Asia and all 7 found there
IV. Lopezieae	4. *Lopezia*	22	Mostly Mexican, 1 S. to Guatemala and a second S. to Panama
V. Hauyeae	5. *Hauya*	2	Mexico to Costa Rica
VI. Onagreae	6. *Gongylocarpus*	2	1 endemic to 2 islands off W. coast of Baja California, 1 widespread in Mexico and Guatemala but not common
	7. *Gayophytum*	9	W. North America, 1 endemic to temperate W. South America and 1 common to both
	8. *Xylonagra*	1	Central Baja California
	9. *Camissonia*	61	W. North America, 1 endemic to temperate W. South America
	10. *Calylophus*	6	Central United States S. to central Mexico
	11. *Gaura*	21	Southwestern and central United States E. to Atlantic Coast and S. to Mexico and Guatemala; centers in Texas

not in Cuba

Table 4.1   Continued.

Tribe	Genus	Estimated number of species	Geographical distribution
	12. *Oenothera*	123	70 North American, 49 South American, plus 4 common to both, 1 of European origin; far more diverse in North America, especially in W. and in Mexico
	13. *Stenosiphon*	1	Great Plains of central United States
	14. *Clarkia*	44	W. North America, all but 1 in California; 1 endemic to temperate W. South America
VII. Epilobieae	15. *Epilobium*	162	Cosmopolitan at high altitudes and high latitudes; 77 spp. in Eurasia (9 common to North America, 6 common to Africa), 46 in Australasia (all endemic), 35 in North America (26 endemic), 12 in South America (10 endemic), 10 in Africa (4 endemic)
	16. *Boisduvalia*	6	W. North America, 1 in W. temperate South America, 1 common to both continents

tion, the family should become an even more instructive example in the future. In this chapter, some of the ways in which this promise has been expressed, and the promise it holds for the future are outlined.

4.2   RELATIONSHIPS WITH OTHER FAMILIES

Onagraceae are members of the clearly defined order Myrtales (Dahlgren and Thorne, 1984; Johnson and Briggs, 1984), sharing all the critical characteristics of this order, including a distinctive set of embryological features (Tobe and Raven, 1983; Schmid, 1984), which, taken together, characterizes the order and is shared by all its members. Myrtales includes about 14 families, the exact number depending on the circumscription of the families and not on the inclusion or exclusion from the order. Cronquist (1984) has argued that Thymelaeaceae ought to be included in Myrtales, but few other botanists agree with him; his view is supported by studies of the *N*-terminal sequences of amino acids of ribulose bisphosphate carboxylase small subunit (Martin and Dowd, 1986), however, which suggest that not only Thymelaeaceae, but also Euphorbiales belong within Myrtales. This matter requires further study, but the overwhelming weight of available evidence contradicts a placement of either Euphorbiales or Thymelaeaceae within Myrtales. Myrtales are themselves clearly related to Rosales (Dahlgren and Thorne, 1984), and clearly belong within the rosid alliance as defined by Cronquist in various publications.

The distributions of the various families of Myrtales are clearly associated with the continents of the Southern Hemisphere at present, and they are likely to have originated in West Gondwanaland (South America plus Africa) or Australasia in the middle Cretaceous Period (perhaps as much as 100 million years ago) or earlier (Raven and Axelrod, 1974). Both the pattern of current distributions and what is known of the fossil record provide evidence in support of this hypothesis, and the case for it appears to be strengthened as more information accumulates.

4.3　AGE AND DISTINCTIVENESS OF ONAGRACEAE

Within Myrtales, Onagraceae are distinctive in a number of features including:

1. A distinctive 4-nucleate embryo sac (summary in Tobe and Raven, 1983);
2. The abundant presence of raphides in the vegetative cells (Gulliver, 1864; Metcalf and Chalk, 1950, p. 1354; Carlquist, 1961, 1975);
3. 'Paracrystalline beaded' pellen ektexine (Skvarla, Raven and Praglowski, 1975; Skvarla and Raven, 1976); and
4. Viscin threads in the pollen (Skvarla *et al.*, 1978).

The combination of these clearly synapomorphic features with the standard morphological ones makes the constitution of Onagraceae as certain as it is for any family of flowering plants, a relationship that underlies the validity of comparisons within the family.

The pollen of Onagraceae is so distinctive that it can easily be recognized in the fossil record, from the Maestrichtian Age (73–65 million years ago) at the end of the Cretaceous Period (Drugg, 1967; Pares Regali, Uesugui and Santos, 1974a,b). It is possible that earlier records of Onagraceae pollen will be discovered, but the pollen is so distinctive that such discoveries are much less likely than for some other groups. Consequently, an origin of the family in the latest Cretaceous Period may be assumed as likely for the present. The oldest records are from Brazil, which is consistent with an origin in West Gondwanaland, and California, which is consistent both with an early dispersal to North America across a broad water gap and with the high degree of diversification of the family in North America at present.

4.4 PATTERNS OF RELATIONSHIP IN ONAGRACEAE

The 16 genera of Onagraceae are grouped unevenly into seven tribes, five of which are monogeneric, one (Epilobieae) includes two genera, and one (Onagreae) includes nine (Table 4.1). Despite having studied relationships within the family for some 30 years, in extensive collaboration with many other botanists, I am able to say relatively little about relationships between its genera, other than the relationships implied by the grouping of genera into tribes. In this connection, the genera have been constructed broadly, to reflect relationships within themselves, and many of them include distinctive elements that have themselves been regarded as separate genera in the past. Obviously, if we had continued to recognize many monotypic or other small genera, these would be seen as having direct relationships with others, e.g. *Dantia* and *Oocarpon* with *Ludwigia*, *Semeiandra* and *Reisenbachia* with *Lopezia*, *Godetia*, *Eucharidium*, and *Heterogaura* with *Clarkia*, and so forth. By including these 'satellite' groups with the other species to which they are directly related, however, we have set up genera within which significant evolutionary lineages can be traced, but between which the connections are often not clear.

Do Onagraceae differ from other plant families in the distinctiveness of their genera? One has the impression that orchids and papilionoid legumes, for example, consist of series of closely related genera than can readily be linked into evolutionary lines. Because of the overall subjectivity of classification, it is difficult to be sure whether the apparent differences in pattern are an artifact of the ways in which genera are recognized in these different families, or whether they reflect differences in the ages or evolutionary patterns of the respective groups. At any rate, Onagraceae are a family of relatively great age, and it is not surprising to find that they include a number of very distinctive groups.

The major feature of relationships within Onagraceae that has been established as a result of our previous studies is that *Ludwigia* (including the formerly recognized genera *Jussiaea* and *Isnardia*) is the 'sister group' of all other Onagraceae (Eyde, 1977, 1979; Raven and Tai, 1979). This hypothesis has recently been confirmed by studies of amino acid sequences of ribulose bisphosphate carboxylase in the genera of Onagraceae (Martin and Dowd, 1986). The sequence of 40 amino acids in the N-terminal end of the small subunit of this molecule resembles that in most plants, whereas the sequence found in all other genera of the family is a unique one that clearly marks them as a separate evolutionary line. One of the most interesting implications of this finding is that, partly because the nectaries of *Ludwigia* are located around the base of the style, inside the anthers, whereas in all other Onagraceae they are located on the floral tube outside the anthers, Eyde (1977, 1979) concluded that epigyny evolved separately in the two main evolutionary lines of the family. The common ancestor of Onagraceae, which possessed all the other distinctive features of the family, had a superior ovary. Differences in vasculature and other details support this conclusion (Eyde, 1981), which could not have been predicted from a less-detailed inspection of the characteristics of the genera of Onagraceae.

Within the second evolutionary line of Onagraceae, it is evident on the basis of floral and vegetative anatomy, as well as chromosome number and morphology, that *Fuchsia*, *Circaea*, *Lopezia*, and *Hauya* are more generalized in their features than are Onagreae (nine genera) or Epilobieae (two genera). At a detailed level, however, it is very difficult to establish relationships between any of these tribes with confidence. Formal cladistic analysis and comparisons of nucleic acids, both underway at present, should eventually reveal some patterns of affinity among the four genera that have the most generalized features.

One of the more perplexing problems involved has to do with the mutual relationships of the nine genera of Onagreae. Of these, the ditypic, Mexican genus *Gongylocarpus*, despite its unique ovaries, seems to be the least specialized overall; it is probably a true relic, retaining among other plesiomorphic features the probable original basic chromosome number of the family, $n = 11$, as well as genetic self-incompatibility in the less specialized of its two surviving species. *Gongylocarpus* resembles the monotypic genus *Xylonagra*, which is also endemic to Baja California, in its capitate stigmas; but *Xylonagra*, perplexingly, shares the reduced chromosome number $n = 7$ with all other genera of Onagreae. The large genus *Camissonia* (61 species, all but one in western North America) could plausibly be allied with *Xylonagra* and *Gongylocarpus*, but the affinities of the very reduced plants that make up the genus *Gayophytum* (nine species) are not at all obvious.

Within the remainder of the tribe, *Calylophus* and *Gaura* are obviously related to one another, but share some apparently apomorphic embryological features with Hauya, which has a basic chromosome number of $x = 10$, whereas they have $x = 7$ – ostensibly a synapomorphy linking them with *Xylonagra*, *Camissonia*, and the other genera of the tribe, unless the independent derivation of $x = 7$ in various lines can be argued as plausible. The monotypic genus *Stenosiphon* is clearly closely related to *Oenothera*, and the single species that was formerly assigned to the monotypic genus *Heterogaura* ought to be regarded as a species of *Clarkia*, as will be discussed below. The relationships of these three lines to one another, or to any of the genera discussed in the preceding paragraph, are unclear. One of the difficulties in deducing affinities in this group lies in the reduced leaves that are characteristic of these plants, which afford few characteristics that might be useful in indicating patterns of relationship. If *Calylophus* and *Gaura* are actually directly related to *Hauya*, it would suggest that the genera of Onagreae have had multiple origins from diverse, mainly extinct ancestors – a pattern of evolution that might in part be unraveled by cladistic and macromolecular studies. *Clarkia*, as judged by any standards, is absolutely distinct within the family.

There is no doubt of the close relationship between *Epilobium* and *Boisduvalia*, in fact it is possible that the two sections of *Boisduvalia* originated as separate evolutionary lines from *Epilobium* (see Section 4.7). However, what the relationships of *Epilobium* and *Boisduvalia* might be with other genera is obscure on the basis of currently available information. They certainly share a number of specialized features, but are not clearly directly related to any other specific group.

4.5 GEOGRAPHICAL DISTRIBUTIONS AND HISTORY

The least specialized species of *Ludwigia*, and all of the genetically self-incompatible species of that genus, occur in South America. In addition, most species of *Fuchsia* are South American, although many distinctive outliers occur elsewhere. In a number of respects, such as leaf anatomy, venation, wood anatomy, and chromosome number and morphology, *Fuchsia* might be regarded as the least-specialized genus of the second evolutionary line of Onagraceae, despite its bird-pollinated flowers and bird-dispersed fruits, which are unlikely to have originated before the Eocene Period (Sussman and Raven, 1978). One of the undoubtedly plesiomorphic features of these two genera, which is also shared with *Hauya*, is the absence of interxylary phloem, a derived characteristic that has evidently originated independently in some species of *Ludwigia*, and also occurs in *Lopezia*, Epilobieae, and Onagreae (Carlquist, 1975, 1977, 1982). One of the lines of evidence that *Fuchsia* probably originated in South America is that it had reached New Zealand by the uppermost Oligocene Period (Mildenhall, 1980), but the relatively rich representation of diverse species, comprising four endemic sections (of a total of 10 in the genus), in Mexico and Central America (Breedlove, Berry and Raven, 1982), implies a relatively great antiquity there also.

It would be difficult to envision an origin for *Ludwigia* anywhere except South America, although members of this genus reached both North America and Europe by Paleogene time, judged from the abundant reports of pollen in the fossil record. All species of the three other groups of Onagraceae that retain many primitive features – *Lopezia*, *Circaea* and *Hauya* – are confined to the Northern Hemisphere, with only *Circaea* among them represented in the Old World. In addition, the greatest diversity of both Onagreae and Epilobieae is found in North America, although some Onagreae also occur in South America and *Epilobium* is cosmopolitan at relatively high latitudes and altitudes. All of these relation-

[handwritten margin note: via island-hopping of the proto. Antilles?]

ships suggest that Onagraceae have been present in North America for a very long time, and that the primary diversification of the evolutionary line that includes all of the genera of the family except *Ludwigia* has taken place there. Although an origin of the family in North America cannot be ruled out on the basis of present evidence, the myrtalean affinities of the family, taken together with the patterns of distribution of *Ludwigia* and *Fuchsia*, still appear to indicate an origin for the family in South America, with early dispersal to North America, where most of its subsequent diversification has taken place.

From the geographical patterns of the individual genera, can be seen that *Ludwigia* and *Epilobium* are very widespread genera that are evidently very easily dispersed. The relatively small seeds of *Ludwigia* are probably spread in mud on birds' feet, but also float readily in most cases and sometimes have extra flotation devices, whereas the plumed seeds of *Epilobium* are blown from place to place, evidently sometimes for great distances. Thus there clearly have been at least two, if not more, independent instances of the dispersal of *Epilobium* species from New Zealand to South America; and one species of the genus, *E. ciliatum* Raf., occurs as a native plant throughout the cooler parts of North America, in eastern Siberia, Korea, Japan, and northeastern China, and over a vast area of South America from the Andes of Chile and Argentina south to Tierra del Fuego and the Falkland Islands (Solomon, 1982). In addition, this species is widespread as an introduced plant in Europe, New Zealand, and some regions of Australia. Despite these very wide ranges, however, it is clear that *Ludwigia* is more diverse in South America than elsewhere, and that *Epilobium* is most diverse in western North America (Raven, 1976).

Most of the sections of *Ludwigia* are primarily South American. The genus is currently divided into 22 sections, of which 11 clearly center in South America, although some have species, such as the widespread *L. octovalvis* Jacq., that occur on other continents. Of the remaining sections, three – *Microcarpium*, *Dantia* and *Ludwigia* – are primarily temperate North American, and the closely related, monotypic *Miquelia* is temperate east Asian. Four additional sections, with a total of five species, are African; two monotypic sections are Asian; and one additional monotypic section is common to Asia and Africa. Many of the evolutionary lines within *Ludwigia* are very distinctive, as Eyde (1977, 1979, 1981) has demonstrated clearly, and as would be expected in a genus that has existed throughout Cenozoic time.

For the other very widespread genus of Onagraceae, *Epilobium*, the pattern of distribution is very different. Of the six sections, one – sect. *Chamaenerion*, with seven species – is mainly Eurasian, whereas four of the remaining five sections are confined to western North America (Raven, 1976). These four sections, however, consist of a total of eight species; the remaining 147 species of the genus all belong to sect. *Epilobium*, which is found on all continents except Antarctica. Sect. *Epilobium* has its most distinctive species in western North America, and can most simply be assumed to have spread from there to other parts of the world. Only 25 species of sect. *Epilobium* occur in North America; one of them is common to South America, where 11 additional native species occur, one common to Australia and New Zealand. In Eurasia, there are 75 native species of sect. *Epilobium*, of which seven are also found in North America and six are also found in Africa; the remaining 62 species are endemic. Australasia has 46 species, all endemic, which seem to have evolved from Eurasian stock (Raven, 1973; Raven and Raven, 1976). Pollen that probably represents sect. *Epilobium* has recently been discovered in the late Oligocene of New Zealand (Daghlian *et al.*, 1984), so the genus appears to have been in Australasia for more than 25 million years. In general, the pattern of distribution of *Epilobium* is that of a highly mobile group, which has produced many, usually quite widespread species in different parts of the world.

The third large genus with a complex pattern of distribution is *Fuchsia*, which consists of approximately 102 species of outcrossing shrubs and some trees. In terms of species numbers, *Fuchsia* is centered in South America, where sect. *Fuchsia* has approximately 60 species (two additional ones occur on Hispaniola), and sects. *Hemsleyella* (14 species), *Quelusia* (8 species), and the monotypic sections *Kierschlegeria* and an undescribed one from northern Peru are endemic. The remaining 16 species of the genus occur in Mexico and Central America (4 sections, 12 species), and in New Zealand and Tahiti (one section, 4 species).

Within South America, sects. *Fuchsia* (Berry, 1982) and *Hemsleyella* (Berry, 1985) have clearly evolved rapidly and become speciose as the Andes have been uplifted to their present heights during the past few million years. The eight species of sect. *Quelusia*, all polyploid, include the most commonly cultivated member of the genus, *F. magellanica* Lam., a native of Chile and Argentina; the seven remaining species of the section occur in the coastal mountains of Brazil, in a pattern that occurs not infrequently among various groups of plants that evidently once had a wider distribution in temperate South America. The fourth and final section of the genus in South America, *F. lycioides* Andr., the only member of sect. *Kierschlegeria*, is unique in the family in being a summer-deciduous shrub; it is confined to central Chile (Atsatt and Rundel, 1982). Whether its tetraploidy supports its geographical distribution and other evidence in indicating a relationship to sect. *Quelusia* is uncertain, but possible.

Of the four small sections of *Fuchsia* that occur as native plants in Mexico and Central America, all are endemic, and two – sect. *Encliandra* (Breedlove, 1969; Arroyo and Raven, 1975) and *Schufia* – are wholly or partly dioecious or subdioecious. The four sections of Mexico and Central America are so distinctive that it is difficult to relate them to their South American relatives, and impossible on the basis of present evidence to rule out the possibility that the genus may actually have originated in North America, although this does seem less likely than a South American origin for it, for reasons we have discussed briefly above.

Fuchsia also occurs in the Old World, where it is represented by the four species of sect. *Skinnera*, three in New Zealand and one in Tahiti. Although these species are diverse, including a large tree, a scandent shrub, and a creeping, almost herbaceous endemic of northern New Zealand, they are certainly closely related to one another. Exactly how they are related to the sections found in the New World is uncertain; they have been separated from them for more than 25 million years, as we have seen. Since Tahiti is no more than a few million years old, the ancestors of the single species there clearly must have reached the island during this period of time; but the exact relationship of the Tahitian species, which has hermaphroditic flowers, to the gynodioecious and dioecious ones of New Zealand, is disputed.

The patterns of distribution in the remainder of the family are less complex than those in *Ludwigia*, *Epilobium* and *Fuchsia*, and can thus be dealt with more briefly. Among the groups that have many generalized features, *Circaea* (seven species; Boufford, 1983) has spread around the Northern Hemisphere, and has its greatest concentration of species in eastern Asia at the present time. A majority of plant genera that occur in the temperate forests of the Northern Hemisphere are better represented in eastern Asia than elsewhere at present; the fossil record indicates that this is because of the presence of more equable conditions, favorable for survival, than because they necessarily originated there. One other group of Onagraceae, *Epilobium* sect. *Chamaenerion*, with seven species, is also best represented in Eurasia; as in *Circaea*, all the species occur there, and two also occur in North America. *Epilobium* as a whole clearly has spread easily and been dispersed many times between North America and Eurasia.

A similar pattern that now concerns only one Asian species concerns the temperate North American groups of *Ludwigia*, sects. *Microcarpium* and *Dantia*; *Ludwigia ovalis* Miq., the only species of sect. *Miquelia*, which occurs at scattered stations in Japan, Korea, and China, is directly related to the North American groups. In addition, *Ludwigia palustris* (L.) Ell., a member of sect. *Dantia*, occurs in Eurasia as well as in North America, and locally in South America and Africa, where it may be introduced.

In summary, *Ludwigia*, *Epilobium*, *Circaea* and *Fuchsia* are the only genera of Onagraceae that occur in the Old World; each of them is known on the basis of definite fossil records to have been there since Paleogene time, i.e. for more than 25 million years. The remaining genera of Onagraceae with many relatively generalized features, *Lopezia* and *Hauya*, are confined to Mexico and Central America, which reasonably may be taken as an indication of great antiquity for the family in that region.

Finally, the tribe Onagreae, with nine genera and about 270 species, occurs primarily in western and central North America. Five of these genera – *Oenothera* (51 species), *Camissonia* (1 species), *Clarkia* (1 species), *Gayophytum* (2 species, 1 common to North America) and *Gaura* (1 species, possibly introduced by man, – also occur in South America. All of them clearly reached South America from North America, judged by the relationships of the species involved, and none, including *Oenothera* (Dietrich, 1978), appears to have been in South America for more than a few million years, judged from the relationships of the species involved. Although the South American species of *Camissonia* and *Clarkia*, and one of the two South American species of *Gayophytum*, are considered to be endemics, they are closely related to individual North American species, and sometimes separable from them only with difficulty. They exhibit the characteristic disjunction in range of mediterranean-climate adapted plants between California and Chile, and their disjunct ranges have almost certainly come about as a result of long-distance dispersal within the last million years (Raven, 1963; Seavey and Raven, 1977). A similar pattern appears to be characteristic of *Boisduvalia* (tribe Epilobieae), with six species, one of which is endemic in Chile and Argentina and one additional species that is common to North and South America. The other genera of Onagreae are restricted to North America, and have regional patterns of distribution consistent with those of many other plant and animal groups.

4.6 REPRODUCTIVE BIOLOGY

As modified from figures summarized a few years ago, about 296 of the approximately 652 species of Onagraceae are modally outcrossed (Raven, 1979) – about 45% of the total. Notable among these are the 102 species of *Fuchsia*, all of which are outcrossed, even though they are self-compatible and self-pollination occurs in some species under certain circumstances. Among the larger genera of the family, about two-thirds of the species of *Lopezia*, *Clarkia* and *Gaura* and about one-third of the species of *Camissonia* and *Oenothera* are outcrossed. Only about 25 (15%) of the 162 species of *Epilobium* are modally outcrossed; very extensive evolution of largely self-pollinating species has occured within sect. *Epilobium*. In some sense, therefore, *Fuchsia*, which consists exclusively of outcrossing species, mostly shrubs, and *Epilobium*, which consists mostly of self-pollinating herbs, stand at opposite poles in this respect. Notwithstanding this difference, both genera have produced large numbers of species in relatively cool habitats, and hybridization between differentiated entities appears to have played an important role in the ability of both to respond rapidly to the demands of their respective changing environments.

In *Ludwigia*, *Circaea* and many species of *Lopezia* and Onagreae, pollinators such as butterflies, flies, and polylectic bees play the predominant role in pollination. More specialized bee pollination, involving many species of oligolectic bees, is characteristic of *Clarkia* and *Camissonia* (summary in Raven, 1979), and *Epilobium* sect. *Chamaenerion*, together with a few species of sect. *Epilobium*, has characteristic bumblebee (*Bombus*)-pollinated flowers.

One of the recurrent themes of pollination within Onagraceae is the frequency of hummingbird-pollinated flowers. These occur in nearly all species of *Fuchsia*, with the New Zealand and Tahitian species pollinated by native birds and some small-flowered Mexican and Central American species apparently pollinated by bees and flies. Hummingbird flowers also characterize the most primitive species of *Lopezia*, and have originated again within the genus from fly-pollinated ancestors. The single species of *Xylonagra* is hummingbird-pollinated, as is one subspecies of *Oenothera* and the two species of *Epilobium* sect. *Zauschneria*. In most of these instances, the shift to hummingbird pollination appears to have been associated with drastic changes in the morphology of the flowers involved, so that the hummingbird-pollinated species of *Epilobium* and *Lopezia* have often been assigned to different genera, thus overemphasizing some of their characteristics and tending to conceal their overall relationships.

Hawkmoth pollination in Onagraceae is characteristic of most of the outcrossing species of *Oenothera*, although three species of the genus in which the flowers open in the morning are bee-pollinated, and one is apparently pollinated by noctuids and other small moths. Hawkmoth pollination likewise is characteristic of *Hauya* and some species of its possible relatives *Calylophus* and *Gaura*; most of the outcrossing species of the latter genus are pollinated by small moths.

In summary, there have apparently been about 22 shifts in pollination system during the course of evolution of the outcrossing species of Onagraceae, but the switch from outcrossing to self-pollination has been much more frequent. This is the case in flowering plants generally. With the exception of *Fuchsia* and the small genera *Hauya*, *Xylonagra* and *Stenosiphon*, the switch to self-pollination has been a characteristic feature of the evolution of all genera of the family, and has usually accompanied the origin of distinct species within the respective groups involved, often with dramatic shifts in their morphology. In highly autogamous species of Onagraceae, for example, there has been a tendency not only for a great reduction in flower size, but also for a loss of flower parts (Raven, 1979, p. 590). No taxon of Onagraceae is wholly cleistogamous, although a few species are frequently so.

The derivation of the species formerly regarded as the monotypic genus *Heterogaura* from species probably assignable to *Clarkia* sect. *Peripetasma* affords a spectacular example of the sorts of alteration in morphology that can accompany an evolutionary shift to autogamy. Using an analysis of restriction-site variation in chloroplast DNA, Systsma and Gottlieb (1986a) showed that *Heterogaura* was more similar to *Clarkia dudleyana* Abrams in the overall set of mutations in its chloroplast DNA (based on an analysis of the 119 mutations that they identified in a set of 605 restriction sites), than either was to any other species. These results confirm morphological and chromosomal similarities in indicating the derivation of the single species formerly assigned to *Heterogaura* from *Clarkia*, and suggest that the two genera should be merged. Similar approaches have clarified the relationships between other species of *Clarkia* sect. *Peripetasma*, of which *C. dudleyana* is a member (Systsma and Gottlieb, 1986b). Macromolecular analyses have likewise shed much light both on relationships and on the course of evolution among other groups of *Clarkia* (e.g. Gottlieb, 1974, 1977; Gottlieb and Weeden, 1979).

4.7 CHROMOSOMAL EVOLUTION

The available information about the chromosomes of Onagraceae is listed in Table 4.2. The most striking feature about the chromosomes of the family is their morphological diversity (Kurabayashi, Lewis and Raven, 1962). Briefly, three distinctive kinds of chromosomes are represented in Onagraceae:

1. Medium-sized to large chromosomes that contract more or less evenly along their entire length during the course of mitosis, and which are not visible in interphase by normal staining methods. Chromosomes of this sort are characteristic of *Fuchsia*, *Circaea* and *Lopezia*, and, of course, are found in most kinds of plants.
2. Dot-like chromosomes that remain heteropycnotic throughout the mitotic cycle, and which, in favorable material, can be counted even in mitotic interphase. Such chromosomes are characteristic of *Ludwigia*, *Epilobium* and *Boisduvalia*.
3. Medium-size chromosomes that are heteropycnotic near the centromere, and have relatively long, lightly staining euchromatic 'tails' that contract during mitosis. Chromosomes of this sort are found in *Hauya* and the nine genera of Onagreae.

The correlation of what is obviously the least-specialized chromosome morphology with a basic chromosome number of $n = 11$ (in *Fuchsia*, *Circaea*, and some species of *Lopezia*, as well as in *Gongylocarpus*, which is in many respects the least specialized genus of the tribe Onagreae) unam-

Table 4.2   Cytology of Onagraceae.

	Total species	Known cytologically	Basic chromosome number	Diploid*	Polyploid†	Aneuploid	Habit‡
1. *Ludwigia*	82	80	$x = 8$	27	53	0	P,WP,A
2. *Fuchsia*	102	77	$x = 11$	60(1)	1(2)	0	T,S,P(2)
3. *Circaea*	7	7	$x = 11$	7	0	0	P
4. *Lopezia*	22	20	$x = 11$	17(1)	2(1)	17§	WP,A
5. *Hauya*	2	2	$x = 10$	1	1	0	T
6. *Gongylocarpus*	2	2	$x = 11$	2	0	0	P,A
7. *Gayophytum*	9	9	$x = 7$	6	3	0	A
8. *Xylonagra*	1	1	$x = 7$	1	0	0	S
9. *Camissonia*	61	59	$x = 7$	38(5)	16(5)	0(1)‖	P,A
10. *Calylophus*	6	6	$x = 7$	5(1)	0(1)	0	P(A)
11. *Gaura*	21	21	$x = 7$	18(1)	2(1)	0	P,A
12. *Oenothera*	123	118	$x = 7$	108(5)	5(5)	0	P,A
13. *Stenosiphon*	1	1	$x = 7$	1	0	0	P
14. *Clarkia*	44	44	$x = 7$	33	10	32¶	A
15. *Epilobium*	163	138	$x = 9$	0	138	7**	WP,P,A
16. *Boisduvalia*	6	6	$x = 10$	3	3	4††	A
Total	652	591	—	327(14)	234(15)	60(1)	—

* Partly diploid, partly polyploid species listed in parentheses.
† Partly polyploid, partly diploid species listed in parentheses.
‡ T = tree, S = shrub, WP = woody perennial, P = perennial, A = annual.
§ Including 2 polyploids based on aneuploid diploids.
‖ Aneuploid plants in some populations of the South American *Camissonia dentata* (Cav.) Reiche only.
¶ Including 9 polyploids based on aneuploid diploids.
** All are polyploids based on hypothetical aneuploid diploids.
†† Including 3 polyploids based on aneuploid diploids.

biguously indicates that the original basic chromosome number for Onagraceae is $n = 11$. However, since *Ludwigia* is the only member of one of the two major evolutionary lines in the family, and has a basic chromosome number of $x = 8$ heteropycnotic chromosomes, it must be assumed that it acquired its distinctive chromosome number and morphology early in the course of its evolution.

One question concerns the possibility of a relationship between *Ludwigia* and Epilobieae. Not only are the chromosomes of these two groups similar in some respects, their pollen is very difficult to distinguish. Many species of both groups shed their mature pollen as tetrads (Skvarla, Raven and Praglowski, 1975), and the individual grains also resemble one another in their morphological detail – so much so that the genera are difficult to distinguish in the fossil record (Daghlian *et al.*, 1984). Nevertheless, the basic floral morphology of Epilobieae is like that of the genera of Onagraceae other than *Ludwigia*, and such morphology seems clearly to represent a synapomorphy. Since that is the case – it would be exceedingly difficult to construct a reasonable hypothesis by virtue of which *Epilobium* would be derived from *Ludwigia* – it clearly seems necessary to assume that the chromosome morphology of *Ludwigia* was derived independently from that of *Epilobium*, and from a starting point that resembled the chromosomes of *Fuchsia* or *Circaea* both in morphology and in number. Furthermore, if *Ludwigia* sect. *Oligospermum* (including sect. *Oocarpon*), which consists of 5-merous species that shed their pollen singly, represents the sister group of other species of *Ludwigia*, as seems probable on the basis of its overall features (cf. Eyde, 1977, 1979, 1981), then pollen tetrads have been derived within *Ludwigia* and cannot be used as a basis of comparison between that genus and *Epilobium*, in which the habit of shedding the mature pollen as tetrads is clearly a primitive feature (Raven, 1976).

The chromosome morphology of the nine genera of Onagreae is distinctive within the family, and is correlated with the occurrence of reciprocal translocations as a regular part of the adaptive system of these genera. *Hauya* apparently has a similar chromosome morphology, although more detailed investigations, which would reveal the structure of the different kinds of chromosomes that occur in Onagraceae, would be highly desirable. At any event, reciprocal translocations do occur frequently in natural populations of all genera of Onagreae, and in no other members of the family. The heteropycnotic portions of the chromosomes of Onagreae, near the centromeres, are apparently not normally recombined during meiosis; but the occurrence of transcribed genes in these regions needs to be investigated. Most genes that have been studied in *Oenothera* occur in the more distal euchromatic regions of the chromosomes.

Aneuploidy has clearly been involved in the evolution of genera of Onagraceae, with a reduction from $n = 11$ to $n = 8$ in the origin of *Ludwigia* and a reduction from $n = 11$ to $n = 10$ in the origin of *Hauya*. *Circaea* and some species of *Lopezia* and Onagreae have retained the original basic chromosome number of $n = 11$. The pattern of chromosomal evolution in Epilobieae will be discussed subsequently.

Within Onagreae, *Gongylocarpus* has $n = 11$, whereas the other genera except *Clarkia* have $x = 7$ as the only basic chromosome number. The occurrence of $x = 7$ in these seven genera is somewhat puzzling, as discussed above, because they do not seem necessarily to be related directly to one another. For example, *Calylophus* and *Gaura* display some embryological features in common with *Hauya*, such as the divided sporogenous tissue in their anthers, in which thick, parenchymatous partitions occur (Tobe and Raven, 1986) and also in features of their integuments (Tobe and Raven, 1985). If these genera are really directly related to *Hauya*, then it would be difficult to imagine how they could also be related to the other genera of

Onagreae; formal cladistic analysis, and the development of additional macromolecular evidence, will apparently be necessary to resolve this problem.

The relationships of the other genera of Onagreae to one another are likewise uncertain, and need further work. Whether their common possession of $x = 7$ represents a synapomorphy, or instead has resulted from parallel evolution, cannot be determined with the available information. Within the genus *Clarkia*, which has no clear relatives, there is extensive aneuploidy, so that species with $n = 5, 6, 7, 8, 9$, and multiples of these numbers occur. Nevertheless, the original basic chromosome number in the genus appears to have been $n = 7$ (Lewis, 1953a,b, Lewis and Lewis, 1955), with both ascending and descending aneuploid changes in chromosome number. The distribution of the different chromosome numbers among the increasingly well-understood species of *Clarkia* generally has served to substantiate this conclusion, but that only deepens the mystery of why $x = 7$ should be ubiquitous in the Onagreae except for *Gongylocarpus*. The former monotypic genus *Heterogaura*, which has $n = 9$, is best regarded as a derived species of *Clarkia*, a conclusion that will be discussed in more detail below.

Chromosomal relationships in Epilobieae are also difficult to interpret. Table 4.3 enumerates the sections and some of the species of this tribe, with range and chromosome number indicated. All species of *Epilobium* are perennial, with the exception of the two species of sect. *Crossostigma* and the single species of sect. *Xerolobium*, whereas all six species of *Boisduvalia*

Table 4.3 Some features of the sections of Epilobieae.

	Chromosome number $(n =)$	Number of species	Geographical distribution
1. *Epilobium*			
A. sect. *Cordylophorum*	15	3	Interior W. United States
B. sect. *Xerolobium*	12	1	W. United States
C. sect. *Zauschneria*	15,30	2	W. United States, N.W. Mexico
D. sect. *Crossostigma*	16,13	2	W. United States
E. sect. *Epilobium*	18	147	Cosmopolitan
F. sect. *Chamaenerion*	18,36,54	7	N. Hemisphere
2. *Boisduvalia*			
A. sect. *Boisduvalia*	10,9,19	4	N. and S. America
B. sect. *Currania*	15	2	N. and S. America

are annual. The remaining species of *Epilobium* regularly flower during their first year, but keep growing and become perennial.

How did the impressive array of chromosome numbers characteristic of Epilobieae originate? In their habit and other features, the members of sects. *Cordylophorum* and *Zauschneria*, with $n = 15$, seem most generalized, and *Epilobium brachycarpum* Presl (formerly known as *E. paniculatum* Nutt. ex Torr. and A. Gray), the only species of sect. *Xerolobium*, seems clearly to be directly related to the members of sect. *Cordylophorum*, from which it differs primarily in being an annual. The relationships of the two annual species of sect. *Crossostigma*, one with $n = 16$ and the other with $n = 13$, are problematical, as are the relationships of any of these groups to sect. *Epilobium* or sect. *Chamaenerion*, or to one another. It seems highly improbable that the two species of *Boisduvalia* with $n = 15$ are related to the species of *Epilobium* with $n = 15$, and the relationships of the species of *Boisduvalia* with $n = 9$ and $n = 10$ are likewise problematical, although the single South American endemic species of the genus, *B. subulata* (Ruiz and Pav.) Raimann, has $n = 19$, a chromosome number almost certainly derived by polyploidy from $n = 9$ and $n = 10$.

Stebbins (1971, p. 193) has hypothesized that the original basic chromosome number may have been $n = 9$, with the basic number in sects. *Epilobium* and *Chamaenerion* ($x = 18$) having been derived by polyploidy. For the origin of other species with $n = 15, 12, 16$, and 13, it is necessary to hypothesize either aneuploid changes linking these chromosomes numbers directly, or descending aneuploidy from $n = 9$ to $n = 6$, and then $n = 12$, 13, 15, and 16 having originated from these postulated now-extinct ancestors by polyploidy (Stebbins, 1971, p. 193; Raven, 1976). Certainly the very close relationship between the two species of sect. *Crossostigma*, with $n = 13$ and $n = 16$ respectively, would seem to argue for the existence of extinct diploid relatives in the group. It is difficult to specify these relationships further, or to understand the possible linkages between the two very distinct sections of *Boisduvalia*, which could certainly have originated independently as annual derivatives from the *Epilobium* complex, as mentioned above, in the absence of formal cladistic analysis and macromolecular evidence, but it should be possible ultimately to resolve these relationships, and consequently the pattern of chromosomal evolution among the groups involved, by these means. The morphological and chromosomal features are so few that it does not appear possible adequately to assess the possible interrelationships within Epilobieae – for example, the possible independent derivation of the two sections of *Boisduvalia* from *Epilobium* – without macromolecular evidence. Here there is a clear case of the importance of modern methodology in elucidating phylogeny, but also one of the importance of comprehensive prior studies to allow the proper formulation of the questions.

Aneuploidy as a regular feature of evolution within genera is best developed in *Clarkia*, as we have seen, and in *Lopezia*. Within *Lopezia*, a complex of four bird-pollinated, somewhat shrubby species with $n = 11$ constitute the sections *Diplandra* (in which both anthers are still functional) and *Jehlia* (with only one functional anther, as in the rest of the genus) and present the most generalized array of characteristics (Plitmann, Raven and Breedlove, 1973; Eyde and Morgan, 1973). In the remainder of the genus, in which fly pollination combined with a snapping mechanism in the flower predominates, there has been an aneuploid reduction of chromosome number in various lines from $n = 10$ to $n = 9, 8,$ and 7, partly correlated, as it is in *Clarkia* and Epilobieae, with the annual habit (Plitmann, Raven and Breedlove, 1975). *Lopezia* sect. *Semeiandra*, which consists of one shrubby, bird-pollinated species and one self-pollinated annual, linked by the obvious synapomorphy of a vertical partition within the flower, has $n = 9$, and seems to have reverted to bird-pollination from fly-pollinated ancestors.

As shown in Table 4.2, polyploidy is characteristic of about two-thirds of the species of *Ludwigia*, where it is a regular feature of the evolution of a number of sections. The highest chromosome number in Onagraceae, $n = 64$ (16-ploid), occurs in some populations of *Ludwigia peruviana* (L.) Hara, a widespread and polymorphic species with at least four different chromosome numbers. Polyploidy is also characteristic of about a quarter of the species of *Clarkia*, where it may play an adaptive role similar to and reinforcing that of aneuploidy, and in *Camissonia*, where aneuploidy is not known to occur. Both polyploidy and aneuploidy seem to play a role in the genera of Onagraceae in which they occur similar to their demonstrated role in other groups of plants, namely in reducing recombination in entities that occur in situations marginal to the range of the genus as a whole. Polyploid species are likewise known in *Fuchsia*, in which two sections and a few other species are exclusively polyploid; *Hauya*, *Gayophytum*, *Gaura*, *Lopezia*, *Boisduvalia*, and *Epilobium* sect. *Chamaenerion*. Only four of the 125 species of *Oenothera* are polyploid, with an additional three species including some polyploid populations; neither polyploidy nor aneuploidy therefore appears to have played an important role in the evolution of that genus. All species of *Epilobium* might be regarded as polyploid, of course, but the list above is restricted to polyploids of obvious derivation within the group involved.

When modern methods are applied to the interpretation of the three (or more) distinctive patterns of chromosomal organization and morphology in Onagraceae, it seems certain that significant results will be forthcoming. These may then provide some indication of the meaning of the changes in chromosome number that have occurred in different parts of the family; for example, the reasons that polyploidy predominates in some groups, and aneuploidy in others, need to be explored in detail.

4.8 PATTERNS OF EVOLUTION IN ONAGRACEAE

In summary, therefore, what can be said about the patterns of evolution in different parts of the family Onagraceae? Can the differing modes of speciation that occur be related to features of the environment or history of the groups involved, or is the sample too small to make such comparisons meaningful? Let us examine this pattern from several different points of view.

First, does the pattern of evolution in the tribe Onagreae, which is the only group of the family in which reciprocal translocations occur as a regular feature of meiosis in naturally occurring populations of all outcrossing species, differ in its mode from other tribes? Not surprisingly, reciprocal translocations (and sometimes chromosomal interchanges also) differentiate species of the different genera of Onagreae, just as they differentiate populations and even individuals in some of genera of this tribe. Such reciprocal translocations are exceedingly rare in wild individuals of other genera, but chromosomal rearrangement has occurred in the evolution of species of *Circaea* (Seavey and Boufford, 1983) and *Boisduvalia* sect. *Boisduvalia* (species with $n = 9$ and $n = 10$; Seavey, 1977). In *Epilobium*, translocations occur rarely, but different chromosomal types tend to characterize whole series of species, whose phylogeny can therefore be traced by studying their distribution (e.g. Seavey and Raven, 1978). In contrast, reciprocal translocations and interchanges have not been observed to occur in hybrids between species of *Ludwigia*, *Fuchsia*, and *Lopezia*.

The reasons for these differences are not evident, but they are not obviously associated with any differences in mode of speciation or evolutionary pattern. *Clarkia* is the only genus of the family in which the classical, 1930s 'biosystematic' species definition may be considered partly applicable. Individual species of *Clarkia* generally consist of individuals and populations that are interfertile, whereas the species that are recognized are generally separated by strong sterility barriers, involving either a failure to form hybrids or the production of highly infertile F_1 individuals. In all other genera of the family, as far as known, the barriers that separate species are essentially the same as those that separate populations within the species; many species that are highly differentiated morphologically, for example, form fertile hybrids in *Fuchsia*, *Ludwigia*, and *Epilobium*.

It is not clear why the pattern of barriers that result in the maintenance and apparently also the origin of species in *Clarkia* apparently differ from those in the rest of the family, but perhaps the reciprocal translocations themselves play a role in this process. Not enough is known about the chromosomal differences that exist within and between species of *Camissonia*, another sizeable (61 species, versus 44 in *Clarkia*) genus of Onagreae that occurs in the same area, to permit a direct comparison, but about a quarter of the species of each genus are polyploid. Of the genera of Onagreae, aneuploidy is found only in *Clarkia*, and may indicate a greater degree of chromosomal lability there than in the others; but the problem of comparing so few examples and attempting to draw generalities from them is a very difficult one. Taking the three large genera of Onagreae – *Camissonia* (61 species), *Clarkia* (44 species) and *Oenothera* (125 species) – one notes that the proportion of self-pollinating species in *Oenothera* is about 57% and that in *Camissonia* about 59%, but that only about 39% of the species of *Clarkia* are self-pollinated. This difference is presumably correlated with the lack of an advantage for chromosomal rearrangements in isolating adaptive genetic combinations in basically self-pollinating populations, but the case certainly cannot be demonstrated rigorously.

The occurrence of complex structural heterozygosity in *Oenothera* (many species), *Gaura* (two species), *Gayophytum* (one species) and *Calylophus* (one species) is certainly an interesting evolutionary phenomenon, highly correlated with self-pollination and the establishment of relatively uniform entities with particular genetic constitutions that are well adapted for their individual areas; but it does not appear to have played any distinctive role in the pattern of speciation of these groups, beyond that of self-pollination in general. The major exception to this generality appears to be *Oenothera* sect. *Oenothera* (including *Euoenothera*), where most of the species are complex structural heterozygotes, and complex plastid-compatibility interactions also affect the viability of particular hybrid combinations.

When the larger genera of Onagraceae are compared in terms of their patterns of speciation, one finds that the woody genus *Fuchsia* is entirely outcrossing and *Clarkia* about 60% outcrossing, whereas about two-thirds of the species of *Oenothera* and *Camissonia* and about three-quarters of the species of *Ludwigia* are modally self-pollinating. In fact, about a quarter of the species of the entire family Onagraceae – more than 150 species – are self-pollinating members of these three large genera. Polyploidy is frequent in *Ludwigia* and *Camissonia*, and, as might have been expected, relatively rare in the outcrossing *Fuchsia* and the largely outcrossing *Clarkia*. Faced with these relationships, one must conclude that despite the obvious importance of particular features in the evolution of individual genera of Onagraceae, the ways in which the individual genera have evolved, their histories, and their present-day habitats are so diverse that a further analysis of these factors must depend on comparisons outside the family rather than comparisons within. Nevertheless, the detailed base of knowledge concerning Onagraceae will form a valuable resource to use in such comparisons, for an improved understanding of variation and evolution in plants in the future. For the 20th century, Ledyard Stebbins has played the most central role in the elucidation of these factors.

4.9 REFERENCES

Arroyo, M. T. K. and Raven, P. H. (1975) The evolution of subdioecy in morphologically gynodioecious species of *Fuchsia* sect. *Encliandra* (Onagraceae). *Evolution*, 29, 500–11.

Atsatt, P. R. and Rundel, P. W. (1982) Pollinator maintenance vs. fruit production: Partitioned reproductive effort in subdioecious *Fuchsia lycioides*. *Ann. Missouri Bot. Gard.*, 69, 199–208.

Berry, P. E. (1982) The systematics and evolution of *Fuchsia* sect. *Fuchsia* (Onagraceae). *Ann. Missouri Bot. Gard.*, 69, 1–198.

Berry, P. E. (1985) The systematics of the apetalous fuchsias of South America, *Fuchsia* sect. *Hemsleyella* (Onagraceae). *Ann. Missouri Bot. Gard.*, 72, 213–51.

Boufford, D. E. (1983) The systematics and evolution of *Circaea* (Onagraceae). *Ann. Missouri Bot. Gard.*, 69, 804–994.

Breedlove, D. E. (1969) The systematics of *Fuchsia* section *Encliandra* (Onagraceae). *Univ. Calif. Publ. Bot.*, 53, 1–69.

Breedlove, D. E., Berry, P. E. and Raven, P. H. (1982) The Mexican and Central American species of *Fuchsia* (Onagraceae) except for sect. *Encliandra*. *Ann. Missouri Bot. Gard.*, 69, 209–34.

Carlquist, S. (1961) *Comparative Plant Anatomy*. Holt, Rinehart, and Winston, New York.

Carlquist, S. (1975) Wood anatomy of Onagraceae, with notes on alternative modes of photosynthate movement in dicotyledon woods. *Ann. Missouri Bot. Gard.*, 62, 386–424.

Carlquist, S. (1977) Anatomy of Onagraceae: Additional species and concepts. *Ann. Missouri Bot. Gard.*, 64, 627–37.

Carlquist, S. (1982) Wood anatomy of Onagraceae: Further species; root anatomy; significance of vestured pits and allied structures in dicotyledons. *Ann. Missouri Bot. Gard.*, 69, 755–69.

Cronquist, A. (1984) A commentary on the definition of the order Myrtales. *Ann. Missouri Bot. Gard.*, 71, 780–2.

Daghlian, C. P., Raven, P. H., Skvarla, J. J. and Pocknall, D. T. (1984) *Epilobium* pollen from Oligocene sediments in New Zealand. *NZ J. Bot.*, 22, 285–94.

Dahlgren, R. and Thorne, R. F. (1984) The order Myrtales: Circumscription, variation, and relationships. *Ann. Missouri Bot. Gard.*, 71, 633–99.

Dietrich, W. (1978) The South American species of *Oenothera* sect. *Oenothera* (*Raimannia, Renneria*; Onagraceae). *Ann. Missouri Bot. Gard.*, 64, 425–626.

Drugg, W. S. (1967) Palynology of the Upper Moreno Formation (Late Cretaceous-Paleocene), Escarpado Canyon, California. *Palaeontographica Abteilung B*, 120, 1–71.

Eyde, R. H. (1977) Reproductive structures and evolution in *Ludwigia* (Onagraceae). I. Androecium, placentation, merism. *Ann. Missouri Bot. Gard.*, 64, 644–55.

Eyde, R. H. (1979) Reproductive structures and evolution in *Ludwigia* (Onagraceae). II. Fruit and seed. *Ann. Missouri Bot. Gard.*, 66, 656–75.

Eyde, R. H. (1981) Reproductive structures and evolution in *Ludwigia* (Onagraceae). III. Vasculature, nectaries, conclusions. *Ann. Missouri Bot. Gard.*, **68**, 470–503.

Eyde, R. H. and Morgan, J. T. (1973) Floral structure and evolution in Lopezieae. *Am. J. Bot.*, **60**, 771–87.

Gottlieb, L. D. (1974) Genetic confirmation of the origin of *Clarkia lingulata*. *Evolution*, **28**, 244–50.

Gottlieb, L. D. (1977) Evidence for duplication and divergence of the structural gene for phosphoglucose isomerase in diploid species of *Clarkia. Genetics*, **86**, 289–307.

Gottlieb, L. D. and Weeden, N. F. (1979) Gene duplication and phylogeny in *Clarkia. Evolution*, **33**, 1024–39.

Gulliver, G. (1864) On Onagraceae and Hydrocharitaceae as elucidating the value of raphides as natural characters. *J. Bot.*, **2**, 68–70.

Johnson, L. A. S. and Briggs, B. G. (1984) Myrtales and Myrtaceae – A phylogenetic analysis. *Ann. Missouri Bot. Gard.*, **71**, 700–56.

Kurabayashi, M., Lewis, H. and Raven, P. H. (1962) A comparative study of mitosis in Onagraceae. *Am. J. Bot.*, **49**, 1003–26.

Lewis, H. (1953a) The mechanism of evolution in the genus *Clarkia. Evolution*, **7**, 1–20.

Lewis, H. (1953b) Chromosome phylogeny and habitat preference in *Clarkia. Evolution*, **7**, 102–9.

Lewis, H. and Lewis, M. E. (1955) The genus *Clarkia. Univ. Calif. Publ. Bot.*, **10**, 241–392.

Martin, P. G. and Dowd, J. M. (1986) Phylogenetic studies using protein sequences within the order Myrtales. *Ann. Missouri Bot. Gard.*, **73**, 441–8.

Metcalf, C. R. and Chalk, L. (1950) *Anatomy of the Dicotyledons.* 2 vols. Clarendon Press, Oxford.

Mildenhall, D. C. (1980) New Zealand Late Cretaceous and Cenozoic plant biogeography: A contribution. *Palaeogeogr. Palaeoclimatol. Palaeoecol.*, **31**, 197–233.

Odrzykoski, I. J. and Gottlieb, L. D. (1984) Duplications of genes coding 6-phosphogluconate dehydrogenase in *Clarkia* (Onagraceae) and their phylogenetic implications. *Syst. Bot.*, **9**, 479–89.

Pares Regali, M. d. S., Uesugui, N. and Santos, A. d. S. (1974a) Palinologia dos sedimentos meso-cenozoicas do Brasil (I). *Bol. Tecn. Petrobras*, **17**, 177–9.

Pares Regali, M. d. S., Uesugui, N. and Santos, A. d. S. (1974b) Palinologia dos sedimentos meso-cenozoicas do Brasil (II). *Bol. Tecn. Petrobras*, **17**, 263–301.

Plitmann, U., Raven, P. H. and Breedlove, D. E. (1973) The systematics of Lopezieae (Onagraceae). *Ann. Missouri Bot. Gard.*, **60**, 478–563.

Plitmann, U., Raven, P. H., Tai, W. and Breedlove, D. E. (1975) Cytological studies in Lopezieae (Onagraceae). *Bot. Gaz.*, **136**, 322–32.

Raven, P. H. (1963) Amphitropical relationships in the floras of North and South America. *Quart. Rev. Biol.*, **38**, 151–77.

Raven, P. H. (1973) The evolution of subalpine and alpine plant groups in New Zealand. *NZ J. Bot.*, **11**, 177–200.

Raven, P. H. (1976) Generic and sectional delimitation in Onagraceae, tribe Epilobieae. *Ann. Missouri Bot. Gard.*, **63**, 326–40.

Raven, P. H. (1979) A survey of reproductive biology in Onagraceae. *NZ J. Bot.*, **17**, 575–93.

Raven, P. H. and Axelrod, D. I. (1974) Angiosperm biogeography and past continental movements. *Ann. Missouri Bot. Gard.*, **61**, 539–673.

Raven, P. H. and Raven, T. E. (1976) The genus *Epilobium* in Australasia: A systematic and evolutionary study. *NZ Dept. Sci. Ind. Res. Bull.*, **216**, 1–321.

Raven, P. H. and Tai, W. (1979) Observations of chromosomes in *Ludwigia* (Onagraceae). *Ann. Missouri Bot. Gard.*, **66**, 862–79.

Schmid, R. (1984) Reproductive anatomy and morphology of Myrtales in relation to systematics. *Ann. Missouri Bot. Gard.*, **71**, 832–5.

Seavey, S. R. (1977) Segregation of translocated chromosomes in *Epilobium* and *Boisduvalia* hybrids (Onagraceae). *Syst. Bot.*, **2**, 109–21.

Seavey, S. R. and Boufford, D. E. (1983) Observation of chromosomes in *Circaea* (Onagraceae). *Am. J. Bot.*, **70**, 1476–81.

Seavey, S. R. and Raven, P. H. (1977) Chromosomal differentiation and the sources of the South American species of *Epilobium* (Onagraceae). *J. Biogeogr.*, **4**, 55–9.

Seavey, S. R. and Raven, P. H. (1978) Chromosomal evolution in *Epilobium* sect. *Epilobium* (Onagraceae). III. *Pl. Syst. Evol.*, **130**, 79–83.

Skvarla, J. J. and Raven, P. H. (1976) Ultrastructural survey of Onagraceae pollen. in *The Evolutionary Significance of the Exine* (eds I. K. Ferguson, and J. Muller), Linnean Society of London, pp. 447–79.

Skvarla, J. J., Raven, P. H., Chissoe, W. F. and Sharp, M. (1978) An ultrastructural study of viscin threads in Onagraceae pollen. *Pollen et Spores*, **20**, 5–143.

Skvarla, J. J., Raven, P. H. and Praglowski, J. (1975) The evolution of pollen tetrads in Onagraceae. *Am. J. Bot.*, **62**, 6–35.

Solomon, J. C. (1982) The systematics and evolution of *Epilobium* (Onagraceae) in South America. *Ann. Missouri Bot. Gard.*, **69**, 239–335.

Stebbins, G. L. (1971) *Chromosomal Evolution in Higher Plants*. Addison-Wesley, Reading, MA.

Sussman, R. W. and Raven, P. H. (1978) Pollination by lemurs and marsupials: An archaic coevolutionary system. *Science*, **200**, 731–6.

Sytsma, K. J. and Gottlieb, L. D. (1986a) Chloroplast DNA evidence for the origin of the genus *Heterogaura* from a species of *Clarkia* (Onagraceae). *Proc. Natl. Acad. Sci. USA*, **83**, 5554–7.

Sytsma, K. J. and Gottlieb, L. D. (1986b) Chloroplast DNA evolution and phylogenetic relationships in *Clarkia* sect. *Peripetasma* (Onagraceae). *Evolution*, **40**, 1248–61.

Tobe, H. and Raven, P. H. (1983) An embryological analysis of Myrtales: its definition and characteristics. *Ann. Missouri Bot. Gard.*, **70**, 71–94.

Tobe, H. and Raven, P. H. (1985) The histogenesis and evolution of integuments in Onagraceae. *Ann. Missouri Bot. Gard.*, **72**, 451–68.

Tobe, H. and Raven, P. H. (1986) Evolution of polysporangiate anthers in Onagraceae. *Am. J. Bot.*, **73**, 475–88.

CHAPTER 5

Phylogenetic aspects of the evolution of self-pollination

ROBERT WYATT

5.1 INTRODUCTION

Phylogenetic reconstructions and classifications of flowering plants rely heavily on morphological changes in reproductive characters. For example, 44 of the 58 morphological characters used by Estabrook and Anderson (1978) in their analysis of *Crusea* were reproductive, and Young's (1981) study of primitive dicots employed 27 reproductive characters among the total of 41 studied. Furthermore, discussions of taxonomically useful characters in the angiosperms often seem to support the ontological primacy of features concerned with reproduction (e.g. Briggs and Walters' (1984) discussion of Gregor's (1938) data for *Plantago maritima*).

Stebbins (1970) suggested that the transition from predominant cross-pollination to predominant self-pollination is more widespread and common in the angiosperms than any other evolutionary change. Obligate cross-fertilization based on genetic self-incompatibility has been replaced by self-fertilization in numerous plant species (reviewed by Uphof, 1938; East, 1940; Lewis, 1942; Stebbins, 1950, 1957, 1958, 1970, 1971, 1974; Baker, 1953, 1959; Fryxell, 1957; Lewis and Crowe, 1958; Grant, 1958, 1963, 1971; Wyatt, 1983).

The morphological changes accompanying these shifts to self-pollination are remarkably similar in totally unrelated groups of plants (Table 5.1). Undoubtedly, the evolution of self-pollination is polyphyletic. This raises the possibility that data derived primarily from reproductive characteristics may confound phylogenetic reconstructions. I wish to evaluate the seriousness of this problem by considering two basic questions:

1. How consistent are the morphological changes that accompany a shift from predominant cross-pollination to predominant self-pollination?
2. Can criteria be formulated to recognize these changes, thereby minimizing potential problems brought about by their parallel or convergent evolution?

109

Table 5.1 List of 21 morphological and phenological character changes that are often associated with the evolution of autogamy (modified from Ornduff, 1969). Numbers in parentheses indicate those studies discussed in the text that reported each change (1, *Leavenworthia*; 2, *Arenaria*; 3, *Limnanthes*; 4, *Gilia*; 5, *Clarkia*; 6, *Plectritis*; 7, *Lycopersicon*; 8, *Eichhornia*).

Outcrossing progenitors	Autogamous derivatives
Flowers many	Flowers fewer (1,2,8)
Pedicels or peduncles long	Pedicels or peduncles shorter (1,2)
Sepals large	Sepals smaller (1,2,4)
Corollas rotate	Corollas funnelform, cylindric, or closed (1,2)
Petals large	Petals smaller (1,2,3,4,5,6,7,8)
Petals emarginate	Petals less emarginate (1)
Floral color pattern contrasting	Floral color pattern less contrasting (1,5)
Nectaries present	Nectaries reduced or absent (1,2,6)
Flowers scented	Flower scentless (1,6)
Nectar guides present	Nectar guides absent (1,6,8)
Anthers long	Anthers shorter (1,2,3,6,7)
Anthers extrorse	Anthers introrse (1)
Anthers distant from stigma	Anthers adjacent to stigma (1,2,3,4,5,6,7,8)
Pollen grains many	Pollen grains fewer (1,2,4,8)
Pollen presented	Pollen not presented (1,2)
Pistil long	Pistil shorter (1,2,3,4,5,6,7,8)
Stamens longer or shorter than pistil	Stamens equal to pistil (1,2,3,4,5,6,7,8)
Style exserted	Style included (1,2,3,4,5,6,7)
Stigmatic area well defined, pubescent	Stigmatic area poorly defined, less pubescent (1,2)
Stigma receptivity and anther dehiscence asynchronous	Stigma receptivity and anther dehiscence synchronous (1,2,3,4,5,6)
Many ovules per flower	Fewer ovules per flower (8)

5.2 PHYLOGENETIC RECONSTRUCTION IN *ARENARIA*

One of the best-documented cases of the evolution of self-pollination involves the winter annual species of *Arenaria* endemic to granite outcrops in the southeastern United States. Taxonomic revisions by two groups working independently in the early 1970s produced similar conclusions regarding these plants: *A. alabamensis* (McCormick, Bozeman, and Spongberg) Wyatt is a self-pollinating derivative of *A. uniflora* (Walt.) Muhl., a more widespread, cross-pollinating species (Weaver, 1970; McCormick, Bozeman and Spongberg, 1971; Wyatt, 1977). Originally described from only two localities in northeastern Alabama, *A. alabamensis* subsequently has been reported from outcrops in north-central Alabama, western Georgia, northern South Carolina, and southwestern North Carolina (Wyatt,

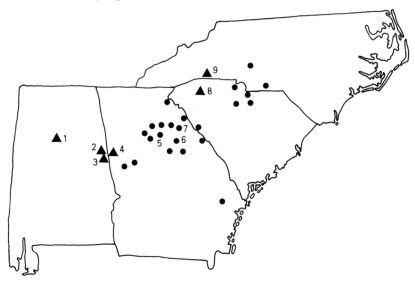

Fig. 5.1   Distribution of cross-pollinating (●) and self-pollinating (▲) populations of *Arenaria uniflora* in the southeastern United States. The self-pollinating populations have been called *Arenaria alabamensis*. Numbers 1–9 indicate field sites selected for intensive study.

1977, 1983, 1984a,b; Wyatt and Fowler, 1977; Fig. 5.1). In contrast, *A. uniflora* occurs from south-central North Carolina to eastern Alabama (Fig. 5.1). Cross-pollinating populations therefore occupy the larger out-crops in the center of the range of exposed rock habitats in the southeast, while self-pollinating ones are restricted to marginal sites (Wyatt, 1986). The two types never occur sympatrically.

Intensive study of nine populations (numbered 1–9 in Fig. 5.1) from throughout the range of these plants showed that all populations are self-compatible (Wyatt, 1984a). In some populations cross-pollination is favored by spatial and temporal separation of male and female functions. Furthermore, because few flowers are open at a time and stems of these exceptionally dense-growing plants (up to 28.8 flowering plants dm^{-2}, Wyatt, 1986) are intertwined, geitonogamy is probably low. Self-pollinating plants lack these spatial and temporal features and show high levels of autofertility under greenhouse conditions (Wyatt, 1984a).

As first pointed out by Weaver (1970) and McCormick, Bozeman and Spongberg (1971), the flowers of *A. uniflora* and *A. alabamensis* differ strikingly. Wyatt (1984a) measured 11 floral characters, including sepal length (SL) and width (SW), petal length (PL) and width (PW), lengths of the antisepalous (ASSL) and antipetalous (APSL) stamens and their associated anthers (ASAL and APAL, respectively), style (STL) and ovary length (OVL), and ovule number (OVNO). In general, these features divided the populations into two groups: populations 5, 6 and 7 from central Georgia were characterized by large petals, stamens and styles, whereas populations 1, 2, 4, 8 and 9 from Alabama and the Carolinas were small-flowered. There was very little overlap between the groups for characters such as petal length and width, stamen lengths and style length: the populations from different outcrop 'islands' were highly distinctive. The correlation structure for the 11 floral characters showed strong, mostly positive relationships.

Wyatt (1984a) also measured morphological variation in seven vegetative characters: rosette diameter (RD), stem number (STNO), stem length (STML), internode length (INTL), pedicel length (PEDL), leaf length (LL), and leaf width (LW). In contrast to the floral characters, the vegetative characters showed greater overlap in population means and less clear separation into groups. Rather than dividing populations on the basis of their breeding systems, characters such as rosette diameter and stem, internode, and pedicel lengths showed clinal patterns that correlated with geographical distance. There was a gradual decrease in size along a southwest–northeast axis (i.e., plants from sites 1, 2, and 3 were largest and plants from sites 8 and 9 were smallest). Similar to the floral characters, correlations among vegetative characters were strong and positive: the larger the plant, the larger its component parts.

Cluster analyses were performed using both the floral and vegetative data sets. The phenogram using floral characters (Fig. 5.2) suggests the presence of three reasonably distinct groups: (1) populations from central Georgia (sites 5, 6, and 7), (2) populations from central and eastern Alabama and South Carolina (sites 1, 2, 8), and (3) populations from eastern Alabama, western Georgia, and North Carolina (sites 3, 4, 9). These clusters reflect

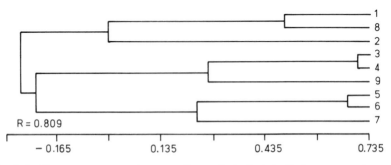

Fig. 5.2 Phenogram expressing overall morphological similarity among nine populations of *Arenaria uniflora* based on 11 floral characters. Population numbers correspond to those in Fig. 5.1.

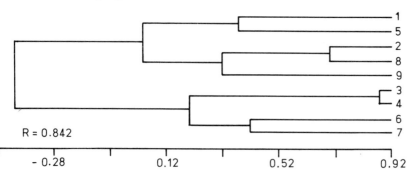

Fig. 5.3 Phenogram expressing overall morphological similarity among nine populations of *Arenaria uniflora* based on seven vegetative characters. Population numbers correspond to those in Fig. 5.1.

similarity in breeding system, rather than geographical proximity. In contrast, the phenogram using vegetative characters (Fig. 5.3) groups populations by their geographical proximity.

The pattern seen in vegetative characters is also evident in the results of experimental cross-pollinations (Wyatt, unpublished). A phenogram summarizing the levels of cross-compatibility in more than 4000 crosses between plants from different populations of *Arenaria* shows that plants from geographically contiguous sites are most cross-compatible (Fig. 5.4). For example, populations 8 and 9 from South Carolina and North Carolina, respectively, are highly compatible. Crossability does not seem to be associated with breeding system. Populations 5, 6 and 7 are equally as compatible with nearby populations as they are among themselves. In contrast, self-pollinating populations from Alabama and the Carolinas, which are similar in terms of floral morphology, appear to be genetically widely divergent on the basis of their low cross-compatibility. This suggests that self-pollinating

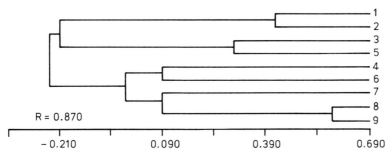

Fig. 5.4 Phenogram expressing overall levels of cross-compatibility among nine populations of *Arenaria uniflora* based on results of reciprocal cross-pollinations. Population numbers correspond to those in Fig. 5.1.

populations on granite outcrops near the geographical margins may have originated independently.

The hypothesis of independent origins can be tested by examining electrophoretically detectable genetic variation. Measures of genetic variability such as percentages of loci polymorphic and mean numbers of alleles per locus proved to be lower for self-pollinating populations from the margins of the range than for the central cross-pollinating populations (Wyatt, unpublished). Furthermore, self-pollinating populations in Alabama and the Carolinas were often fixed for different alleles that were polymorphic in cross-pollinating populations. This is reflected in a phenogram, resulting from a cluster analysis using Rogers's (1972) genetic similarities (Fig. 5.5). On the basis of genetic similarity at 14 putative gene loci, populations 1 and 2 from central and eastern Alabama, respectively, were allied with population 5 from Atlanta, Georgia. In contrast, population 9 from North Carolina was linked to population 6 from Loganville, Georgia, while population 8 from South Carolina was joined with population 7 from Pendergrass, Georgia. This suggests at least two, and possibly more, independent origins of self-pollinating derivatives from originally cross-pollinating populations on the large outcrops in central Georgia.

Wyatt (1983) hypothesized that similarity in floral features of populations in Alabama and the Carolinas was the result of convergent evolution driven by similar selective forces in the two areas. More recently, Wyatt (1986) showed that competition for pollinators may have stimulated the evolution of self-pollination in *Arenaria uniflora*. *Arenaria glabra* Michx., another winter annual endemic to rock outcrops in the southeast, is pollinated by the same syrphid flies and andrenid and halictid bees as *A. uniflora* but has a larger, showier floral display. Wherever *A. glabra* is sympatric with *A. uniflora* (which is only on the wetter outcrops at the margins of the range

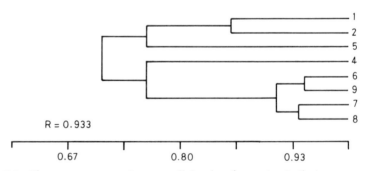

Fig. 5.5 Phenogram expressing overall levels of genetic similarity among eight populations of *Arenaria uniflora* based on Rogers' (1972) coefficient of genetic similarity using 14 putative gene loci. Population numbers correspond to those in Fig. 5.1.

in Alabama and the Carolinas), only self-pollinating plants of the latter are found.

To test the efficacy of cladistic analysis to resolve phylogenetic relationships within *A. uniflora*, I carried out tests using floral characters only, vegetative characters only, or a combined set of floral and vegetative characters. The resulting data matrix is shown in Table 5.2. Character compatibility analysis made use of the program CLIQUE (version 2.2) in Felsenstein's (1983) Phylogeny Inference Package (PHYLIP). Wagner parsimony analysis was done using Felsenstein's (1983) WAGNER program (version 2.3) and his PENNY program (version 2.5), which uses a branch-and-bound algorithm to find all of the most parsimonious trees.

Character compatibility analysis found the 11 floral characters to be highly compatible, forming a clique that included 10 of the 11 characters. Only ovule number, which Wyatt (1984a, 1984c) had noted earlier as a character with high intraplant variation, was excluded from this largest clique. The resulting tree (Fig. 5.6) is very similar to the phenogram based on floral characters (Fig. 5.2). Plants from the large-flowered populations in central Georgia are placed together as are the small-flowered, self-pollinating populations from Alabama and the Carolinas. Plants from site 3 in eastern Alabama constitute a third, intermediate group. This pattern is entirely consistent with current taxonomy of the group (Weaver, 1970; McCormick, Bozeman and Spongberg, 1971): the large-flowered populations (5, 6, 7) would be called *A. uniflora sensu stricto*, while the small-flowered populations (1, 2, 4, 8, 9) would be placed in *A. alabamensis*. On the basis of electrophoretically detectable genetic variation and cross-compatibility, however, we suspect that this phylogeny is incorrect. Floral characters yield a misleading phylogeny because the self-pollinating morphs were probably derived independently. The phylogeny resulting from character compatibility analysis is misleading.

Four largest cliques, all including 4 of the 7 vegetative characters, were found by character compatibility analysis of the data set for vegetative

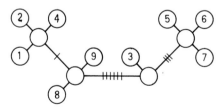

Fig. 5.6 Unrooted phylogenetic tree constructed by character compatibility analysis using 11 floral characters of *Arenaria uniflora*. Population numbers correspond to those in Fig. 5.1.

Table 5.2 The taxon × character state matrix used in the phylogenetic analyses of *Arenaria uniflora*. Population numbers correspond to those shown in Fig. 5.1. Abbreviations for the characters are explained in the text (p. 111). The first 11 characters are floral features; the last 7 are vegetative features.

Popu-lation	SL	SW	PL	PW	ASSL	APSL	ASAL	APAL	STL	OVL	OVNO	RD	STNO	STML	INTL	PEDL	LL	LW
1	1	1	1	1	1	1	1	1	1	1	1	0	0	0	0	0	0	1
2	1	1	1	1	1	1	1	1	1	1	0	1	0	0	1	0	0	1
3	0	0	0	1	0	1	0	0	1	0	0	1	0	1	1	0	0	0
4	1	1	1	1	1	1	1	1	1	1	1	1	1	1	1	0	0	1
5	0	0	0	0	0	0	0	0	0	0	1	1	0	0	1	0	1	1
6	0	0	0	0	0	0	0	0	0	0	0	1	0	0	1	1	0	1
7	0	0	0	0	0	0	0	0	0	0	0	1	0	0	1	0	0	1
8	1	1	1	1	1	1	1	1	1	1	0	0	1	1	1	1	1	1
9	1	0	1	1	1	1	1	1	1	1	1	1	1	1	1	1	1	1

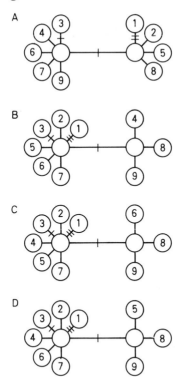

Fig. 5.7 Unrooted phylogenetic trees constructed by character compatibility analysis using seven vegetative characters of *Arenaria uniflora*. Population numbers correspond to those in Fig. 5.1.

features (Fig. 5.7). Rosette diameter, internode length, and leaf width were included in each of the four cliques. The small number of characters analyzed leaves the detailed structure of the trees largely unresolved. Nevertheless, it is apparent that the populations form different groups than those based on breeding systems. The trees, like the phenogram for vegetative characters (Fig. 5.3), suggest that self-pollinating populations from Alabama and the Carolinas are polyphyletic.

Character compatibility analysis applied to the combined data set (Table 5.2) found 13 of 18 characters to be compatible in the two largest cliques. The resulting trees are very similar, differing only in the placement of population 4 from western Georgia (Fig. 5.8). As might be expected, the larger number of floral than vegetative characters results in a tree that is more similar to that generated by floral characters alone (Fig. 5.6). This is compounded by the fact that the floral characters tend to be more compatible among themselves. Again, the phylogeny is consistent with current

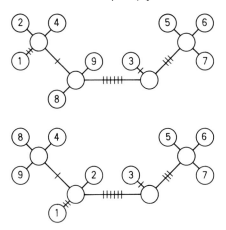

Fig. 5.8 Unrooted phylogenetic trees constructed by character compatibility analysis using a combined set of 18 floral and vegetative characters of *Arenaria uniflora*. Population numbers correspond to those in Fig. 5.1.

taxonomy but at odds with evidence concerning degrees of cross-compatibility and genetic similarity.

The branch-and-bound procedure discovered 20 trees with a minimum of 13 character changes in the 11 floral characters. Many of these were identical to trees found in multiple runs of the WAGNER program, including the tree shown in Fig. 5.9. This most parsimonious tree would suggest that populations 5, 6 and 7 from central Georgia are closely related and that populations 1, 2, 4, 8 and 9 from Alabama, western Georgia, and the Carolinas form a monophyletic group. The pattern is entirely consistent with that resulting from character compatibility analysis (Fig. 5.6) and that resulting from phenetic analysis (Fig. 5.2).

Using the seven vegetative characters, the branch-and-bound procedure found 20 trees with a shortest distance of 10 steps. The WAGNER program failed to find any trees with this minimum number of character state

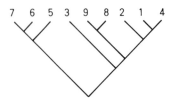

Fig. 5.9 Unrooted phylogenetic tree constructed by Wagner parsimony analysis using 11 floral characters of *Arenaria uniflora*. Population numbers correspond to those in Fig. 5.1.

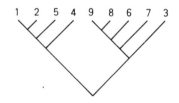

Fig. 5.10 Unrooted phylogenetic tree constructed by Wagner parsimony analysis using seven vegetative characters of *Arenaria uniflora*. Population numbers correspond to those in Fig. 5.1.

changes. The most parsimonious tree shown in Fig. 5.10 is remarkably similar to the phenogram in Fig. 5.5 based on genetic similarity. It is also consistent in broad outline with the phenogram based on vegetative characters (Fig. 5.3) and the trees resulting from character compatibility analysis (Fig. 5.7). It suggests that the self-pollinating populations in Alabama and western Georgia originated from outcrossing stock similar to population 5 from a site south of Atlanta, Georgia. In contrast, self-pollinating populations from South and North Carolina appear to be more closely related to outcrossing populations 6 and 7 from sites north and east of Atlanta, Georgia. This pattern is concordant with that based on crossability (Fig. 5.4), as well as genetic similarity (Fig. 5.5).

The branch-and-bound procedure found 12 trees requiring a minimum of 26 character state changes for the combined data set of 18 characters. The WAGNER program did not discover any trees with so few changes. Again, the tree shown in Fig. 5.11 is nearly identical to that produced using only floral characters (Fig. 5.9). It suggests that the self-pollinating populations form a monophyletic group distinct from the outcrossing populations in central Georgia.

It is apparent, therefore, that cladistics fails to discover what we suspect are the true phylogenetic relationships among the nine populations of *A. uniflora*. Both major approaches within cladistics, character compatibility and parsimony, constructed phylogenies that included groups which, on

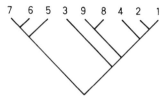

Fig. 5.11 Unrooted phylogenetic tree constructed by Wagner parsimony analysis using a combined set of 18 floral and vegetative characters of *Arenaria uniflora*. Population numbers correspond to those in Fig. 5.1.

the basis of other evidence, are polyphyletic. Convergent evolution in floral morphology has led to incorrect phylogenies.

I believe that similar problems arise commonly in cladistic analyses of angiosperms, since such studies typically place heavy emphasis on floral morphology, and changes in floral morphology associated with the evolution of self-pollination are common and widespread. Although the *Arenaria* example involved populations of a single, highly variable species, there are numerous examples of similar situations at higher taxonomic levels (e.g. Grant and Grant's (1965) studies of Polemoniaceae, Ornduff and Crovello's (1968) analyses of Limnanthaceae, and Eckenwalder and Barrett's (1986) research on Pontederiaceae).

5.3 PHYLOGENETIC STUDIES IN OTHER GROUPS

The evolution of autogamy in *Leavenworthia* (Cruciferae) constitutes one of the best-documented examples of the derivation of self-compatible and self-pollinating races and species from self-incompatible progenitors (Rollins, 1963; Lloyd, 1965, 1967; Solbrig, 1972; Solbrig and Rollins, 1977). Lloyd (1965) concluded that the evolution of selfing in *L. crassa* and *L. alabamica* was based on reproductive assurance. Populations of these species that invaded small cedar glades with shallow soil were selected to flower earlier, in order to set seed before the glades dried out. In some cases this led to very early flowering that preceded the emergence of the native bees that pollinate *Leavenworthia*. These drier sites were also less favorable for nesting by the bees, so that pollinator abundances were diminished. The paucity of bees selected for self-pollination in races of *L. crassa* and *L. alabamica* occupying drier sites. Lloyd's (1965) data suggested that the advantage of having a greater percentage of flowers pollinated at these sites more than compensated for decreases in the number and quality of seeds resulting from each self-pollination.

Rollins (1963) and Solbrig and Rollins (1977) reached similar conclusions for other species of the genus. They found that populations of the self-incompatible species were much larger than populations of the self-compatible species and that the self-compatible species typically grew in drier glades. Like Lloyd (1965), they concluded 'that the selective force that accounts for the switch from self-incompatibility to self-compatibility and the distribution of these two types is the timing of potential pollinator emergence in relation to the drying out of the glade.'

The loss of self-incompatibility in some of the races of *L. crassa* and *L. alabamica* has been accompanied by a whole complex of correlated changes in other characters. Among these Lloyd (1965) included an increase in autofertility, a change from extrorse to introrse anthers, and a decrease in the pollen—ovule ratio.

Solbrig (1972) commented that in *Leavenworthia* 'all seven species are very similar in their morphology, being almost indistinguishable in the vegetative state.' On the other hand, differences among the species and races in floral characteristics are often quite striking and associated with the breeding system. The evolution of self-pollination in *Leavenworthia* has clearly been polyphyletic (Rollins, 1963; Lloyd, 1965), prompting Lloyd (1965) to remark therefore that 'characters associated with the evolution of self-compatibility have a very limited use as indicators of relationships between the races of *L. crassa* and *L. alabamica*.' In fact, the primitive, self-incompatible races of the two different species are so similar that they can only be distinguished on the basis of fruit characters. The derived, self-compatible races likewise are identical in many features. Nevertheless, Lloyd (1965) was able to construct a phylogeny of the races using characters that were not associated with the change in the breeding system, such as leaf lobing and curvature. Had he not been so careful, he would surely have been misled by the striking parallel evolution in floral traits.

Limnanthes (Limnanthaceae) includes a number of North American annual herbs that have been the subject of detailed study with respect to their breeding systems (Mason, 1952; Ornduff and Crovello, 1968; Ornduff, 1969; Arroyo, 1973a,b, 1975; Jain, 1976, 1978; Brown and Jain, 1979; McNeill and Jain, 1983; Kesseli and Jain, 1984, 1985). As in *Leavenworthia*, reproductive assurance was advanced as the most likely explanation for the evolution of autogamy in *Limnanthes floccosa*. Arroyo (1973b, 1975) suggested that periodic droughts reduced population sizes of the usual bee pollinators, as well as the plants, and that even if cross-pollination occurred, inadequate soil moisture would prevent seed maturation. Autogamy was therefore advantageous in insuring reproduction and, because such plants were free to grow and flower earlier in the spring, in avoiding drought. Presumably, marginal populations of outcrossing *L. alba* served as the progenitors of *L. floccosa* (Arroyo, 1975). Unfortunately, Arroyo (1973a,b, 1975) presented no data bearing directly on the questions of pollinator scarcity and the effects of drought on seed abortion. More recently, Brown and Jain (1979) challenged Arroyo's (1975) observations and hypotheses regarding the role of drought.

Arroyo (1973b) summarized the major morphological differences between *L. alba* and *L. floccosa* by noting that the latter had shorter petals and pistils, produced less pollen, and held the anthers at the level of (rather than below) the stigma. The self-pollinating *L. floccosa* also was only weakly protandrous and showed high autofertility. It also lacked the fragrant flowers with well-developed nectaries and nectar guides, which characterized *L. alba*. Brown and Jain (1979) found that the two species differed significantly in 9 of 15 floral characters.

Arroyo (1973a) carried out a cluster analysis on 26 populations of

Limnanthes floccosa. On the basis of the resulting phenogram, she described five subspecies, differentiated primarily on the basis of degree of autogamy. Her groups cut across traditional taxonomic boundaries established by Mason (1952). Arroyo (1973b) explained this lack of concordance by stating that 'Mason (1952) emphasized pubescence differences, whereas I emphasize differences in floral morphology related to levels of autogamy as well as pubescence differences.' Brown and Jain's (1979) electrophoretic data showed that the different populations of *L. floccosa* and *L. alba* are genetically heterogeneous and suggested that the subspecies might be polyphyletic. Using a more complete data set, McNeill and Jain (1983) concluded that Arroyo's (1973b) subspecific phylogeny for *L. floccosa* was probably correct. They also argued, however, that the high degree of genetic divergence between *L. alba* and *L. floccosa* is inconsistent with the hypothesis that *L. floccosa* was recently derived from *L. alba*.

Earlier, Ornduff and Crovello (1968) had carried out a more extensive numerical taxonomic analysis of the entire Limnanthaceae. Their work detected 'parallelisms in the evolution of floral characters associated with increasing autogamy,' and they concluded that 'analyses emphasizing floral characters will place evolutionarily remotely related outcrossing OTUs [operational taxonomic units] together and unrelated, autogamous OTUs together'. Their analysis using 17 vegetative characters agreed better with Mason's (1952) phylogeny of the group than did an analysis using 18 reproductive characters or an analysis using all 35 characters. With respect to *L. floccosa* and *L. alba*, Ornduff and Crovello (1968) showed that *L. floccosa* var. *pumila* falls closer to the varieties of *L. alba*, rather than to Mason's (1952) other two varieties of *L. floccosa*, on the basis of vegetative characters. These findings raise further questions as to the validity of Arroyo's (1973b) conclusions regarding relationships among these taxa, since her analysis relied very heavily on floral characteristics known to be associated with breeding system shifts within the family.

Gilia achilleifolia (Polemoniaceae) is a diploid, entomophilous, annual herb endemic to hillsides and canyons in the central Coast Range of California. Grant (1954) postulated that the species is composed of xenogamous and autogamous races. Schoen (1977, 1982a, 1982b, 1983) showed that outcrossing rates differed between these two races and that autogamous populations are common in the northern portion of the range east of San Francisco. He suggested that a northward migration in the past had brought the species outside the range of its normal pollinators (beeflies and halictid and andrenid bees). Schoen (1982a) hypothesized that 'outside the range of normal pollinators and, in small, unattractive and newly founded populations, the more highly autogamous variants (i.e., those with low degrees of protandry) may have been favored over their comparatively xenogamous counterparts via fecundity selection.' Thus, as in *Leavenwor-*

thia and *Limnanthes*, the stimulus for the evolution of autogamy was paucity of pollinators.

The self-pollinating race differs from the cross-pollinating race in a number of features. Grant (1954) emphasized that the outcrossers occupied sunny hillsides and produced dense inflorescences of large, showy flowers, while the selfers occupied shady places and produced inflorescences with fewer, smaller flowers. Schoen (1977) quantified these differences, showing that plants of the self-pollinating race had smaller corollas, stigmas maturing below the anthers, low pollen production, low pollen–ovule ratios, homogamy, many pollen grains per stigma, and high seed-set. Further research showed that selfers were weakly, if at all, protandrous, had weakly exserted stigmas and low flower weights, and set fruit abundantly in the absence of pollinators (Schoen, 1982a).

Cluster analysis based on Nei's (1972) genetic distance showed two well-defined groups that corresponded to the northern, selfing populations and the southern, outcrossing populations (Schoen, 1982b). Although only seven populations were included in this survey, the phenogram strongly suggests that populations of the self-pollinating race are monophyletic. The unusually large genetic distance between the races further suggests that the two groups might deserve taxonomic rank as distinct species. Grant and Grant (1965) earlier accorded subspecific rank to the autogamous races as *G. achilleifolia* ssp. *multicaulis*.

Clarkia (Onagraceae) comprises about 43 annual species of western North and South America. All species of *Clarkia* are self-compatible, but the floral structure favors outcrossing, with bees being the usual pollinators (Lewis and Lewis, 1955). Self-pollination, however, has evolved at least 11 times in the genus (Moore and Lewis, 1965). It is believed that the stimulus for the evolution of self-pollination was 'reduction of the population to extremely small size during a growing season truncated by water stress, and that the genotypes surviving this catastrophic selection were those produced by the earliest flowering individuals. In a very small population genes promoting self-pollination would be at an advantage.' The importance of this mode of evolution in *Clarkia* has been discussed by Lewis (1953, 1962, 1966, 1973).

One species that has been investigated intensively in terms of the evolution of self-pollination is *Clarkia xantiana*. Over most of its range in the southern Sierra Nevada foothills of California, it has large lavender-pink flowers and is outcrossing because the mature style overarches the stamens and pollen is shed before the stigma becomes receptive (Moore and Lewis, 1965). Derived from this 'pink outcrosser' are two populations with small flowers that are normally self-pollinated. One of these, the 'pink selfer', has flowers the color of the outcrosser; the other, the 'white selfer', has white flowers. Moore and Lewis (1965) showed that means for petal length, petal width, style length, and anther height were approximately 2–3 times larger for the pink outcrosser than for the selfers.

On the basis of geographical and cytogenetic evidence, Moor and Lewis (1965) proposed that the white selfer was derived from the pink selfer and that, in turn, the pink selfer was derived from the sympatric population of pink outcrossers. To test this phylogeny, Gottlieb (1984) analyzed electrophoretic variation at 40 gene loci. He found that the pink selfer and the white selfer are identical at 32 loci and that the pink selfer was not more similar to allopatric populations of the pink outcrosser than to the sympatric one. He also concluded that the white selfer probably originated from the pink one and not independently, thus upholding Moore and Lewis's (1965) view.

5.4 GENERAL CONSIDERATIONS

Ornduff (1969), expanding upon Lloyd's (1965) list for *Leavenworthia*, noted 32 character changes that often accompany a shift from predominant outcrossing to predominant selfing. From these, I have extracted a shorter list of 21 strictly morphological or phenological characters (Table 5.1), from the studies described above and several others that have been reported by researchers. Among the characters that appear nearly always to change are petal size, anther size, anther distance from stigma, pistil length, stamen length relative to the pistil, style exsertion, and synchrony of stigma receptivity and anther dehiscence. Seed-set and fruit-set, which appear in Ornduff's (1969) list, are also usually increased in autogamous derivatives but can be viewed simply as direct consequences of the shift in breeding system (and often as the reason for the shift). Dichogamy (character 20) and herkogamy (13, 17, 18) are common means of preventing intrafloral selfing that are generally available to all hermaphroditic plants and may have a simple genetic basis (e.g. Gleeson, 1982). Selection to decrease flower life span (to reduce dichogamy) might also cause reduced development of petals, stamens, pistils, and distances separating anthers and stigmas. Many of these changes that are so frequent and widespread in autogamous taxa, therefore, may simply reflect developmental correlations among the component characters (Lord and Hill, 1987).

A number of characters rarely change, such as corolla orientation, petal margins, floral color pattern, nectar guides, and anther dehiscence (Table 5.1). Other characters, such as flower number, pedicel length, pollen presentation, and stigmatic surface, probably do change often but have seldom been examined. Although Ornduff (1969) listed a reduced number of ovules per flower as characteristic of selfers, only Barrett (1985) reported such a trend. Most researchers have found no consistent differences (e.g.

Lloyd, 1965; Wyatt, 1984d). Rollins (1963), however, reported an increase in ovule number in interspecific comparisons of *Leavenworthia*.

On the basis of his observations of *Leavenworthia*, Lloyd (1965) divided the character changes associated with the selection of autogamy into three groups. The first included loss of features associated with the attraction of pollinators, such as the decrease in petal lengths, the decrease in contrast between yellow and white portions of yellow-centered flowers, and the decrease in brightness of nectar guides. He speculated that the loss of adaptations to insect pollination could be viewed as saving on 'energetic costs' in the unfavorable habitats of the self-compatible races. Garnock-Jones (1976, 1981) reached similar conclusions with respect to autogamous subspecies and species of *Parahebe* (Scrophulariaceae) in New Zealand. Patterns of variation were continuous and correlated with mode of pollination in four characters principally concerned with floral display (peduncle length, pedicel length, corolla lobe width, anther length). These patterns were thought to result from selection for 'metabolic economies', counteracted strongly by selection for enhanced display in cross-pollinating populations, but less so in self-pollinating populations, which occurred in energy-limited, upland habitats.

A second category of changes included those which contributed positively to enhanced levels of self-pollination. These included the evolution of introrse anthers, the location of stigmas and anthers at the same height within the flower, and the closing of the petals in late afternoon. Decreases in anther lengths and pollen—ovule ratios were also viewed in this context, although the latter would appear to be more a result, rather than a cause, of increased self-pollination (cf. Wyatt, 1984d).

Lloyd's (1965) third category included those changes that appear to be simple responses to the specific conditions under which the self-compatible races grow. Among such shifts were decreases in the lengths of branches, numbers of flowers, silique lengths, and average seed weights. These were presumed to reflect the reduced growth potential of the self-compatible races on drier cedar glades.

Because he had studied a number of different races that appeared to have evolved to varying degrees toward autogamy, Lloyd (1965) was able to speculate about the order and rate of change of characters associated with this shift in the breeding system of *Leavenworthia*. He concluded that the evolution of these characters had not followed a completely constant pattern.

Unfortunately, the selective forces driving the evolution of autogamy (reviewed by Jain, 1976) are often unknown and the character state changes often are not fully documented (Table 5.1). Most studies to date have proposed reproductive assurance as the stimulus for the evolution of autogamy. For comparative purposes, it would be most enlightening to have additional examples in which other selective forces were responsible. It is not possible to determine at this time if the great similarity in morphological changes in widely divergent taxonomic groups is due to similar selective pressures.

It would also be most informative to learn if the morphological consequences of the evolution of autogamy depend on the mode of self-fertilization that is involved. Lloyd (1979) concluded that competing selfing is less often favored than prior selfing and that delayed selfing is most often advantageous. As he noted, however, many plants may not be able to select freely among these modes because of features of their floral biology. The resulting morphological changes in features related to removal of pollen by pollinators, such as flower number, pedicel length, and pollen presentation, may range from no change at all with delayed selfing to complete reduction or loss with prior selfing.

Lloyd's (1979) models for the evolution of self-fertilization are fully reversible. Most authors, however, have assumed that the evolution of autogamy is irreversible (Jain, 1976). Stebbins (1974), for example, has argued that 'self-fertilizing species are usually the ends of evolutionary lines, and rarely if ever contribute to major evolutionary trends.' Bull and Charnov (1985) discussed the evolution of complete selfing in a hermaphroditic population as an example of irreversible evolution. Like Lande and Schemske (1985), they reasoned that 'inbreeding depression is apparently due to largely recessive, deleterious genes maintained by mutation; if selfing begins to evolve, the frequencies of these deleterious genes are greatly reduced, inbreeding depression is lessened, and the advantage of selfing is intensified.' The dichotomy in breeding systems suggested by this view was supported by Schemske and Lande's (1985) tabulation of outcrossing rates for 55 plant species.

Aide (1986) argued, however, that this strongly bimodal pattern was artifactual and resulted from inclusion of a large number of studies of wind-pollinated grasses and pines. Waller (1986) also strongly criticized Schemske and Lande's (1985) tabulation and interpretation of the empirical data regarding outcrossing rates in plants. He further attacked Lande and Schemske's (1985) model, questioning whether disruptive selection for self-fertilization necessarily occurs. In fact, Uyenoyama (1986) showed that models incorporating relatedness among mating pairs can lead to stable, intermediate rates of outcrossing. She concluded that 'development of experimental programs directed toward the detection and estimation of frequency-dependent costs and benefits associated with selfing would permit an informed assessment of the possible adaptive mechanisms for the maintenance of mixed mating systems.'

The likelihood of reversibility of an evolutionary trend to self-pollination depends in part on the degree to which self-pollination predominates. Stebbins (1957) noted that 'no species of plant has been studied extensively throughout its entire range of distribution and has been found to be exclusively self pollinated everywhere.' For example, Adams and Allard (1982) found that *Festuca microstachys*, a nearly obligate selfer, still maintained considerable genetic variation and achieved outcrossing rates as high as 6.7%. Barrett and Shore (1987) discussed several examples of apparent re-evolution of outcrossing mechanisms in derived homostylous varieties of *Turnera ulmifolia*. Direct tests of the assumption of irreversibility, involving artificial selection experiments to increase and decrease outcrossing rates, would be most useful in determining the likelihood of such reversals. Such studies would also reveal which characters change in association with a shift in the breeding system. Artificial selection experiments attempting to derive self-pollinating plants from an originally cross-pollinating population and to derive cross-pollinating plants from originally self-pollinating stock are currently in progress using *Arenaria* (Wyatt and Fone, unpublished).

5.5 CONCLUSION

Are there any general criteria to avoid making errors in phylogenetic reconstructions due to parallel or convergent evolution in floral characteristics? First, if all the species under consideration have identical breeding systems and pollination syndromes, there is likely to be no problem in using floral characters. On the other hand, if there are breeding system or pollinator transitions within the group under study, caution should be used in including floral characters. Perhaps the best approach would be one similar to that used above in the analysis of *Arenaria* or that used by Ornduff and Crovello (1968) in their analysis of Limnanthaceae. In those studies characters were divided into subsets consisting of vegetative characters only and reproductive characters only. These subsets were then examined carefully for congruence. Of course, one must also be aware of the possibility of parallel or convergent evolution affecting vegetative characters. For example, Eckenwalder and Barrett (1986) concluded that vegetative characters of the Pontederiaceae 'provide a poor measure of evolution in the character suite as a whole . . . since vegetative characters in aquatic plants are often evolutionarily labile and highly homoplasious.' Interestingly, their phylogenetic trees failed to align homostylous derivatives with their presumed tristylous ancestors in *Eichhornia*, suggesting instead that the homostyles form a monophyletic group.

Finally, as I attempted to do in each of the cases discussed above, it is useful to test critical predictions of the phylogeny using evidence from multiple sources. Most useful in this regard are data regarding cytological differences, chemical differences, cross-compatibility, and genetic distance data from proteins and DNA.

5.6 ACKNOWLEDGEMENTS

I thank Ann Stoneburner for valuable field assistance and discussions, David M. Lane and Derek Holcomb for help with morphological measurements, and Leslie D. Gottlieb for encouraging me to think about the subject of this chapter. I also thank Spencer C. H. Barrett, Leslie D. Gottlieb, and Jeffrey P. Hill for comments on an earlier draft of the manuscript. Support for my research on *Arenaria* was provided by NSF Grant No. DEB-8001986.

5.7 REFERENCES

Adams, W. T. and Allard, R. W. (1982) Mating system variation in *Festuca microstachys*. *Evolution*, **36**, 591–5.

Aide, T. M. (1986) The influence of wind and animal pollination on variation in outcrossing rates. *Evolution*, **40**, 434–5.

Arroyo, M. T. K. de (1973a) A taximetric study of infraspecific variation in autogamous *Limnanthes floccosa* (Limnanthaceae). *Brittonia*, **25**, 177–91.

Arroyo, M. T. K. de (1973b) Chiasma frequency evidence on the evolution of autogamy in *Limnanthes floccosa* (Limnanthaceae). *Evolution*, **27**, 679–88.

Arroyo, M. T. K. de (1975) Electrophoretic studies of genetic variation in natural populations of allogamous *Limnanthes alba* and autogamous *Limnanthes floccosa* (Limnanthaceae). *Heredity*, **35**, 153–64.

Baker, H. G. (1953) Race formation and reproductive method in flowering plants. *Symp. Soc. Exp. Biol.*, **7**, 114–31.

Baker, H. G. (1959) Reproductive methods as factors in speciation in flowering plants. *Cold Spring Harbor Symp. Quant. Biol.*, **24**, 177–91.

Barrett, S. C. H. (1985) Ecological genetics of breakdown in tristyly. in *Structure and Functioning of Plant Populations. II. Phenotypic and Genotypic Variation in Plant Populations* (eds J. Halck and J. W. Woldendorp), North-Holland, Amsterdam.

Barrett, S. C. H. and Shore, J. S. (1987) Variation and evolution of breeding systems in the *Turnera ulmifolia* L. complex (Turneraceae). *Evolution*, **41**, 340–54.

Briggs, D. and Walters, S. M. (1984) *Plant Variation and Evolution*, 2nd edn, Cambridge University Press, Cambridge, UK.

Brown, C. R. and Jain, S. K. (1979) Reproductive system and pattern of genetic variation in two *Limnanthes* species. *Theor. Appl. Genet.*, **54**, 181–90.

Bull, J. J. and Charnov, E. L. (1985) On irreversible evolution. *Evolution*, **39**, 1149–55.

East, E. M. (1940) The distribution of self-sterility in the flowering plants. *Proc. Am. Phil. Soc.*, **82**, 449–518.

Eckenwalder, J. E. and Barrett, S. C. H. (1986) Phylogenetic systematics of Pontederiaceae. *Syst. Bot.*, **11**, 373–91.

Estabrook, G. F. and Anderson, W. R. (1978) An estimate of phylogenetic relationships within the genus *Crusea* (Rubiaceae) using character compatibility analysis. *Syst. Bot.*, **3**, 179–96.

Felsenstein, J. (1983) *Phylogeny Inference Package*, Version 2.3. Seattle, WA.

Fryxell, P. A. (1957) Mode of reproduction in higher plants. *Bot. Rev.* (Lancaster), **23**, 135–233.

Garnock-Jones, P. J. (1976) Breeding systems and pollination in New Zealand *Parahebe* (Scrophulariaceae). *NZ J. Bot.*, **14**, 291–8.

Garnock-Jones, P. J. (1981) Change of adaptations from entomophily to autogamy in *Parahebe linifolia*. *Pl. Syst. Evol.*, **137**, 195–201.

Gleeson, S. K. (1982) Heterodichogamy in walnuts: Inheritance and stable ratios. *Evolution*, **36**, 892–902.

Gottlieb, L. D. (1984) Electrophoretic analysis of the phylogeny of the self-pollinating populations of *Clarkia xantiana*. *Pl. Syst. Evol.*, **147**, 91–102.

Grant, V. (1954) Genetic and taxonomic studies in *Gilia*. IV. *Gilia achilleaefolia*. *Aliso*, **3**, 1–18.

Grant, V. (1958) The regulation of recombination in plants. *Cold Spring Harbor Symp. Quant. Biol.*, **23**, 337–63.

Grant, V. (1963) *The Origin of Adaptations*. Columbia University Press, New York.

Grant, V. (1971) *Plant Speciation*. Columbia University Press, New York.

Grant, V. and Grant, K. A. (1965) *Flower Pollination in the Phlox Family*. Columbia University Press, New York.

Gregor, J. W. (1938) Experimental taxonomy. 2. Initial population differentiation in *Plantago maritima* in Britain. *New Phytol.*, **37**, 15–49.

Jain, S. K. (1976) The evolution of inbreeding in plants. *Ann. Rev. Ecol. Syst.*, **7**, 469–95.

Jain, S. K. (1978) Breeding system in *Limnanthes alba*: Several alternative measures. *Am. J. Bot.*, **65**, 272–5.

Kesseli, R. V. and Jain, S. K. (1984) New variation and biosystematic patterns detected by allozyme and morphological comparisons in *Limnanthes* sect. *Reflexae* (Limnanthaceae). *Pl. Syst. Evol.*, **147**, 133–65.

Kesseli, R. V. and Jain, S. K. (1985) Breeding systems and population structure in *Limnanthes*. *Theor. Appl. Genet.*, **71**, 292–9.

Lande, R. and Schemske, D. W. (1985) The evolution of self-fertilization and inbreeding depression in plants. I. Genetic models. *Evolution*, **39**, 24–40.

Lewis, D. (1942) The evolution of sex in flowering plants. *Biol. Rev. Cambridge Philos. Soc.*, **17**, 46–7.

Lewis, D. and Crowe, L. K. (1958) Unilateral interspecific incompatibility in flowering plants. *Heredity*, **12**, 233–56.

Lewis, H. (1953) The mechanism of evolution in *Clarkia*. *Evolution*, **7**, 1–20.

Lewis, H. (1962) Catastrophic selection as a factor in speciation. *Evolution*, **16**, 257–71.

Lewis, H. (1966) Speciation in flowering plants. *Science*, **152**, 167–72.

Lewis, H. (1973) The origin of diploid neospecies in *Clarkia*. *Am. Naturalist*, **107**, 161–70.

Lewis, H. and Lewis, M. E. (1955) The genus *Clarkia*. *Univ. Calif. Publ. Bot.*, **20**, 241–392.

Lloyd, D. G. (1965) Evolution of self-compatibility and racial differentiation in *Leavenworthia* (Cruciferae). *Contr. Gray Herb.*, **195**, 3–134.

Lloyd, D. G. (1967) The genetics of self-incompatibility in *Leavenworthia crassa* Rollins (Cruciferae). *Genetica*, **38**, 227–42.

Lloyd, D. G. (1979) Some reproductive factors affecting the selection of self-fertilization in plants. *Am. Naturalist*, **113**, 67–79.

Lord, E. M. and Hill, J. P. (1987) Evidence for heterochrony in the evolution of plant form. in *Development as an Evolutionary Process*. Alan R. Liss, Inc.

McCormick, J. F., Bozeman, J. R. and Spongberg, S. (1971) A taxonomic revision of granite outcrop species of *Minuartia* (*Arenaria*). *Brittonia*, **23**, 149–60.

McNeill, C. I. and Jain, S. K. (1983) Genetic differentiation and phylogenetic inference in the plant genus *Limnanthes* L. section *Inflexae*. *Theor. Appl. Genet.*, **66**, 257–69.

Mason, C. T. (1952) A systematic study of the genus *Limnanthes*. *Univ. Calif. Publ. Bot.*, **25**, 455–512.

Moore, D. M. and Lewis, H. (1965) The evolution of self-pollination in *Clarkia xantiana*. *Evolution*, **19**, 104–14.

Nei, M. (1972) Genetic distance between populations. *Am. Naturalist*, **106**, 283–92.

Ornduff, R. (1969) Reproductive biology in relation to systematics. *Taxon*, **18**, 121–33.

Ornduff, R. and Crovello, T. J. (1968) Numerical taxonomy of Limnanthaceae. *Am. J. Bot.*, **55**, 173–82.

Rogers, J. S. (1972) Measures of genetic similarity and genetic distance. *Univ. Texas Publ.*, **7213**, 145–53.

Rollins, R. C. (1963) The evolution and systematics of *Leavenworthia* (Cruciferae). *Contr. Gray Herb.*, **192**, 3–198.

Schemske, D. W. and Lande, R. (1985) The evolution of self-fertilization and inbreeding depression in plants. II. Empirical observations. *Evolution*, **39**, 41–52.

Schoen, D. J. (1977) Morphological, phenological, and pollen-distribution evidence of autogamy and xenogamy in *Gilia achilleifolia* (Polemoniaceae). *Syst. Bot.*, **2**, 280–6.

Schoen, D. J. (1982a) The breeding system of *Gilia achilleifolia*: Variation in floral characteristics and outcrossing rate. *Evolution*, **36**, 352–60.

Schoen, D. J. (1982b) Genetic variation and the breeding system of *Gilia achilleifolia*. *Evolution*, **36**, 361–70.

Schoen, D. J. (1983) Relative fitnesses of selfed and outcrossed progeny in *Gilia achilleifolia* (Polemoniaceae). *Evolution*, **37**, 292–301.

Solbrig, O. T. (1972) Breeding system and genetic variation in *Leavenworthia*. *Evolution*, **26**, 155–60.

Solbrig, O. T. and Rollins, R. C. (1977) The evolution of autogamy in species of the mustard genus *Leavenworthia*. *Evolution*, **31**, 265–81.

Stebbins, G. L. (1950) *Variation and Evolution in Plants*. Columbia University Press, New York.

Stebbins, G. L. (1957) Self-fertilization and population variability in the higher plants. *Am. Naturalist*, **41**, 337–54.

Stebbins, G. L. (1958) Longevity, habitat, and release of genetic variability in higher plants. *Cold Spring Harbor Symp. Quant. Biol.*, **23**, 365–78.

Stebbins, G. L. (1970) Adaptive radiation in angiosperms. I. Pollination mechanisms. *Ann. Rev. Ecol. Syst.*, **1**, 307–26.

Stebbins, G. L. (1971) *Chromosomal Evolution in Higher Plants*. Arnold, London.

Stebbins, G. L. (1974) *Flowering Plants: Evolution Above the Species Level*. Belknap Press, Cambridge, MA.

Uphof, J. C. T. (1938) Cleistogamous flowers. *Bot. Rev. (Lancaster)*, **4**, 21–49.

Uyenoyama, M. (1986) Inbreeding and the cost of meiosis: The evolution of selfing in populations practicing biparental inbreeding. *Evolution*, **40**, 388–404.

Waller, D. M. (1986) Is there disruptive selection for self-fertilization? *Am. Naturalist*, **128**, 421–6.

Weaver, R. E. (1970) The *Arenarias* of the southeastern granitic flat-rocks. *Bull. Torrey Bot. Club*, **97**, 40–52.

Wyatt, R. (1977) *Arenaria alabamensis*: A new combination for a granite outcrop endemic from North Carolina and Alabama. *Bull. Torrey Bot. Club*, **104**, 243–4.

Wyatt, R. (1983) Pollinator–plant interactions and the evolution of breeding systems. in *Pollination Biology* (ed. L. Real), Academic Press, New York.

Wyatt, R. (1984a) The evolution of self-pollination in granite outcrop species of *Arenaria*. I. Morphological correlates. *Evolution*, **38**, 804–16.

Wyatt, R. (1984b) The evolution of self-pollination in granite outcrop species of *Arenaria*. IV. Correlated changes in the gynoecium. *Am. J. Bot.*, **71**, 1006–14.

Wyatt, R. (1984c) Intraspecific variation in seed morphology of *Arenaria uniflora* (Caryophyllaceae). *Syst. Bot.*, **9**, 423–31.

Wyatt, R. (1984d) Evolution of self-pollination in granite outcrop species of *Arenaria* (Caryophyllaceae). III. Reproductive effort and pollen-ovule ratios. *Syst. Bot.*, **9**, 432–40.

Wyatt, R. (1986) Ecology and evolution of self-pollination in *Arenaria uniflora* (Caryophyllaceae). *J. Ecol.*, **74**, 403–18.

Wyatt, R. and Fowler, N. (1977) The vascular flora and vegetation of the North Carolina granite outcrops. *Bull. Torrey Bot. Club*, **104**, 245–53.

Young, D. A. (1981) Are the angiosperms primitively vesselless? *Syst. Bot.*, **6**, 313–30.

CHAPTER 6

Evolution of mating systems in cultivated plants

CHARLES M. RICK

6.1 INTRODUCTION

In the course of plant domestication and improvement, many genetic changes have been wrought by selection and other techniques to transform wildlings to biotypes that are adapted for cultivation and useful productivity. During this process the mating system of various cultigens has been altered either intentionally or unintentionally. In investigations of the tomato species we have encountered much evidence of changes in mating systems both intra- and interspecifically. This chapter will discuss primarily modifications during and prior to domestication of the cultivated tomato (*Lycopersicon esculentum*) and other crop species. Various aspects of this subject have been treated by Frankel and Galun (1977), de Nettancourt (1977), Simmonds (1979), and others.

Many aspects of the evolution of domesticates differ from those of wild species, even from their respective ancestral taxa. At least in modern terms, cultivated plants live in a much more highly regimented and controlled system. One of the major differences is the extreme genetic uniformity sought and often achieved by plant breeders. Another feature, often misunderstood, is the vastly higher harvest index (proportion of total plant weight constituted by the harvested part) in cultigens. As in the majority of instances (the context of this chapter), the part harvested consists of seeds or fruit, the enhancement of which leads to severe drains on the plant carbohydrate sink. The stresses resulting from this intense drain often render the plant more susceptible to disease, arthropod attack, and adverse effects of the environment. It is therefore no surprise that our modern cultivars require the sheltered environment of cultivation for their best performance or even survival. Consequently, populations of modern cultivars could rarely survive more than one or a very few generations if left to fare for themselves.

133

Other impacts of domestication deal with modifications of the breeding system. In seed-propagated crops, the desired uniformity can be achieved in various ways, the simplest being enforced autogamy, but also in the F_1 hybrid system, by virtue of the pure-line nature of parental lines and of high homoeostasis of the hybrid progeny. The demands of modern agriculture lay increasing demands on uniformity to render cultivation more efficient, to meet market standards, to facilitate harvest, and to provide the crop at the correct stage of maturity for such exacting operations as freezing and canning, in which the tolerances of variable maturity are exceedingly low. Little wonder then that the desiderata for survival differ vastly between wild and cultivation situations.

The genus *Lycopersicon* and closely related species of *Solanum* comprise a wide diversity of mating systems. Table 6.1 classifies these species according to mating system, as determined by the work of many investigators. In several of these species, the situation is dynamic in the respect that the individual species does not breed in a uniform manner throughout its range. As an example *L. pimpinellifolium* (considered by some authorities as an ancestor of the cultivated tomato), although all forms are capable of self-pollinating, is outcrossed to the extent of 40% in the center of its elongated distribution, the level decreasing gradually through intermediate territory, reaching values approaching zero at the north and south periphery. These clines are associated with decrease in flower size, extent of stigma exsertion, and genetic variability as measured by allozymes – indices of polymorphy and percentage heterozygosity (Rick, Fobes and Holle, 1977). A similar pattern is found in *L. hirsutum*, except that the central, highly variable populations are self-incompatible (SI), in contrast to the marginal populations, which are self-compatible (SC) and virtually monomorphic. These differences are also associated with similar changes in flower size and structure (Rick, Fobes and Tanksley, 1979). In the more exclusively SI species *L. pennellii* and *L. peruvianum*, SC is rare and found only in the periphery. These are examples of the evolution of self-fertilization as derived from ancestral allogamous forms as successful pioneering elements, conforming to patterns delineated and assessed by Stebbins (1957).

Table 6.1 Tomato species listed according to mating system.

Autogamous: *L. cheesmanii, esculentum, parviflorum*
Facultative
   Self-compatible: *L. chmielewskii, pimpinellifolium*
   Self-compatible/incompatible: *L. hirsutum, pennellii, peruvianum*
Allogamous: (entirely self-incompatible): *L. chilense, S. juglandifolium, lycopersicoides, ochranthum, rickii*

6.2 INTENSIFIED INBREEDING IN NATURALLY SELF-POLLINATED SPECIES

The cultivated tomato, as known earlier in Europe and North America, was remarkably depauperate in genetic variability. Although offering investigators variation in fruit size, shape and color and in leaf shape and plant habit, these differences are relatively simply inherited and of relatively limited use. Attempts to select for early maturity, increased yield, and pest resistance seldom succeeded. The rate of progress manifest in yield statistics (USDA, 1922–1980) shows remarkably little improvement prior to the 1940s. Diminution of genetic variation in the early tomato stocks is also demonstrated by a survey of allozymes (Rick and Fobes, 1975): more than 80% of the tested European and North American cultivars are monomorphic for the same genotype at all tested loci. In contrast, the rate of progress since 1940 has been phenomenal, yield per acre of processing tomatoes having increased fourfold. Whilst better cultural methods – improved use of fertilizers, planting distances, pest control, etc. – have no doubt affected performance, comparisons of vintage cultivars in demonstration plots clearly reveal a major genetic input.

Disease resistance ranks high among the many facets of recent genetic improvement of tomatoes. Inherited disease resistance not only benefits production by preventing decreased yields and production of inferior harvests, but also permits production in infested areas where the crop could otherwise not be grown. In the history of tomato breeding, no strong effective resistance was known to major diseases until the discovery in *L. pimpinellifolium* by Bohn and Tucker (1940) of a single dominant gene I for fusarium wilt resistance, followed by introduction of a resistant cultivar by Porte and Wellman (1941). From that time research in this specialty has accelerated until the present, when resistance to at least thirty important diseases has been detected in wild *Lycopersicon* spp., and resistance to at least 16 has been bred into important commercial cultivars (survey by Rick, *et al.*, 1987). In some cases a single gene confers resistance to a broad spectrum of pathogenic races and species, as in the case of *Mi* for resistance to root-knot nematodes and Tm-2^2 for resistance to tobacco mosaic virus. Resistance to as many as six diseases has been combined in single, pure-breeding cultivars and up to 10–12 in F_1 hybrid cultivars. Genes thus introduced have also improved yield, facilitated mechanized harvest, and improved various factors affecting fruit quality, as well as other attributes of the crop. The tomato may therefore have benefited from such exploitation of wild relatives to a greater extent than any other cultivated plant.

It is to the circumstances leading to such depletion of genetic variation in the cultivated tomato that I would like to call attention. I conjecture that a combination of factors accounts for this depletion. The wild form and generally accepted immediate predecessor of the cultivated tomato is the cherry tomato, *L. esculentum* var. *cerasiforme*, which is self-fertile and normally largely self-pollinated throughout its range. According to all available evidence – archeological, ethnographic, linguistic, genetic, and other – domestication took place in Mesoamerica (Jenkins, 1948). Introduction of the domesticated forms to southern Europe took place shortly after the conquest of Mexico (Matthiolus, 1544), and a return of the cultivated forms to North America took place in the 18th and 19th centuries. Undoubtedly through all of these migrations genetic variability was diminished by founder events and natural and artificial selection. Nearly all Mesoamerican *cerasiforme* and early European and North American cultivars are identical in allozyme genotype (Rick and Fobes, 1975). New allozymic variants appear in some of the more recently bred cultivars, but some of these (ex. $Aps-1^1$) can be traced to interspecific hybridizations made to incorporate desirable genes not present in older cultivars. Thus with genetic diversity at a near-zero level, the older tomato cultivars contrast sharply with related wild *Lycopersicon* spp. whose variability is many magnitudes greater (Rick, 1984).

Concomitant with these developments, a progressive change in flower form can be traced (Rick, 1976). For many reasons it is likely that the primordial *Lycopersicon* stock was SI, a condition prevailing in extant *L. chilense, hirsutum, pennellii* and *peruvianum* as well as in the closely allied *Solanum* spp. (Table 6.1) (Rick, 1982). Characteristic of these species is a strongly exserted stigma, obviously positioned to facilitate the outcrossing enforced by SI. Stigmas of many accessions of var. *cerasiforme* and older Mexican cultivars are also exserted to varying degrees. The older European cultivars (as well as the tomato flowers figured in early herbals) have stigmas either exserted or positioned at the mouth of the anther tube, doubtless because automatic self-pollination is essential to good fruit set, particularly in lines intended for greenhouse production. The same condition prevails in the older North American cultivars, probably because they descended directly from the older European stocks. A more recent development has been a further foreshortening of the style to place the stigma within the anther tube several mm below its orifice. This change further improves fruiting by guaranteeing self-pollination, and by the same token prohibits outcrossing (Rick and Dempsey, 1969). This relationship was demonstrated by a relatively high negative correlation between stigma level and fruitfulness in the F_2 of a cross between parents with high and low stigma levels. In the evolution of the cultivated tomato, the stigma has thus progressively shifted from a well-exserted position through intermediate positions to its present inserted level with drastic effects on the pollination system, grading from obligate outcrossing to strict autogamy.

It is noteworthy that a somewhat similar evolution has occurred in the domestication of the eggplant, of which the probable immediate ancestor, the allogamous *S. incanum*, has strongly exserted stigmas, in contrast to stigmas at about the same level as anther tips in the cultivated, largely self-pollinated *S. melongena* (D. Zohary, personal communication). Since these events transpired in Asia, probably India, they must have occurred independently of the changes in *Lycopersicon*. Other examples of transitions from semi-outcrossed SC forms to largely selfed cultigens are found in the domestication of *Capsicum* peppers, flax, and others (Simmonds, 1979).

Although intentionally or unintentionally selected, these changes have led to rapid establishment of a pure-line condition. Further, they must have served also to restrict genetic variability in the manner described above for the tomato.

The aforementioned contrasts in mating system between cultigens and their respective wild progenitors are not universal in seed-reproduced crops. Unfortunately, satisfactory data exist for very few related wild species, but the self-pollinating barley (*Hordeum sativum*) is a notable, well-documented exception. Extensive tests in the generally accepted wild parent, *H. spontaneum*, failed to demonstrate any appreciable outcrossing, despite much higher natural variability than in *H. sativum* (Nevo *et al.*, 1986). Zohary (personal communication) relates this apparent anomaly to a high premium for self-fertilization among desert annuals of the eastern Mediterranean.

6.3 SWITCH FROM SELF-INCOMPATIBILITY TO SELF-COMPATIBILITY

It is likely that all SC crop species have descended from primordial SI ancestors. In many instances the line of descent would have to be relatively long such as cases in which SI is unknown in related species, in some instances, within the same botanical family. But examples are known in which the change has occurred within a species or between rather closely related species. One example is *Brassica oleracea*, to which the astonishingly varied group of cabbage, collards, sprouting broccoli, kohlrabi, leafy and marrow stemmed kales, cauliflower and Brussels sprouts belongs. Now, whilst the wild form of the species, a leafy rosette plant native to the coastline of west central Europe and the Mediterranean, is highly SI, SC has ·been found in all of the derived cultigens, and nearly all of the temperate summer cauliflowers are self-fertile (Hoser-Krauze, 1979). Thus, unquestionably the substitution of SC for SI has been advantageous for the improvement of this cultigen, and it can be assumed that considerable inbreeding occurs, even under conditions of open pollination, without adverse effect.

Another SI crop in which SC has been repeatedly detected is rye (*Secale cereale*) (Geiger, 1982), where such variants are exploited for various purposes, including the rapid derivation of inbred lines for hybrid breeding programs. As far as I can determine, however, exclusively SC lines suffer inbreeding depression, and no SC cultivars have sufficient vigor to gain acceptance.

Another example, dealing with two closely related species, is endive (*Cichorium endivia*), which is SC (Ernst-Schwarzenbach, 1932), in contrast to its probable progenitor, chicory (*C. intybus*), which is SI (Stout, 1916). Thus, whether intentional or not, artificial selection for the special features in endive included the transformation from SI to SC, together with increased fasciation and other desired characteristics (Rick, 1953). Other examples of the same theme are found in the ornamental *Antirrhinum majus*, the only SC exception to a genus of SI species (Stebbins, 1957), and in alfalfa.

6.4 CHANGES IN MONOECIOUS SPECIES

A fascinating case is presented by the cultivated cucurbits. The domesticated species of *Citrullis* (1 species), *Cucumis* (2 species), *Cucurbita* (4 species), and other genera are monoecious or variants thereof, hence classified as outcrossers. Nevertheless, contrary to the situation in allogamous species, they suffer little or no degeneration from inbreeding. This anomaly, verified by experiments in each of the aforementioned major cultivated species (cf. reviews by Robinson *et al.*, 1976; Whitaker and Davis, 1962), permits unlimited use of inbred lines, even for direct use as cultivars and as parents of commercial F_1 hybrids. In contrast to this lack of inbreeding depression, heterosis has been repeatedly demonstrated and is actively exploited by breeders and seedsmen. Thus, at the present time, the majority of cucumber cultivars and a large proportion of squash cultivars in the US are F_1-hybrid type.

This anomaly still lacks a satisfying explanation. Perhaps the most widely accepted is the old concept that during the evolution of cucurbits, the crops, by virtue of their large plant size and generally small populations, had to adapt their breeding systems accordingly. In primitive agriculture in tropical and subtropical America, for example, the dooryard garden might have had only a few plants, sometimes only one of each crop. Under these circumstances, generation after generation, the species clearly would not have survived if subject to severe inbreeding depression as known in maize, potatoes, and other outcrossers. Evidently deleterious genes were successfully eliminated in the course of their evolution.

Since these features appear to be universal for cultivated cucurbit species, the law of parsimony of assumptions would suggest that adaptation to inbreeding is an ancient trait, evolved once in a primordial stock, presumably in the wild, not under cultivation.

With regard to breeding systems, extent of genetic variation in cultivated vs. wild forms, etc., the available data are extremely limited. The strictly monoecious species are highly outcrossed, but the andromonoecious melons are selfed to a surprising extent. Zink (unpublished data), for example, has measured outcrossing via monogenic seedling markers in several seasons at Davis and found rates in the range of 4%, which would hardly qualify the crop as an outcrosser.

Whatever the causes, the cultivated cucurbits are examples of inbreeding or inbreeding-tolerant cultigens derived from various wild allogamous species and constitute another example of the general trend toward inbreeding in cultivated plants.

6.5 CONVERSION OF SELFING SPECIES TO ALLOGAMY

This section, as well as the theme of this contribution, is concerned with long-term changes of mating systems; therefore, transitory modifications will not be considered. As an example, genetic male sterility has been exploited effectively to facilitate large-scale hybridization in many crop plants, but since it usually operates in only one generation, it does not qualify for consideration here. Again we shall consider SI as the main device to promote allogamy. The problems of converting selfing species to a workable system of SI have been considered and evaluated by de Nettancourt (1977). The prime motive for changes of this unusual nature, apart from the stimulation of academic challenge, would be to facilitate large-scale production of F_1 hybrid cultivars. Lines thus converted could be allowed to intercross economically via natural vectors without the need for emasculation of pistillate parents and hand pollination. As also indicated by de Nettancourt, other purposes that might be served by the conversion are to suppress fructification, when desired, and to promote the production of seedless fruits in situations where natural parthenocarpy does not occur. He also discusses possible alternative procedures for implementing this change in the mating system:

1. Inducing the necessary mutations.
2. Reconstructing SI via genetic combinations within the SC species (Larson *et al.*, 1973). This proposal assumes that SC species are derived from SI ancestors via mutations to loss of activity, particularly at different S loci (as in the *Gramineae*). It is consequently conceivable that a workable SI system might be reconstituted by hybridizations between genotypes of the SC species either in F_1 or in subsequent segregating generations.
3. Transfer by hybridization of the genes necessary for a SI system. This alternative, having received more support than the others, was investigated by Martin (1961, 1968) and Mather (1943), and actively advocated by Denna (1971).

All these proposals suffer from difficulties as acknowledged by their advocates. A very low success rate renders alternative (1) unrealistic. Induced S alleles that are fully functional are extremely rare. The problem is further compounded by the need to obtain at least three different alleles for a functional gametophytic S system, thus the task for this locus alone is increased by this factor. Proposal 2, although logical, may be based on faulty assumptions, and no progress has been reported yet on this approach. Regarding proposal 3, severe problems were encountered by Martin (1968) and Mather (1943) in attempts to implement the change in interspecific combinations of *Lycopersicon* and *Petunia*. Backcrosses in the former reveal the activity of S and another major locus, both necessary for functional SI; furthermore a slow, progressive loss of SI in later backcross generations suggests the need for additional polygenic control for adequate expression – a problem also discovered by Hardon (1967) in a different *Lycopersicon* interspecific hybridization. Difficult as it may seem, this approach has been advocated more seriously than the other two alternatives.

Serious as these obstacles are, other problems exist in the implementation of a SI system in entomophilous species, as recognized by de Nettancourt (1977) and by Frankel and Galun (1977). This issue was faced in experiments between a tiny-flowered accession of the SC *L. pimpinellifolium* and large-flowered SI accessions of *L. hirsutum* and *L. pennellii* (Rick, 1982). In backcrosses to the former, inheritance of SI was studied, along with that of other floral features essential to a functional system. As in the earlier research on *Lycopersicon*, two major genes appear to determine SI. The other essential features – exposure of the stigma and components of floral attractivity – are inherited quantitatively. With few exceptions SI and the various floral traits were inherited independently or correlated to an extent no greater than expected for random linkages and allometric relationships. As discussed below, interspecific transfers of monogenic traits can be fraught with difficulties; therefore the task of introgressing not one, but several, semi-independent metric traits would become one of a magnitude difficult to contemplate.

The complexity of this undertaking is further compounded if the required characters must be transferred by wide crosses. Even the transposal of a single major gene can be thwarted by tight linkages with undesirable traits from a wild parent. For example, in breeding root-knot nematode resistance (*Mi*) from *L. peruvianum*, Gilbert and his colleagues were frustrated by such problems, and it was 14 years from the time of the initial interspecific hybridization until the first acceptable F_1 hybrid cultivars were introduced (Gilbert, McGuire and Tanaka, 1960), and still more time was required before commercially acceptable pure-breeding cultivars were obtained. However, clean transfer of desired major genes may soon be routinely accomplished via genetic engineering techniques; but, until other short cuts are devised, we must still rely on conventional plant breeding for introgressing metric traits.

An additional complication engendered by the proposed transfer concerns the product of hybridization between converted lines. If a workable system could be achieved, by definition the products of crossing between converted SI lines would be SI and have floral traits adapted for outcrossing, not selfing. They could yield satisfactorily only in the presence of adequate pollinator activity – a rare situation in most temperate areas of tomato cultivation and never encountered in greenhouses or other enclosures. Problems of the same magnitude would probably be encountered in attempts to convert other self-pollinated crops to an effective SI system.

The evidence thus indicates that a workable implementation of the SI system requires many formidable genetic changes and that the system is vastly more easily disrupted than established. It also follows that experimental conversion to an effective SI system could be an extremely difficult, if not impossible, task. The undertaking would presumably be less fraught with difficulties in anemophilous species, which would not need as many structural changes.

The reader must be advised that these considerations are related to a shift from SC to SI in agricultural crops. The task should be theoretically simpler in other plant species in which productivity, product quality, and other considerations would not apply. Further, the problems would be less complex if the necessary features could be derived within rather than between species. In the extreme situation of very recently derived SC, as in the example of *Brassica oleracea*, temperate summer cauliflowers could be converted successfully back to SI merely be substitution of the appropriate *S* alleles.

6.6 OTHER CROP GROUPS

The above examples are largely drawn from seed-reproduced annuals or biennials, which in many respects are the most advanced of cultivated plants (greater progress in improved yields, quality, etc.). The clonally propagated groups (fruits, ornamentals and certain forage crops) differ in several respects. Clonal reproduction ensures extreme uniformity without the necessity of breeding for homozygosity – in most respects, a great boon to the plant breeder. It follows that mating systems seldom need to be manipulated. Thus, in general, examples parallel to those presented above are rare or lacking in certain clonal groups. One example of the incentive for change of mating system is found in certain almond, cherry, and plum cultivars, in which SI obstructs fruit setting unless the grower provides an intercompatible genotype. It is standard practice in such crops to interplant with, or topgraft to, intercompatible cultivars; in the same vein, much effort must often be expended to augment pollinating bee populations.

In other instances the problem has been solved more effectively by selecting SC cultivars. Spontaneous SC variants have been described and thus exploited in apples, pears (Crane and Lawrence, 1947), *Feijoa* guava, etc. In the absence of naturally occurring SC, investigators have successfully induced such mutations in sweet cherries (*Prunus avium*) Lapins, 1971), pineapple (*Ananus*) (Marr, 1964), and other fruit crops (Lapins, 1983). It must be observed, however, that monoculture of single SI genotypes can serve an important purpose. Thus, in commercial pineapple production, since the development of seeds is highly undesirable and fruits can develop parthenocarpically, it is illegal in certain districts to plant any other line than the standard cultivar – for example, 'Cayenne' in Hawaii.

Similar considerations apply to clonal dioecious species. In the case of strawberries (*Fragaria* spp.) most cultivars of the 19th century, like their wild progenitors, were dioecious, requiring complex interplanting of pistillate and staminate lines. Accordingly, small wonder that, as soon as hermaphroditic variants were discovered, they soon predominated in new cultivars (Bringhurst and Voth, 1984). The same scenario occurred in the history of grape cultivation, dioecious types now being rare and retained only because equivalent hermaphroditic cultivars have not yet been found (Janick and Moore, 1975). A similar situation exists in papaya (*Carica papaya*).

The standards for breeding clonal ornamentals differ radically in respect to the nature of the product. Thus, with a premium on high production of showy flowers, fruit and seed production can often be undesirable because it tends to reduce the extent of flowering. Consequently, in contrast to clonal fruits, a single SI genotype can be highly advantageous, and no gain would result from meddling with the mating system.

Although, as previously argued, vegetative propagation offers great advantages, it suffers such defects as the risk of harboring devastating virus diseases. For this and other reasons, enterprising researchers have sought other avenues of propagation. For example, Bobisud and Kamemoto (1982) have succeeded in deriving amphidiploids of *Dendrobium* spp. (*Orchidaceae*), which are fertile, true-breeding, and extraordinarily showy and productive. Thus, in the history of this group, nursery practices have changed from the original seed reproduction to clonal, as the current standard mode of prapagation, but are now reverting to seed, notwithstanding the resistance of traditional attitudes of orchid fanciers.

Cloned forage crops are generally less advanced than the aforementioned horticultural groups, i.e. cultigens are more similar to wild progenitors; in fact, they can be essentially the same. They resemble clonal ornamentals to the extent that fruit and seed production may not be desirable; in fact, may decrease yield of fodder by competing for the plants' carbohydrate sink.

The rates of changes in mating system during plant domestication are highly speculative, but sufficient data exist to warrant the following considerations. The switch from SI to SC is one of the least complex changes. As mentioned above, the SI system, requiring the integration of many complexly inherited characters for its proper functioning in natural populations, is easily disrupted by simple genetic changes. The mere allele substitution at the *S* locus and others can convert *Nicotiana alata* to SC. Alleles with similar effects have been reported in many other SI plant species (de Nettancourt, 1977). The addition of a centric chromosome fragment bearing the *S* locus can effect the same change in *Petunia inflata* (Brewbaker and Natarajan, 1960) and in *N. alata* (van Gastel and de Nettancourt, 1975). It is therefore not surprising that the shift from SI to SC has occurred repeatedly in wild populations and during domestication and that, as in the case of *Brassica oleracea*, this transition took place within a relatively short time span. The recently documented cases of mutagen-induced changes were effected in very few generations.

Conversion from SI to SC does not *a priori* lead to autogamy, although it is an essential step in the process. The attractivity characters that ensure adequate cross-pollination in a workable SI system will continue to do so in newly derived SC lines. In the case of *B. oleracea*, the SC and SI genotypes differ little if any for flower size, elevation of the raceme, number of flowers per inflorescence, and other display features. In the case of *Lycopersicon* spp. exserted stigma, always associated with SI, not only aids insect pollination, but may also, in fact, lead to unfruitfulness of SC derivatives in the absence of pollinators. The evolution of autogamy in *L. pimpinellifolium*, *cheesmanii*, and *parviflorum* required reduction in style length so that the stigma could readily receive pollen from anthers of the same flower. Now, such a change might be assumed to be simply inherited, and, in fact, a mutant *ex* for exserted stigma is known (Rick and Robinson, 1951). Nevertheless, quantitative control by an estimated nine genes was reported by Ruttencutter and George (1975), and in a study of linkages with allozymes, Tanksley, Medina-Filho and Rick (1982) found that quantitative trait loci for stigma position could be identified on at least eight chromosomes.

The attractivity traits asssociated with SI are inherited in similar complex fashion and are largely independent of each other and of S (Rick, 1982). Thus, although the change from SI to SC is relatively simply determined, the change from facultative outcrossing to autogamy is far more complex and would require a correspondingly greater time to evolve. The relevance of the display traits in this context might be challenged, but if, as in the *B. oleracea*, pollen vectors are abundant, floral display could counter efforts to breed complete autogamy.

Transitions from dioecy, which guarantees cross-fertilization, to monoecy or hermaphroditism are of a similar nature. These changes in floral structure are generally controlled by alleles at the sex-determining locus alone or in concert with an independent locus (Westergaard, 1958). Thus, the basic change in sex expression is rather simply determined and progress in changes of this nature might be expected to ensue rapidly following muta-tion, but transitions to enforced inbreeding systems as in SI → SC are more complex and progress consequently likely to be much slower.

6.8 CONCLUSIONS

From the preceding discussion we can see that the transition from allogamy to autogamy, well documented in wild species, is often intensified in domesticated plants. In certain species, like the cultivated tomato, the trend is from a facultative selfer to a total selfer. In others (cucurbits), the change is from facultative outcrossers to variable selfers. In many others, devices for promotion of outcrossing (for example, self-incompatibility) have been eliminated, and the cultigen thereby rendered amenable to more rapid selection and achievement of uniformity. No doubt certain of these changes were effected by primitive man (tomatoes, cucurbits). Thus, the simplest selection procedures applied for uniformity of a desired phenotype could have no doubt abetted breakdown of devices that promote allogamy. Others are more modern developments in which breeders intentionally select for changes toward self-fertilization with a rather good understanding of the underlying basic principles.

Reversals of this trend – i.e. from autogamy to allogamy as the primary reproductive system in a crop via interspecific hybridization – are virtually unknown. Experiments reveal the magnitude of problems engendered by attempting such changes and indicate that, at least in entomophilous crops, such transitions are extremely difficult, if not impossible, to achieve.

A considerable divergence in mating system between wild and cultivated taxa has been emphasized but, such changes accompanying domestication constitute only one of the many classes of modification effected in the evolution of cultivated plants. In this regard, Schwanitz (1966) enumerates a series of some 12 classes of change, many of which are losses of wild characters (for example, loss of dormancy, mechanical protection, toxic compounds and dispersal devices). Altogether, in the highly evolved cate-gory of seed-produced annuals, the amount of such change may be so great that certain comparisons between wild ancestors and derived cultigens become difficult, if not absurd.

Finally, we must acknowledge that much of our understanding of these events owes to Stebbins' analysis and conclusions on the evolutionary aspects of autogamy in plants, as communicated by him in 1957.

6.9 REFERENCES

Bobisud, C. A. and Kamemoto, H. (1982) Selection and inbreeding in amphidiploid *Dendrobium* (Orchidaceae). *J. Am. Soc. Hort. Sci.*, **107**, 1024–7.

Bohn, G. W. and Tucker, C. M. (1940) Studies on *Fusarium* wilt of the tomato. I. Immunity in *Lycopersicon pimpinellifolium* Mill. and its inheritance in hybrids. *Missouri Agr. Expt. Sta. Res. Bull.*, **311**, 1–82.

Brewbaker, J. L. and Natarajan, A. T. (1960) Centric fragments and pollen-part mutation of incompatibility alleles in *Petunia*. *Genetics*, **45**, 699–704.

Bringhurst, R. S. and Voth, V. (1984) Breeding octoploid strawberries. *Iowa State J. Res.*, **58**, 371–81.

Crane, M. B. and Lawrence, W. J. C. (1947) *The Genetics of Garden Plants*. MacMillan, London.

Denna, D. W. (1971) The potential use of self-incompatibility for breeding F_1 hybrids of naturally self-pollinated vegetable crops. *Euphytica*, **20**, 542–8.

Ernst-Schwarzenbach, M. (1932) Zur Genetik und Fertilität von *Lactuca sativa* L. und *Cichorium Endivia* L. *Arch. Julius Klaus-Siftung. Zürich*, **7**, 1–35.

Frankel, R. and Galun, E. (1977) *Pollination Mechanisms, Reproduction and Plant Breeding*. Springer, Berlin.

Gastel, A. J. G. Van, and de Nettancourt, D. (1975) The effects of different mutagens on self-incompatibility in *Nicotiana alata* Link and Otto II. Acute radiation with X-rays and fast neutrons. *Heredity*, **34**, 381–92.

Geiger, H. H. (1982) Breeding methods in diploid rye (*Secale cereale* L.). *Tagunsber. Akad. Landwirtschatswiss. DDR*, **198**, 305–32.

Gilbert, J. C., McGuire, D. C. and Tanaka, J. (1960) Indeterminate tomato hybrids with resistance to eight diseases. *Hawaii Farm Sci.*, **9**(3), 1–3.

Hardon, J. J. (1967) Unilateral incompatibility between *Solanum pennellii* and *Lycopersicon esculentum*. *Genetics*, **57**, 795–808.

Hoser-Krauze, J. (1979) Inheritance of self-incompatibility and the use of it in the production of F_1 hybrids of cauliflower (*Brassica oleracea* var. *botrytis* L. subvar. *cauliflora* D.C.). *Genet. Polon.*, **20**, 341–67.

Janick, J. and Moore, J. N. (eds) (1975) *Advances in Fruit Breeding*. Purdue University Press. W. Lafayette, IN.

Jenkins, J. A. (1948) The origin of the cultivated tomato. *Econ. Bot.*, **2**, 379–92.

Lapins, K. O. (1971) 'Stella', a self-fruitful sweet cherry. *Can. J. Plant Sci.*, **51**, 252–3.

Lapins, K. O. (1983) Mutation breeding. in *Methods in Fruit Breeding* (eds J. N. Moore and J. Janick), Purdue University Press, W. Lafeyette, IN, pp. 74–99.

Larsen, J., Larsen, K. Lundqvist, A. and Osterbye, U. (1973) Complex self-incompatibility systems within the angiosperms and the possibility of reconstructing a self-incompatibility system from different forms within a self-fertile species. *Incomp. Newslett. Assoc. EURATOMITAL, Wageningen*, **3**, 79–80.

Marr, G. S. (1964) Study of self-incompatibility in the pineapple. *Agr. Res. Dept. Agr. Tech. Serv. Rot. S. Africa*, **2–6**: 561.

Martin, F. W. (1961) The inheritance of self-incompatibility in hybrids of *Lycopersicon esculentum* Mill. × *L. chilense* Dun. *Genetics*, **46**, 1443–54.

Martin, F. W. (1968) The behavior of *Lycopersicon* incompatibility alleles in an alien genetic milieu. *Genetics*, **60**, 101–9.

Mather, K. (1943) Specific differences in *Petunia*. I. Incompatibility. *J. Genet.*, **45**, 215–35.

Matthiolus, P. A. (1544) *Di Pedacio Dioscoride Anazsarbeo libri cinque della historia, et materia medicinale trodotti in lingua volgare Italiana*. Venetia.

Nettancourt, D. de (1977) *Incompatibility in Angiosperms*. Springer, Berlin.

Nevo, E., Zohary, D., Beiles, A., Kaplan, D. and Storch, N. (1986) Genetic diversity and environmental associations of wild barley, *Hordeum spontaneum*, in Turkey. *Genetica*, **68**, 203–13.

Porte, W. S. and Wellman, H. B. (1941) The Pan America tomato, a new red variety highly resistant to fusarium wilt. *USDA Circ.*, **611**, 1–6.

Rick, C. M. (1953) Hybridization between chicory and endive. *Proc. Am. Soc. Hort. Sci.*, **61**, 459–66.

Rick, C. M. (1976) Tomato. in *Evolution of Crop Plants* (ed. N. W. Simmonds), Longman, London, pp. 268–73.

Rick, C. M. (1982) Genetic relationships between self-incompatibility and floral traits in the tomato species. *Biol. Zbl.*, **101**, 185–98.

Rick, C. M. (1984) Evolution of mating systems: evidence from allozyme variation. *Proc. XV Int. Congr. Genetics (New Delhi)*, **3**, 215–21.

Rick, C. M. and Dempsey, W. H. (1969) Position of the stigma in relation to fruit setting of the tomato. *Bot. Gaz.*, **130**, 180–6.

Rick, C. M., DeVerna, J. W., Chetelat, R. T. and Stevens, M. A. (1987) Potential contributions of wide crosses to improvement of processing tomatoes. *Scient. Hort.*, **200**, 45–55.

Rick, C. M. and Fobes, J. F. (1975) Allozyme variation in the cultivated tomato and closely related species. *Bull. Torrey Bot. Club*, **102**, 376–84.

Rick, C. M., Fobes, J. F. and Holle, M. (1977) Genetic variation in *Lycopersicon pimpinellifolium*: evidence of evolutionary change in mating systems. *Pl. Syst. Evol.*, **127**, 139–70.

Rick, C. M., Fobes, J. F. and Tanksley, S. D. (1979) Evolution of mating systems in *Lycopersicon hirsutum* as deduced from genetic variation in electrophoretic and morphological characters. *Pl. Syst. Evol.*, **132**, 279–98.

Rick, C. M. and Robinson, J. (1951) Inherited defects of floral structure affecting fruitfulness in *Lycopersicon esculentum*. *Am. J. Bot.*, **38**, 639–52.

Robinson, R. W., Munger, H. M., Whitaker, T. W. and Bohn, G. W. (1976) Genes of the *Cucubitaceae*. *Hort. Science*, **11**, 554–68.

Ruttencutter, G. E. and George, W. L. (1975) Genetics of stigma position. *Rept. Tomato Genetics Coop.*, **25**, 20–1.

Schwanitz, Fr. (1966) *The Origin of Cultivated Plants*. Harvard University Press.

Simmonds, N. W. (1979) *Principles of Crop Improvement*. Longman, London.

Stebbins, G. L. (1957) Self fertilization and population variability in the higher plants. *Am. Nat.*, **41**, 337–54.

Stout, A. B. (1916) Self- and cross-pollinations in *Cichorium intybus* with reference to sterility. *Mem. NY Bot. Gard.*, **6**, 333–454.

Tanksley, S. D., Medina-Filho, H. and Rick, C. M. (1982) Use of naturally occurring enzyme variation to detect and map genes controlling quantitative traits in an interspecific backcross of tomato. *Heredity*, **49**, 11–25.

USDA (1922–1980) *Agricultural Statistics*, US Dept. Agric. US Government Printing Office.

Westergaard, M. (1958) The mechanism of sex determination in dioecious flowering plants. *Adv. Genetics*, 9, 217–81.

Whitaker, T. W. and Davis, G. N. (1962) *Cucurbits*. Leonard Hill, London.

Editors' commentary on Part 2

One major goal in the study of plant evolutionary biology is to determine phylogenetic relationships. This interest need not reflect aesthetic concern nor an attempt to obtain more natural classifications. Rather, correct phylogeny provides the framework necessary for many particular inquiries of the form 'How did A evolve?' where A is understood to be a new species or other taxon, or an attribute, be it a morphological trait, an adaptation, or a gene sequence. Once it becomes possible to state, with precision and certainty, that 'A evolved from B', then pathway and manner of evolution become amenable to analysis. Otherwise A and B are merely similar or not and, compared to C, more or less so.

At what level can phylogeny be known? From the standpoint of experimental taxonomy (biosystematics), the level is defined by the ability to carry out crosses to study compatibility, chromosome pairing relationships and hybrid fertility. The results generally identify a genetic distance characteristic of species. At greater levels of divergence, the extent and type of morphological resemblance is the primary evidence for relationships, but uncertainty remains because convergence may not be recognized and novel characters, for example, anatomical details of vegetative and floral structures or perhaps an unusual secondary metabolite, frequently suggest competing phylogenetic alignments.

In recent years, it has become evident that phylogenetic relationships among plant species, genera and, perhaps, even at higher ranks can be determined by comparison of the structure and sequence of their DNAs. Two approaches are used. The first is based on comparisons of nucleotide sequences of genes or the size pattern of fragments cut from DNA by restriction enzymes, and leads to cladograms of shared derived mutations or of the extent of overall similarity. The second approach makes use of genetic changes that are likely to have occurred only once within a particular lineage so that those taxa that exhibit them can be considered to have descended from a single common ancestor, no matter what their present morphological divergence. Such uniquely derived attributes include large inversions of the chloroplast genome and duplications of specific nuclear

149

genes. Thus the two approaches differ in that one makes use of DNA sequence and the other DNA structure.

The importance of molecular evidence for inferring phylogenetic relationships in plants is explicitly acknowledged in Peter Raven's review of the Onagraceae (Chapter 4). This family of flowering plants, comprising 16 genera and about 652 species, is the best known family of its size and has been studied from many standpoints including cytology, breeding system, morphology, vegetative and floral anatomy, as well as biochemistry. The family is characterized by a number of derived features including certain anatomical details of its pollen that are so distinctive that the pollen is readily recognized in the fossil record starting about 70 million years ago.

Although the family is clearly monophyletic, Raven points out that the relationships among its genera have not been resolved except that *Ludwigia* appears to represent a lineage that is distinct from all other Onagraceae. The problems of identifying relationships among the other 15 genera reflect their apparent antiquity (the putative intermediate taxa are extinct) and the lack of concordant character variation that might serve to group them. As a consequence, he concludes that molecular evidence is necessary to reconstruct the generic phylogeny. That such an approach is likely to be productive seems assured because it has already provided significant information in *Clarkia*, a well studied genus in the family (see below).

We believe molecular evidence will be applied more and more often to answer phylogenetic questions in plants because the evidence is precise and objective and can reveal a detailed account of relationships. The molecular evidence also provides a new and unexpected view of the structure and organization of genetic information. We now know that significant change takes place within the genome as sequences, particularly multigene families, expand by duplication, diverge by mutation (substitutions as well as short deletions and additions), and become subject to various processing mechanisms including gene correction. DNA sequences (transposons) are also moved within the genome and when they insert within genes or their control regions their presence modifies gene expression. Small functional pieces of genes (exons) can be mobilized and recombined to code for novel proteins with often useful functions. Thus DNA is manipulated within cells by a variety of mechanisms and the consequence of change in its structure, organization and coding properties defines a context for evolution that must be specified and understood just as the more familiar processes of divergence and adaptation between organisms and the external environment ('Nature') must be understood.

Therefore, from the standpoint that molecular genetic information is essential to understand plant evolution, we have juxtaposed in this section two intensive and rich reviews of the organelle and nuclear genomes (Chapters 2 and 3) with three chapters on the broad field of systematics

(Chapters 4, 5 and 6). In Chapter 2 Birky describes the chloroplast and mitochondrial genomes and emphasizes how their structure, coding properties and mode of inheritance affect their evolution. He points to a number of factors that distinguish the organelle genomes from each other and from the nuclear genome and suggests how the factors might interact to cause different patterns of variation during evolution. In Chapter 3 Tanksley and Pichersky describe the organization and structure of the nuclear genome as well as their own intensive studies on the molecular genetics and evolution of the small subunit of RubisCo and chlorophyll a/b binding proteins in tomato. They show that a significant proportion of nuclear genes is duplicated and that frequently the duplicated copies are unlinked. How this comes about remains uncertain.

The other three chapters in the section cover more familiar ground. In addition to Raven's review of the Onagraceae, two chapters are presented that examine the morphological correlates and genetic basis of the evolution of self-pollinating breeding systems from outcrossing ones. In Chapter 5 Wyatt's intensive analysis of the diminutive annual *Arenaria uniflora*, native to granite outcrops in the southeastern United States, emphasizes that the morphological changes in the flower associated with the switch to self-pollination are remarkably similar whenever they occur. He shows how the use of multiple lines of evidence can reveal the occurrence of convergent changes at an early level of overall divergence, and warns others to be on the lookout for similar problems when related taxa have different breeding systems. In Chapter 6 Rick reviews the frequent examples of the evolution of self-pollination in cultivated plants which facilitated establishment of pure lines, but also restricted genetic variability. Of interest is the parallel morphological change in tomato (*Lycopersicon*) and eggplant (*Solanum incanum*), both in the Solanaceae, that resulted in shortening stigma height to or below the level of the anthers, as well as the apparent genetic simplicity underlying changes in the compatibility system.

Most of the phylogenetically interesting studies of DNA in plants have involved the chloroplast genome and have been extensively reviewed (see Chapter 2). Two remarkable results stand out. The first illustrates the use of evidence derived from DNA fragment patterns obtained by cutting chloroplast DNA with restriction enzymes. Sytsma and Gottlieb (1986) examined the relationship of the monotypic genus *Heterogaura* (Onagraceae) to the related and speciose genus *Clarkia* (with about 43 species). *Heterogaura heterandra* is a self-pollinating annual plant native to California and Oregon that has such unusual floral and fruit characters that it has been ranked as a genus since its discovery in the 19th century, although in recent years it has come to be viewed as allied to *Clarkia* (Raven, 1979). The DNA analysis was straightforward, and revealed clearly that *Heterogaura* did not stand in a taxonomically basal position with respect to *Clarkia*, but rather shared a recent common ancestor with a particular species of *Clarkia*, *C. dudleyana*, a member of a phylogenetically advanced section. The two taxa were more closely related to each other than either was to any other species; in fact, they were more closely related than were nearly all other pairs of *Clarkia* species tested. The finding was remarkable because phylogenetic relationships have been the explicit focus of intensive study in Onagraceae and the genera had been considered stable (Raven, 1979). More importantly, the DNA analysis revealed a relationship that was completely obscured by morphological divergence.

The second example that documents the unexpected value of molecular data in plant systematics is the identification of the ancestral stock within the Asteraceae. The relevant evidence was discovered during a restriction enzyme survey of chloroplast genomes in the family. A 22 kbp inversion was found in 57 genera representing all of the tribes, but it was absent from the subtribe Barnadesiinae of the tribe Mutisieae as well as from ten putative out-group families and from Angiosperms in general (Jansen and Palmer, 1987). The absence of the inversion in the Barnadesiinae suggests that this subtribe represents the most primitive lineage in the family.

Unlike the evidence in the previous case which made use of a large number of mutational changes to reveal the relationship between *Heterogaura heterandra* and *Clarkia dudleyana*, the single chloroplast inversion in the Asteraceae has resolving power because its origin is almost certainly unique since it appears to have occurred only once in about 400 million years of land plant evolution. The molecular evidence in Asteraceae appears to solve a long-standing controversy regarding the identification of the most primitive lineage in the family, and also supports suggestions that the family originated in montane South America, the present center of distribution of the Barnadesiinae (Jansen and Palmer, 1987). In addition, the discovery is consistent with previous suggestions that bilabiate flowers and woody habit which predominate in the subtribe are primitive.

The proposition that a single genetic or molecular attribute could reveal phylogenetic relationships above the species level and irrespective of differences in morphological and cytological data was not novel, but was suggested following the discovery of the duplication of the nuclear gene encoding the cytosolic isozyme of phosphoglucose isomerase (PGI) in *Clarkia* (Gottlieb and Weeden, 1979). The PGI duplication was revealed during the course of electrophoretic studies of species relationships in *Clarkia* when it was found that some species had a single cytosolic PGI like most other diploid plants whereas about a dozen *Clarkias* had two PGI isozymes as a result of inheriting a duplication of the coding gene. The value of gene duplications to reveal unsuspected phylogenetic relationships was also shown by recent analysis of duplicated isozymes of 6-phosphogluconate dehydrogenase (6PGD) in *Clarkia* (Odrzykoski and Gottlieb, 1984). The phylogenetic applications of gene duplications are unusual because, like inversions of the chloroplast genome, they have a very high probability of being unique. The hypothesis can be tested directly when appropriate gene sequences become available.

Phylogenetic information is useful for experimentalists because it directs specific research into many worthwhile questions that may have seemed unapproachable. For example, prior to our present realization that *Heterogaura heterandra* and *Clarkia dudleyana* share a relatively recent common ancestor, there was little reason to pay particular attention to differences in their fruits because the expectation was that even intensive analysis would simply provide more details. However, now that we know that the few-seeded indehiscent *Heterogaura* fruit evolved from the many-seeded elongated capsule characteristic of *Clarkia*, an analysis of fruit development might reveal which steps during fruit development were modified. Attempts to hybridize the species also become worthwhile even if it requires heroic steps such as bypassing the normal route of pollination or embryo culture. In addition, since the entire *Heterogaura* fruit is the unit of dispersal, it becomes worthwhile to study demography and colonization in this species and contrast the results directly to that in *Clarkia*. The point is that related species that differ markedly in their morphology and other attributes are excellent experimental objects, but the possibilities depend initially on recognizing a close phylogenetic relationship.

Phylogenetic information at higher taxonomic levels is also valuable. Thus, identification of the primitive lineage within the Asteraceae quickly led to suggestions regarding where the family arose and which morphological characters were likely to be primitive. From the standpoint of molecular biology, phylogenetic information is necessary to understand the evolution of gene sequences and gene families as pointed out by Birky as well as Tanksley and Pichersky. Molecular evidence is beginning to provide a fertile meeting ground from which biologists with different interests can harvest many new and unexpected ideas.

REFERENCES

Gottlieb, L. D. and Weeden, N. F. (1979) Gene duplication and phylogeny in *Clarkia*. *Evolution*, **33**, 1024–39.

Jansen, R. K. and Palmer, J. D. (1987) A chloroplast DNA inversion marks an ancient evolutionary split in the sunflower family (Asteraceae). *Proc. Natl. Acad. Sci. USA*, **84**, 5818–22.

Odrzykoski, I. J. and Gottlieb, L. D. (1984) Duplications of genes coding 6-phosphogluconate dehydrogenase in *Clarkia* (Onagraceae) and their phylogenetic implications. *Syst. Bot.*, **9**, 479–89.

Raven, P. H. (1979) A survey of reproductive biology in Onagraceae. *NZ J. Bot.*, **17**, 575–93.

Sytsma, K. J. and Gottlieb, L. D. (1986) Chloroplast DNA evidence for the origin of the genus *Heterogaura* from a species of *Clarkia* (Onagraceae). *Proc. Natl. Acad. Sci. USA*, **83**, 5554–7.

PART 3

Development and evolution

CHAPTER 7

Ontogeny and phylogeny: phytohormones as indicators of labile changes

TSVI SACHS

... our ignorance of the pathways from genes to characters is one of the chief obstacles to our understanding of evolutionary trends (Stebbins, 1974, p. 102).

... we can postulate that those trends of angiosperm evolution which involve the change in location and *de novo* activity of intercalary meristems are based upon mutations that alter either the locations where growth substances are synthesized, the way in which they move through the plant, or both of these factors (Stebbins, 1974, p. 112).

7.1 THE PROBLEM: COULD ONTOGENY CONSTRAIN THE EVOLUTION OF FORM?

Darwin's evolutionary theory was the unifying concept that brought together evidence from fossils, comparative anatomy and morphology, biogeography and adaptations. More recently, the Neo-Darwinists accounted for the mechanism of evolution in a genetic context. The present Neo-Darwinism does not, however, explain why form evolved in one way rather than another. This could be due to the gap between genes and mature structures. It is often believed that large changes in the phenotype necessarily result from a large number of gene changes. This assumption may be misleading – morphological differences that appear to be large or complex need not depend on many genes (Frankel, 1983). The missing links are ontogenetic processes, and the general question to be considered here is the ways in which these could constrain and limit the consequences of natural selection.

There have been suggestions that major aspects of evolution do not depend on information transmitted by DNA (Goodwin, 1982). A different approach is followed here: the working hypothesis is that form depends on genes and that at the gene level evolution conforms with Neo-Darwinian principles. This would mean that genes change one at a time and that all stages of the evolution of a new structure are subject to selection. Mature form, however, is necessarily the product of ontogeny, and this involves a complex, coordinated action of many genes. It is to be expected, therefore, that the evolution of ontogeny could involve principles that stem from its complexity. These principles need not be specified (nor contradicted) by Neo-Darwinism (Maynard Smith, 1983).

How could the complexity of ontogeny influence the course of evolution? In plants, organs that have varied functions often exhibit structural similarity that reflects their homology. This similarity suggests that the evolution of form is constrained by ontogenetic processes (Stebbins, 1974; Alberch, 1983; Bachmann, 1983). A key question is, therefore, why are some aspects of ontogeny conservative while others are labile. Differences in the likelihood of various changes could depend on traits of networks of many genes (Sander, 1983). The operation of complex networks necessarily involves genes with different functions. Thus control systems could be expected, and their changes could lead to large modifications of many aspects of ontogeny. Furthermore, genes that operate early could be expected to have effects that ramify in development, much more so than genes that operate late in the development of the very same tissue or organ (Stebbins, 1974; Sachs, 1982). Since changes in genes with few effects are more likely to have a selective advantage, some courses of evolution are more likely whereas others are unlikely or perhaps impossible. All this is general enough to be vague, but it does suggest that constraints of evolutionary change reflect not only gene substitutions but also how the particular gene action participates in ontogenetic processes.

7.2 PHYTOHORMONE CHANGES AS CONVENIENT MODELS

The ontogeny of plants and the role of their known phytohormones offer major advantages for studies of specific examples of ontogenetic constraints of the evolution of form. The 'continued embryology' of plants means that organogenesis can be studied in apices of vigorous seedlings and not only in minute, delicate embryos. The absence of cell movement and the relative simplicity of plants are also advantageous for research. Mutations that have large effects on form are viable and not necessarily rare (Gottlieb, 1984). Even in such advantageous organisms, however, a discussion of all evolutionary consequences of ontogenetic complexity would be too ambitious — it would even be impossible in our present state of ignorance concerning the controls of ontogeny. It is, therefore, necessary to concentrate on part of the problem, and one where ontogenetic controls are relatively well known should be a good starting point.

Phytohormones are a class of substances known to act as controlling factors of plant development (Sachs, 1986). Their importance is shown by the results of the incorporation of genes that disrupt their synthesis, as in the crown gall disease. Such incorporation results in the development of tumors or various partial disruptions of organization (Hooykaas, Ooms and Schilperoort, 1982). It follows that the regulated synthesis of phytohormones and limited competence to respond to their effects are essential for organized development (Sachs, 1975, 1986). The original hypothesis, the one that led to the isolation of the known phytohormones, was that each substance controls a specific process. Available evidence, however, shows that this fruitful hypothesis was not correct. Phytohormones appear to be agents of spatial integration of plant development that specify the presence of organs while any specific effect they elicit depends on the developmental history of the responding tissue (Sachs, 1986). This means that their effects need not be proportional to their measureable internal concentrations and it is also not clear that concentrations, rather than gradients, flow and changes with time, are the critical parameters (Sachs, 1986). Further information about phytohormones is discussed below, the important point for the following discussion being that they are correlative signals and could determine the location of developmental processes.

The purpose here is, therefore, to consider quantitative changes of the action of known phytohormones as models of the evolution of form. As could be expected, there are various genes that have quantitative effects on the synthesis of phytohormones and on the sensitivity of tissues to their influence (Phinney, 1985). Adding phytohormones and various inhibitors to developing tissues are ways of studying the influence of genes that are necessary for phytohormone action. A major question must be how such changes could become localized, causing changes of form and not just overall size. It is not assumed, of course, that genes that influence phytohormones are the only controls of development; it is only suggested that they are convenient model systems for considering general principles of the evolution of form.

What we would wish to know is the sort of differences of form that could result from quantitative changes of one key substance at a time. These are the changes that could be expected to be common on the basis of Neo-Darwinian theory. No less important would be a definition of what does not occur under these conditions. This could be a start towards an evaluation of differences between related organisms that could supplement and even go beyond quantitative differences in DNA structure. For the present purposes, therefore, phytohormonal effects should be considered not at the level of the controls of gene activity – where new techniques are yielding important results – but rather at the level of the integration of multigene developmental processes, whose evolutionary consequences have hardly been considered.

Leaf gaps and the evolution of leaves

A first example of the possible role of phytohormones in the evolution of ontogeny deals with internal rather than external form. Vascular strands are the most prominent internal differentiation in plants and their development is closely correlated with the growth of the organs they supply. There are proofs, furthermore, that these correlations are mediated by phytohormones: shoots and roots induce the differentiation of vascular strands by acting as sources and sinks for auxin and, presumably, other developmental signals (Sachs, 1981). Evolutionary changes in phytohormonal relations could therefore be expected to be expressed by internal as well as external structure.

A specific example of such evolutionary differences concerns the presence of leaf gaps. These are regions of parenchyma in the vascular system, immediately above the connection of the strands or traces of a leaf with those of the rest of the shoot (Fig. 7.1). The presence of leaf gaps thus means that there are no direct contacts between vascular strands of a given leaf and the strands leading to the younger leaves above. The occurrence of leaf gaps was found to be correlated with the membership of plants in large phyletic groups and they have therefore been considered to be structures of evolutionary significance. The occurrence of leaf gaps is also correlated with other traits of leaves, especially relative size and the complexity of the internal vascular system. Leaves have therefore been classified into two morphological categories: macrophylls, where leaf gaps occur in the adjoining stem, and microphylls where they do not. It has been often stated that these organs evolved independently and are therefore not homologous; from this it has been concluded that their evolutionary relations could not be considered.

It is thus known that vascular differentiation is induced by phytohormones and at least one aspect of vascular pattern, leaf gaps, is of evolutionary interest. This raises the question whether the presence and absence of leaf gaps can be understood on a basis of phytohormonal relations. The control of the formation of contacts between new and old vascular strands was studied using strands induced by exogenous auxin (Sachs, 1968, 1981). A newly induced strand always contacts older strands that do not lead to either a natural or an artificial source of auxin (Fig. 7.1). When the older strand is supplied with auxin, on the other hand, the formation of contacts is inhibited (though not invariably prevented). These results are in agreement with the natural distribution of contacts at the base of leaves: contacts are present when the leaves are weak and are presumably poor sources of auxin. Experimental weakening of leaves, furthermore, leads to the formation of vascular contacts where they do not occur naturally (Sachs, 1968, 1981). It might be possible to induce gaps in plants where they do not occur, by the addition of auxin to leaf primordia at a very early stage. These experiments have apparently not been performed.

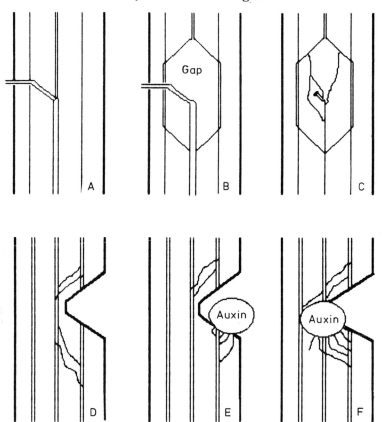

Fig. 7.1 Leaf gaps and the possible role of auxin in their formation. A and B: Contacts of the vascular tissues leading to a leaf. These contacts are either direct (A) or indirect (B). When they are indirect there is a gap of parenchyma, rather than vascular tissue, above the region where the strands connect with the leaf. C: When the leaf was damaged in the primordial stage, its vascular contacts were reduced. When the damage was of a very young primordium, vascular strands differentiated across the leaf gap. D: Wounds in the vascular system led to regeneration. The new strands were directed so they connect to the cut strands below the wound, the ones that did not lead to any leaves. E: The 'directing effect' of the cut strands was reduced or prevented when they were supplied with an exogenous source of auxin. F: A source of auxin induced new strands. These formed preferential contacts with strands that have been cut so they did not lead to any leaves.

Experimental evidence thus suggests that the occurrence of leaf gaps is an expression of the degree to which a leaf dominates the adjacent axial tissues. From an evolutionary point of view it is important that the occurrence of leaf gaps need not be specified by special genes and need not necessarily be correlated with qualitative traits that might characterize homologous organs. Instead, leaf gaps depend on quantitative phenomena: the relative formation of auxin, and presumably other developmental signals, in leaves and adjoining stems. This could mean that evolutionary changes of leaf gaps are not difficult and need not be rare. In agreement with this conclusion, leaf gap formation in angiosperms is not always regular or invariable. For example, in the water plant *Elodea*, where the leaves have limited vascular systems, leaf gaps are absent. Even in peas leaf gaps are not invariable: at the base of the first leaves on seedlings, which are small bracts, their occurrence is sporadic.

The localization of intercalary growth

Tubular structures, such as corollas and ovaries, appeared at various times during the evolution of angiosperms. Generally, if not invariably, the formation of these structures occurred by processes of 'intercalary concrescence' or 'congenital fusion': a continuous ring of intercalary growth at the base of a whorl of newly initiated organs (Cusick, 1966; Sattler, 1978; Stebbins, 1974). The same processes were also the origin of peltate leaves (Fig. 7.2; Stebbins, 1974). The ontogenetic sequence can be recognized in the mature state by small or even minute protrusions above the structure produced by intercalary growth, as well as by the course of the vascular tissues. Changes in the localization of intercalary growth are therefore an important evolutionary process. Since phytohormones are known to influence growth processes, the question to be considered here is whether changes in these substances could change the location of intercalary growth.

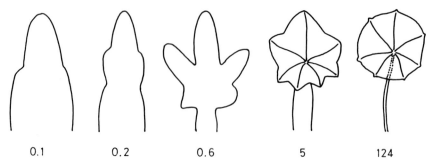

0.1   0.2   0.6   5   124

Fig. 7.2 Development of *Tropaeolum majus* leaves. Leaves of different ages are drawn the same size while the actual length in mm appears below each leaf. This was done so as to emphasize changes of shape during leaf ontogeny. Primordia of leaf lobes appear when the leaf is very small, but most of the leaf surface is formed by intercalary growth at the base of these embryonic lobes. Peltate leaves of *Tropaeolum* and other plants are therefore common examples of 'intercalary concrescence' or 'congenital fusion'.

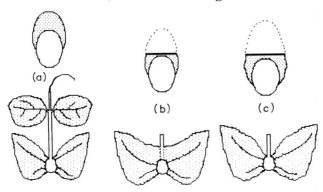

Fig. 7.3 New intercalary growth during pea leaf regeneration. A: Normal primordium and mature leaf. B: All distal parts of the primordium were cut when it was about 0.5 mm long. The increase in the size of the stipules included intercalary growth, and this joined the stipules with the petiole. Such growth did not occur in untreated plants. C: Similar cut, but in a primordium over 1 mm long. The remaining part of the leaf grew to an unusual size, but their shape was similar to that found in untreated plants.

When a young leaf primordium is damaged, the mature size of the remaining parts is larger than in leaves that are left intact (Sachs, 1969). In primordia over 0.2 mm long new forms appear: the regenerative processes do not necessarily replace the missing parts. These new forms include joining by intercalary growth of parts that are normally separate (Fig. 7.3). It is remarkable that this intercalary growth occurs in locations and in species where it is unknown in intact plants. It may be concluded that the capacity for 'intercalary concrescence' may be a common trait of meristematic tissues, apparently those that have undergone the early stages of polarization. This raises the question of the nature of the developmental controls of this capacity for growth.

Soon after the discovery of auxin it was found that its application to shoot apices causes leaf shape deformations (Laibach and Mai, 1936). Similar effects of auxin, and of substances that interfere with auxin transport, have since been reported for many plants. These observations were readily repeated with pea seedlings, a convenient experimental object for work on primordial tissues. Auxin (1% indoleacetic acid, in a lanolin paste) was applied to growing shoot apices. The deformed leaves (Fig. 7.4) show that the influence of auxin increases intercalary growth, often in locations in which it never occurs in untreated plants. This auxin-induced growth can cause 'intercalary concrescence' or 'congenital fusion': it thus leads to 'phenocopies' of evolutionary processes. These observations suggest that auxin could be a control of the occurrence of intercalary growth.

The appearance of intercalary growth during regeneration indicates that the expression of this developmental capacity is normally limited by correlative relations, and these relations are changed when part of the leaf is removed. This was confirmed by an examination of many leaves deformed by auxin treatments (Fig. 7.4): the increase of growth of one region in response to auxin was accompanied by decreased growth in neighboring regions. The most pronounced changes of shape occurred when the lamina grew more, and the rachis less, than in intact primordia. This means that auxin shifted the location of growth rather than merely increasing the size of part of the leaf. It appears, therefore, that there is a control of the overall

Fig. 7.4 Leaf deformations induced by auxin applied to apices of pea seedlings. A normal leaf appears on the upper left. Other leaves are examples of the wide variety of deformations that resulted from auxin treatment of young primordia. These leaves were deformed by the occurrence of intercalary growth in unusual locations. This growth was accompanied by a reduction of elongation in other parts of the leaf. Indoleacetic acid (1%) in lanolin was applied directly to exposed apices of lateral pea buds. These buds grew following the removal of the main, dominant shoot.

size of a leaf primordium that depends on interactions between its parts. The stage at which this control operates and its mechanism are yet to be defined, but an examination of the effects of auxin does suggest one generalization. Growth at the base of the leaf reduces growth further up, towards the tip; effects in the opposite direction were not found.

The deformations that resulted from the application of auxin were varied and unpredictable. The next question was, therefore, how the effects of auxin become localized, so that shape is changed and not only the size of the leaves. A first object of the research was therefore to find the critical parameters for obtaining uniform, repeatable deformations. Auxin paste was applied carefully, under a dissecting microscope, to different parts of young leaf primordia. Differences in the location of the applied auxin were not, however, correlated with specific deformations of shape. On the other hand, the age of the primordium at the time it was treated with auxin, as judged by its shape and size, had a repeatable relation with the deformations found at maturity (Fig. 7.5). The results obtained to date have been variable; Fig. 7.5 therefore presents an hypothesis rather than an analysis of precise observations. It was a general rule, however, that the larger, and therefore older, the primordium at the time it was treated the more apical the region of abnormal growth. The largest primordia to respond showed modifications of the tip of the leaf, a joining of the tendril with the leaflets. Even these primordia, however, were very small (less than 1 mm long). It may be concluded that during the earliest stages of primordial development there is a shift, from the base towards the tip, of the region of *competence to respond* to auxin treatment.

A careful observation of the deformation of leaf primordia by high auxin concentrations thus reveals the following principles, all of them possible determinants of leaf shape.

1. In developing leaf primordia there is a capacity for intercalary growth that is not normally expressed.
2. This growth is limited by correlative effects of tissues that were formed earlier, effects that regulate the size of the leaf.
3. Another, perhaps related, control of this growth is the availability of auxin.
4. In young primordia there is an age-dependent shift in the region of competence to respond to auxin.

The operation of these four processes or controls means that a mere increase in the *supply* of a natural substance that induces development, such as auxin, could result in a shift of the *location* intercalary growth. Neither a pulse nor a spatial control of gene activity are essential; an increase in concentration, an overall effect of genes, could suffice. The influence of the increased concentration is localized by an interaction with the competence

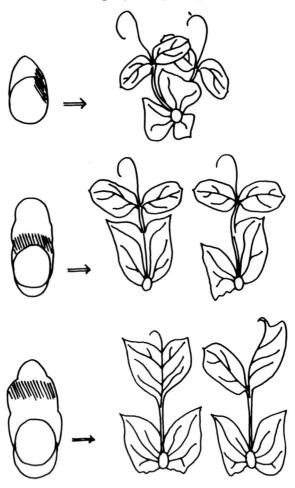

Fig. 7.5 Correlation between the size of primordia and their deformation by auxin. On the left are outlines of apices and leaf primordia, as seen from above. On the right are examples of the leaves that resulted from auxin treatment of apices at the stages represented on the right. The striped areas indicate likely regions of increased intercalary growth, induced by the applied auxin. The location of these regions, the ones most competent to respond to auxin, moves towards the tip as the primordium matures. Treatment of primordia over 1 mm long did not result in leaf deformations.

gradient that depends on the age of the tissue – in reference to a pre-pattern of tissue differences that depends on the apical growth of plants. It is thus possible to understand how quantitative changes, caused by known controls, could lead to *localized* growth and thus to a change in the shape of the mature organ.

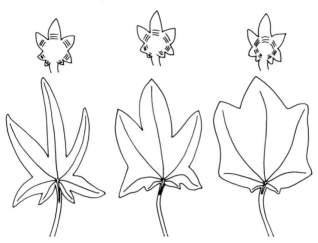

Fig. 7.6 Lobing of leaves of different cotton varieties. The top row shows primordia and the locations (striped) where increased intercalary growth could lead to the various shapes shown in the lower row. The results obtained with pea leaves suggest that an increase in auxin concentration in the leaf primordium could shift intercalary growth, leading to forms with reduced lobes.

This suggests a working hypothesis concerning evolutionary processes that have led to common morphological differences. For example, closely related plants may have either peltate or deeply lobed leaves. Genes that control the relative size of lobes are known in various plants (Hilu, 1983), for example in cotton (Fig. 7.6). These genes could act by changing the supply of phytohormones, such as auxin. This change could also involve the competence of the cells to respond to auxin, or the concentration of auxin in one cellular compartment, so it may not be measurable by available means. The important point concerning the changed location of growth, however, is that cellular change need not be large nor limited to one part of the leaf so as to be expressed by pronounced morphological differences: an increased influence of auxin should lead to a basal *shift of intercalary growth*, and thus to reduced lobing. A similar explanation might apply to the intercalary growth at the base of primordia that results in 'intercalary concrescence' or 'congenital fusion'. Thus the results concerning pea leaves do not allow firm conclusions, but they do suggest the possibility of experimental work.

Correlative inhibition and the origin of new organs

A third, and last, example of the possible role of phytohormonal changes in the evolution of form concerns the differentiation of specialized organs. A potato plant has two types of shoot: vertical ones with broad leaves and

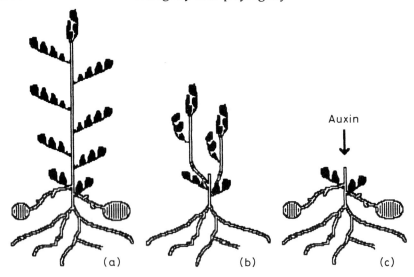

Fig. 7.7 Dominance and hormonal control of lateral shoot differentiation in potato. A: Intact plant. Lateral buds at the base of the plant developed as runners, with tubers at their ends. B: The main shoot was removed at an early stage. The laterals at the base of the plant replaced the removed part and became vertical shoots with broad leaves. C: A source of auxin was placed where the main shoot had been. The auxin imitated growing shoots in inducing the specialized differentiation of the lateral buds.

runners with a storage tuber at their tips. When a vertical shoot of a potato is removed, it is replaced by one or more of the laterals which would have otherwise developed as tuber-bearing runners (Fig. 7.7; Sachs, 1887; Woolley and Wareing, 1972). The result mentioned for potato is, furthermore, but an example of a general rule: when a growing shoot apex is removed the development of one or more of the remaining apices is modified (Snow, 1931; Umrath, 1948). The results of shoot removal thus show that the determination of organ differentiation can be studied experimentally.

The most common modification associated with removal of growing shoots is, however, an increase in growth rate rather than a differentiation or re-differentiation of the remaining apices. The simple experiment of shoot removal is, therefore, a demonstration of the general phenomenon known as 'apical dominance' (Thimann, 1977). It must mean that there are signals that correlate the development of the different parts of a plant and thus have a role in controlling its overall form. In potato and in other plants the influence of these correlative relations is expressed by apical differentiation as well as by growth rate. It thus appears worthwhile to consider the possible role of correlative controls in the morphological evolution of homologous organs.

There is firm evidence that phytohormones act as signals of the correlative relations between shoot apices. The evidence is that phytohormones replace the influence of organs, such as shoots or roots, in which they are known to be formed (Fig. 7.7; Thimann and Skoog, 1933). Though the role of phytohormones in organ correlations has often been questioned, the basic evidence has not been refuted (Sachs, 1975, 1986). The most common arguments have been an unjustified expectation that phytohormones must account for all correlative phenomena and a confusion between the role of hormones in correlative relations and their unknown mechanism of action. It has also been expected that phytohormones should be characterized by a 'typical' influence on a developmental process. As mentioned above, however, at least some phytohormones can be characterized as messengers of organs of one morphological type. The response to their influence, on the other hand, is not characteristic: it varies as a function of the tissue that is observed (Sachs, 1975, 1981).

Thus many facts suggest that auxin can be defined as a major correlative agent of growing shoot tissues (Sachs, 1981, 1986). As such it replaces a main shoot in inhibiting the growth of later buds (Thimann and Skoog, 1933). The important point for the present discussion is that in appropriate plants auxin also replaces vertical, leafy shoots in causing other shoots to differentiate as specialized organs. In potato, for example, auxin causes the lower buds to become runners that develop tubers at their tips (Woolley and Wareing, 1972). Cytokinins, on the other hand, can be defined as major correlative agents of growing roots. As such they have various effects other than the 'typical' one of causing cell divisions. In relation to apical dominance, they replace the roots in promoting the growth of lateral buds, even in the presence of a dominant shoot. In appropriate plants they influence the differentiation of meristems, tending to make shoots grow vertically and bear broad leaves. This is seen in potato, where cytokinins prevent lateral buds from developing as runners (Woolley and Wareing, 1972).

In the discussion above it was suggested that correlative relations depend on specialized, hormonal signals while a role of metabolites essential for growth was not mentioned. Competition for metabolites could also be important in growth rate correlations, but this does not conflict with the statements above, since phytohormonal and competition mechanisms are not mutually exclusive. However, the development of different organs, including ones that are specialized for storage, requires the same metabolites as a dominant shoot. It follows that correlations expressed by the differentiation of apices, rather than by a changed growth rate, can be expected to require the participation of developmental signals, such as phytohormones.

It may be concluded that phytohormones serve as signals, though not necessarily unique signals, determining apical differentiation. The original inequalities between apices could depend on the meristems being of different ages, an invariable consequence of the way plants grow. Age differences, however, have to be expressed physiologically if they are to specify developmental processes. A physiological expression of this type could depend on developmental rates and phytohormonal correlations. This suggests that mutations modifying the activity of phytohormones – their synthesis, transport and the responses of the tissues – would change the quantitative relations between organs of different types. Changes in phytohormone activity could even lead to the appearance or disappearance of specialized organs, such as potato tubers, or to modifications of their relative location on the plant. Mutations could also lead to increased development of a special kind, such as the transverse growth typical of storage organs, in a location specified by phytohormonal relations. All this suggests the working hypothesis that, both in ontogeny and phylogeny, the earliest developmental stages of a specialized organ could be differences in the phytohormonal relations of apices that are becoming specialized.

7.3 DISCUSSION: GENERALIZATIONS CONCERNING THE EVOLUTION OF FORM

As mentioned above, the effects of added phytohormones could be phenocopies of changes caused by point mutations. A first generalization these effects of phytohormones suggest is, therefore, that *changes of single genes could have far-reaching and unexpected consequences*. The modifications of mature form include changes in the location of vascular contacts, the location of intercalary growth and the formation of specialized storage organs. It is perhaps significant that such traits have been used to separate large systematic groups and even morphological categories. Additional examples of large effects of applied phytohormones have been reported. Thus phenocopies of the unusual morphology of the genus *Streptocarpus* were obtained by the application of known substances (Rosenblum and Basile, 1985). The possibility of the changes being so large is clearly dependent on manipulations of control functions. These controls are part of the organization of development and are essential because of its complexity. It follows that even apparent evolutionary 'saltations' of morphological structure, such as unusual structures in the genus *Streptocarpus*, need not be large when viewed as developmental or genetic changes.

A second but no less important generalization is concerned with what does not occur when developmental controls are changed. Though the effects of mutations of one or very few genes can be large, they are still only modifications of existent developmental systems. The new forms result from variations of the relative role, location and timing of different processes, but *the developmental background of the plant does not change readily*. In other words, though they may appear readily, the ontogeny of new forms is constrained and reflects the origin of the plant. Thus, for example, intercalary growth induced in leaf primordia represents a shift in the location of growth, not a new program of leaf development. The differentiation of storage or other specialized organs does not proceed from unique primordia. Instead, some of the primordia of a given category are modified by changes in the relative role and the orientation of growth and differentiation processes that occur in all stems. The presence of leaf gaps, finally, depends only on a shift in the location of vascular contacts, while the contacts themselves are present in all plants.

On theoretical grounds it is possible to suggest a basis for this difference between traits that are labile from an evolutionary point of view and ones that are conservative. Neo-Darwinian principles predict that genetic changes are rare and random. It follows that control functions which may depend on one substance or signal could be labile to changes. The appearance of a new type of growth, primordial organization or differentiation, on the other hand, would constitute new developmental processes and would depend on the coordinated function of many genes. When these genes appear one at a time they generally do not serve an adaptive function, and are therefore likely to be eliminated by selection. When a functional problem or possibility arises, therefore, solutions based on *modifications of controls of development would be the first to appear* while multigene developmental processes could be expected to remain unchanged. Even if these are not ideal solutions, they would compete and thus inhibit the slower appearance of new multigene processes. The economy or efficiency of the individual steps of development is presumably important in itself, but often less so than the various adaptive functions of the mature structures resulting from this development.

It may be appropriate to illustrate the consequences of this discussion by considering the expected evolution of new plant organs. *Sui generis* organs should not occur: they would require many novel and co-ordinated gene activities to appear at the same time. Instead, when adaptation calls for a new structure, some or all of the members of an existent organ type could be modified. Many of the mutations leading to such modifications influence controls of rate, duration, location and relative timing of growth and differentiation, rather than changing the developmental processes that are the 'building blocks' of form (Fig. 7.8). Mutations would thus be expressed by quantitative changes in the expression and localization of existent processes. The locations where these mutations would be expressed would be specified by correlative relations, and thus would be dependent on the position of the modified organs on the plant. Changes of control functions, furthermore, could cause developmental processes originally expressed in one organ to appear in another (Corner, 1958; Zimmermann, 1961; Sachs, 1982). The structures that result from such changes would be 'hybrid'

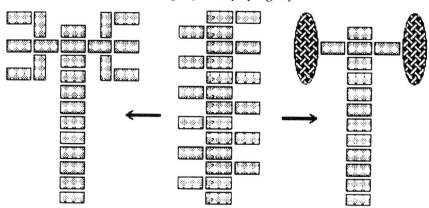

Fig. 7.8 Schematic presentation of two types of ontogenetic changes that could occur during evolution. The hypothetical structure in the center changed in two ways. On the left is a result of a modification of the controls of the number, distribution and orientation of its 'building blocks'. On the right is a result of the appearance of a new type of 'building block'. Both available facts and theoretical considerations suggest that changes of the controls of development are more common than the appearance of new developmental processes that depend on many genes.

between organs with different homologies (Sattler, 1974; 1984). All this is, of course, but a way of stating the basic tenets of plant morphology, discussed by botanists at least since Goethe's 'metamorphosis' of leaves.

Thus though changes of plant form can occur readily, not all changes are possible. The changes that do occur, furthermore, are modifications of existent ontogeny and in this sense evolution is constrained by ontogenetic systems (Goodwin, Holder and Wylie, 1983). A major result of these constraints is that homologies between organs can be recognized, and that they are most apparent in young, primordial stages (Sachs, 1982). It may be concluded that the effects of known controls of development conform with, or even predict, the conclusions of 'classical' plant morphology. This is major support for the working hypothesis mentioned above: there is, at present, no need to assume an evolution of form that is not dependent on genes or contradicts Neo-Darwinian principles.

The separation of development into controls and processes, the latter being the 'building blocks' of Fig. 7.8, is certainly artificial. This separation does, however, express extremes of reality relevant to the likelihood of evolutionary modification. It suggests that a knowledge of the controls of plant ontogeny could eventually provide measures of the relative importance of different morphological and developmental, characters as indicators of evolutionary relations. Ontogeny could thus offer quantitative criteria for the study of phyletic groups (Sachs, 1982). It could account for an apparent paradox: it is often possible to tell a group of plants by details of their structure, but not by the overall shape. Thus idioblasts, such as hairs and glands, are often excellent phyletic guides while the shape and size of organs or the entire plant are not. Different ontogenetic systems could, furthermore, be labile to different evolutionary modifications. This could be one reason for the *phyletic diversity* of plant evolution (Sachs, 1978): for evolution not having proceeded only by adaptive radiations of very few lines, the ones most successful at photosynthetic and other major functions. The various parallel lines could differ, for example, in the potential of their developmental systems for the formation of new secretory and other structures related to the evolving biotic environment.

7.4 THE NEED FOR THE STUDY OF COMPARATIVE MORPHOGENESIS

A key practical question raised by this discussion concerns ways of recognizing and evaluating different types of ontogenetic change. There is no doubt that essential knowledge is bound to come from the rapidly advancing work on molecular mechanisms of gene and hormone action. These, however, could not be the entire story; the examples and discussion above show that important constraints depend on the relations between different processes, on a level of the patterning and organization of ontogeny. Information that one gene or one substance is all that has changed is therefore important, even essential, but it is not a sufficient criterion for an eventual decision whether the outcome of this change is labile or conservative from an evolutionary point of view. This means that intermediate controls, between molecules and the ecological adaptations of mature structure, can not be ignored.

A first, as yet neglected, stage of the study of evolution of ontogeny could be a classification of differences in the course of development – differences considered as changes of ontogenetic controls and not only microscopic events of growth and cell division. The results above show that there should also be specific, quantitative controls of the expression of different processes. In addition, changes in the relative location of developmental processes could be expected, and prime examples of such changes are homeotic mutations. Separate from these would be classes of changes in the relative timing of different processes, changes that could lead to large differences in mature form. All these classes of ontogenetic changes may be either early or late in the development of individual organs and may involve either the formation of developmental signals or the competence to respond to them.

No classification of ontogenetic differences appears to have been documented, though some changes of ontogenetic controls have, of course, been mentioned in the literature, especially the literature dealing with animals (Alberch, 1983). The stress has, however, been on heterochrony, changes in the relative timing of processes that influence the entire life of the organism (Gould, 1977). Genes that shift the timing and hence location of flowering are known, and there have been suggestions that neoteny, one form of heterochrony, has played an important role in plant evolution (Corner, 1958; Zimmermann, 1961; Takhtajan, 1972). The examples of the effects of hormones considered above show, however, that changes in the location in which different processes are expressed may be at least as important modifications of form as changes of relative timing. No less important, these examples show that the attention to the development of the entire organism might have been misleading; changes of tissues and organ primordia must also be considered (Stebbins, 1974).

Thus an analysis of ontogeny based on comparative studies is called for. The most promising differences to be studied are ones that involve single factors and, hopefully, single processes. This comparative aspect could be an important use of knowledge of the genetic controls of form and of experimental manipulations, such as studies of regeneration and the addition of known phytohormones. The study of the structural differences must include comparative descriptions of ontogeny, with a major stress on the neglected primordial stages. Wherever possible, concentrations of critical substances and details of molecular mechanisms would naturally be desirable. The examples above show, however, that consequences of the interdependence of processes may be recognized even when the processes themselves are 'black boxes'. An analysis of the basis of structural differences could be a major contribution towards understanding the controls of patterned development and at the same time it would be an essential step in any causal study of the evolution of form. It is only on the basis of an analysis of this type that the appearance of new multigenic processes could be recognized. Perhaps it is such processes that should be recognized as major steps in evolution, as true evolutionary novelties.

7.5 REFERENCES

Alberch, P. (1983) Mapping genes to phenotypes, or rules that generate form. *Evolution*, 37, 861–3.

Bachmann, K. (1983) Evolutionary genetics and the genetic control of morphogenesis in flowering plants. *Evol. Biol.*, 16, 157–208.

Corner, E. J. H. (1958) Transference of function. *J. Linn. Soc. London*, 56, 33–40.

Cusick, F. (1966) On phylogenetic and ontogenetic fusions. in *Trends in Plant Morphogenesis*, (ed. E. G. Cutter) Longmans, London. pp. 170–83.

Frankel, J. (1983) What are the developmental underpinnings of evolutionary changes in protozoan morphology. *Symp. Br. Soc. Dev. Biol.*, 6, 279–314.

Goodwin, B. C. (1982) Development and evolution. *J. Theor. Biol.*, **97**, 43–55.

Goodwin, B. C., Holder, N. and Wylie, C. C. (1983) Development and Evolution. *Symp. Br. Soc. Dev. Biol.* Vol. 6.

Gottlieb, L. D. (1984) Genetics and morphological evolution in plants. *Am. Naturalist*, **123**, 681–709.

Gould, S. J. (1977) *Ontogeny and Phylogeny*. Harvard University Press, Cambridge, MA.

Hilu, K. W. (1983) The role of single-gene mutations in the evolution of flowering plants. *Evol. Biol.*, **16**, 97–128.

Hooykaas, P. J. J., Ooms, G. and Schilperoort, R. A. (1982) Tumors induced by different strains of *Agrobacterium tumefaciens*. in *The Molecular Biology of Plant Tumors* (eds G. Kahl and J. S. Schell), Academic Press, New York, pp. 373–90.

Laibach, F. and Mai, G. (1936) Über die künstliche Erzeugung von Bildungsabweichungen bei Pflanzen. *Wilhelm Roux' Archiv für Entwicklungsmechanik der Organismen*, **134**, 200–6.

Maynard Smith, J. (1983) Evolution and development. *Symp. Br. Soc. Dev. Biol.*, **6**, 33–57.

Phinney, B. O. (1985) Gibberellin A_1 dwarfism and shoot elongation in higher plants. *Biol. Plant.*, **27**, 172–9.

Rosenblum, I. M. and Basile, D. V. (1985) Hormonal regulation of morphogenesis in *Streptocarpus* and its relevance to evolutionary history of Gesneriaceae. *Am. J. Bot.*, **71**, 52–64.

Sachs, J. von (1887) *Lectures on the Physiology of Plants*. Translated by H. Marshal Ward. The Clarendon Press, Oxford.

Sachs, T. (1968) On the determination of the pattern of vascular tissue in peas. *Ann. Bot.*, **32**, 781–90.

Sachs, T. (1969) Regeneration experiments on the determination of the form of leaves. *Israel J. Bot.*, **18**, 21–30.

Sachs, T. (1975) Plant tumours resulting from unregulated hormone synthesis. *J. Theor. Biol.*, **55**, 445–53.

Sachs, T. (1978) Phyletic diversity in higher plants. *Pl. Syst. Evol.*, **130**, 1–11.

Sachs, T. (1981) The control of the patterned differentiation of vascular tissues. *Adv. Bot. Res.*, **9**, 151–262.

Sachs, T. (1982) A morphogenetic basis of plant morphology. *Acta Biother.*, **31A**, 118–31.

Sachs, T. (1986) Cellular interactions in tissue and organ development. *Soc. Exp. Biol. Symp.* Vol. 40 (*Plasticity in Plants*, ed. D. H. Jennings and A. J. Trewavas) pp. 181–210.

Sander, K. (1983) The evolution of patterning mechanisms: gleanings from insect embryogenesis and spermatogenesis. *Symp. Br. Soc. Dev. Biol.*, **6**, 137–60.

Sattler, R. (1974) A new conception of the shoot of higher plants. *J. Theor. Biol.*, **47**, 367–82.

Sattler, R. (1978) 'Fusion' and 'continuity' in floral morphology. *Notes R. Bot. Gard. Edin.*, **36**, 397–405.

Sattler, R. (1984) Homology – a continuing challenge. *Syst. Bot.*, **9**, 382–94.

Snow, R. (1931) Experiments on growth and inhibition. II New phenomena of inhibition. *Proc. R. Soc. Lond. B*, **108**, 305–16.

Stebbins, G. L. (1974) *Flowering Plants. Evolution Above the Species Level.* The Belknap Press of Harvard University, Cambridge, MA.

Takhtajan, A. (1972) Patterns of ontogenetic alterations in the evolution of higher plants. *Phytomorphology*, **22**, 164–70.

Thimann, K. V. (1977) *Hormone Action in the Whole Life of Plants.* The University of Massachusetts Press, Amherst.

Thimann, K. V. and Skoog, F. (1933) Studies on the growth hormone of plants. III The inhibiting action of growth substance on plant development. *Proc. Nat. Acad. Sci. Wash.*, **19**, 714–16.

Umrath, K. (1948) Dornenbildung, Blattform und Blutenbildung in abhängigkeit von Wuchsstoff und Korrelativer Hemmung. *Planta*, **36**, 262–97.

Woolley, D. J. and Wareing, P. F. (1972) The role of roots, cytokinins and apical dominance in the control of lateral shoot form in *Solanum andigena. Planta*, **105**, 33–42.

Zimmermann, W. (1961) Phylogenetic shifting of organs, tissues and phases in pteridophytes. *Can. J. Bot.*, **39**, 1547–53.

Editors' commentary on Part 3

It seems obvious that to understand how new morphological characters evolve in plants requires integrated genetic and developmental analyses, yet there are few examples. Most of the genes that have been examined that change morphology arose as mutations in cultivated plants and were studied outside an evolutionary context, for example, hooded awn in barley (Stebbins and Yagil, 1966), afila and acacia mutants of pea leaf (Marx, 1974) and numerous others reviewed by Hilu (1983). The study of hooded barley was particularly important because it distinguished the initial changes in timing and orientation of cell divisions brought about by a single mutant gene from subsequent epigenetic consequences—the formation of a primordium-like tissue mass on the upper surface of the awn that led to the differentiation of the unusual hood. The result was important because it convincingly demonstrated that a large difference in morphology need not reflect numerous and/or complex genetic changes.

The conclusion that apparently complex phenotypic changes can be achieved by simple genetic means commonly emerges from developmental genetic studies (Kaiser, 1935; Bachmann, 1983; Garcia-Bellido, 1983), though we do not know how frequent this is in nature. Its significance has been generally neglected by plant evolutionists and others because of an *a priori* expectation that evolution was mostly gradual and resulted from selection of mutations having relatively slight effects on the phenotype; the subject is reviewed by Bachmann (1983) and Gottlieb (1984).

Most plant geneticists are engaged in breeding work to improve crop yields or identify markers for breeding programs. They frequently examine end products of growth such as height, leaf area, fruit size, and seed number or weight. Characters of this type are often responsive to factors that affect general physiological state and, consequently, are influenced by genes that affect numerous processes such as photosynthesis, nutritional state, assimilate partitioning, and other metabolisms. The analysis assesses all genes in the breeding population that contribute to phenotypic variance in yield components without regard to the fact that the contribution of many genes reflects their effects on growth and vigor rather than on specific features of character expression. Thus genetic studies of yield components generally

177

conclude that the continuous variation patterns found in progenies are due to large numbers of segregating factors. The results are valuable for many agronomic purposes such as predicting population response to various selective regimes, but provide little information about the genetic basis of differences between particular traits.

The integration of phenotype and underlying genotype is also ignored by most plant morphologists. Many elegant analyses of morphological issues have been carried out, but few of them have taken advantage of genetical designs. The lesson from hooded barley is put aside. The consequence is that no matter how precise the description of the course of development of different morphs, the extent of the differences that might be attributed to changed heredity cannot be identified. The result is that we don't learn how genetic changes modify developmental programs to produce new traits. If traits that confer fitness were studied, combined developmental and genetic analysis would elucidate how differences in fitness evolve.

Our interest in integrating genetical and developmental analyses to understand morphological evolution in plants was anticipated more than 50 years ago by Edmond Sinnott and his students in their studies of shape change in fruits of squashes and peppers (reviewed in Gottlieb, 1984). Nowadays serious information seems to be coming mostly from the molecular biologists. One of the best examples has to do with the molecular basis of anthocyanin pigmentation patterns on snapdragon (*Antirrhinum majus*) flowers.

The wildtype snapdragon normally has full red flowers. The red colors are caused by anthocyanin pigments, and several genes required for pigment synthesis in snapdragon have been identified. Mutations at the *pallida* locus may result in completely ivory flowers or ivory flowers with red spots or sectors. The latter phenotypes are particularly interesting because they are the consequence of somatic excision of a transposable element from the promoter region upstream of the *pallida* structural gene (Coen, Carpenter and Martin, 1986). These excisions are often imprecise in the sense that they add or subtract small numbers of nucleotides. As a result, they permit partial gene function leading to flowers having unusual spatial patterns or different intensities of pigmentation. Many of the resulting changes are stable once the transposable element has departed.

How these minute changes in gene sequence are translated into pattern differences is still under study. But the implications for population genetics are already intriguing. As suggested by Coen, Carpenter and Martin (1986), when an active transposon attains moderate frequency in a natural population, repeated imprecise excision can generate a burst of allelic variation in a short time period. The inference is that variation patterns need not always reflect uniform and gradual processes of single base-pair substitutions. The results with *pallida* will stimulate molecular studies of floral patterns in many species. Some of these patterns affect pollinator behavior so that we can anticipate not only more frequent connections between molecular biology and development, but also between these disciplines and ecological genetics.

Several methods to identify specific genes involved in flower development were reviewed during a symposium devoted to the molecular aspects of plant development (Goldberg, 1987). The technical problem is to identify genes that 'reorganize' vegetative into floral meristems that produce the sepals, petals, stamens and pistils characteristic of flowers. One approach isolates cDNA clones corresponding to messages present during early stages of meristem reorganization and tests their consequences in tissue cultures that can be induced to undergo specific morphogenesis. A second approach isolates mutants that cause meristems to make one organ in place of another. Such homeotic mutants have been induced in *Arabidopsis thaliana* by chemical mutagenesis. Since this species has the smallest known genome in higher plants, it may prove possible to map the mutant to a particular DNA restriction fragment. More precise localization on the fragment can be carried out by additional procedures, and the actual gene can be identified, at least, in principle, by transforming cells with the wildtype gene. If the correct gene is added, the mutant is expected to be complemented and plants regenerated from transformed cells will be normal. A third method is to transform plants with specific transposons since genes mutated by insertion of a transposon can be obtained by established methodologies (Paz-Ares *et al.*, 1986; Wienand *et al.*, 1986; Cone, Burr and Burr, 1986). Thus, it seems certain that molecular techniques will make available many plant genes in which altered sequences affect morphology. Eventually it will be possible to study the evolution of certain morphological traits by examining homologous genes isolated from related species. The potential for important interactions that take advantage of evidence from molecular biology, development, and evolution is now before us.

The prospects are exciting but it should not be thought that the tough problems are already solved or that gene sequences *per se* will reveal how morphogenesis proceeds. Substantial insight into development in plants has accumulated from traditional studies and is likely to provide important guideposts for molecular approaches. From this standpoint, we present the thoughtful chapter by Tsvi Sachs (Chapter 7) which describes how quantitative changes in the level of auxin, or perhaps in the competence of particular cells or tissues to respond to auxin, changes readily imagined to reflect different types of specific gene action, might result in localized changes in form or shifts in the spatial position of new growth.

A key question for Sachs is why some traits are conservative in evolution whereas others are not. He supposes that answers will reveal the extent of complexity (the number and extent of interactions) of the controlling gene systems. He also notes that mutations in genes that act early in ontogeny are likely to have different and more serious consequences than those that act late because the earlier the change the greater the probability of its ramification during development. He concludes by emphasizing the importance of changes in single genes and points out that even large changes in morphological structure 'need not be large when viewed as developmental or genetic changes'. But, of additional interest is that even though such changes are possible, the changed morphology does not signal changes in 'developmental background'. Controls can be modified but programs (multigene processes) are not readily evolved. From this standpoint, it is 'easier' to modify existing organs to serve new functions than to evolve new organs for novel purposes. Ledyard Stebbins has written extensively about such 'transference' of function and has provided numerous examples (Stebbins, 1970a,b, 1974).

REFERENCES

Bachmann, K. (1983) Evolutionary genetics and the genetic control of morphogenesis in flowering plants. *Evol. Biol.*, **16**, 157–208.

Coen, E. S., Carpenter, R. and Martin, C. (1986) Transposable elements generate novel spatial patterns of gene expression in *Antirrhinum majus*. *Cell*. **47**, 285–96.

Cone, C., Burr, F. A. and Burr, B. (1986) Molecular analysis of the maize anthocyanin regulatory locus *C1*. *Proc. Natl. Acad. Sci. USA*, **83**, 9631–5.

Garcia-Bellido, A. (1983) Comparative anatomy of cuticular patterns in the genus *Drosophila*, in *Development and Evolution* (eds B. C. Godwin, N. Holder and C. C. Wylie), Cambridge University Press, Cambridge, pp. 227–55.

Goldberg, R. B. (1987) Emerging patterns of plant development. *Cell*, **49**, 298–300.

Gottlieb, L. D. (1984) Genetics and morphological evolution in plants. *Am. Naturalist*, **123**, 681–709.

Hilu, K. W. (1983) The role of single-gene mutations in the evolution of flowering plants. *Evol. Biol.*, **16**, 97–128.

Kaiser, S. (1935) The factors controlling shape and size in *Capsicum* fruits: a genetic and developmental analysis. *Bull. Torrey Bot. Club*, **62**, 433–54.

Marx, G. A. (1974) A scheme for demonstrating some dominant genetic principles in the classroom. *J. Hered.*, **65**, 252–4.

Paz-Ares, J., Wienand, U., Peterson, P. A. and Saedler, H. (1986) Molecular cloning of the *c* locus of *Zea mays*: a locus regulating the anthocyanin pathway. *EMBO J.*, **5**, 829–33.

Stebbins, G. L. (1970a) Adaptive radiation in angiosperms. I. Pollination mechanisms. *Ann. Rev. Ecol. Syst.*, **1**, 307–26.

Stebbins, G. L. (1970b) Transference of function as a factor in the evolution of seeds and their accessory structures. *Israel J. Bot.*, **19**, 59–70.

Stebbins, G. L. (1974) *Flowering Plants: Evolution above the Species Level*. Harvard University Press, Cambridge, MA.

Stebbins, G. L. and Yagil, E. (1966) The morphogenetic effect of the hooded gene in barley. I. The course of development in hooded and awned genotypes. *Genetics*, **54**, 727–41.

Wienand, U., Weydemann, U. Niesbach-Klosgen, U. Peterson, P. A. and Saedler, H. (1986) Molecular cloning of the *c2* locus of *Zea mays*, the gene coding for chalcone synthase. *Mol. Gen. Genet.*, **203**, 202–7.

PART 4

Adaptation: two perspectives

CHAPTER 8

Biophysical limitations on plant form and evolution

KARL J. NIKLAS

8.1 INTRODUCTION

The botanist is confronted with a remarkably diverse group of organisms that nonetheless have identical or very similar metabolic requirements. Apart from a relatively few parasitic species, the majority of plants require similar resources to synthesize their primary and secondary metabolites (Farquhar, Caemmerer and Berry, 1980; Mooney and Gulmon, 1979; Troughton, Card and Hendy, 1974). And with a few exceptions, such as CAM and C_4 metabolism, the biosynthetic pathways leading to these metabolites are virtually identical (Pearcy and Ehleringer, 1984; Schmitt and Edwards, 1981). Hence the physiological diversity among many algae, bryophytes, pteridophytes and seed plants is much more limited than is that of the vertebrates. Yet among the land plants, and the embryophytes in particular, the divergence in form and structure often requires that biochemical and ultrastructural evidence be used to prove common ancestry (Stewart and Mattox, 1975). The combination of metabolic homogeneity and structural heterogeneity has produced a dilemma for the functional morphologist. The growth and survival of a green plant is predicated on a photosynthetic imperative, yet form can be highly variable. Four explanations have been offered to reconcile this apparent contradiction:

1. Structural heterogeneity among land plants is less than it appears, and masks essential similarities. This explanation emphasizes the convergence seen in the shape and anatomy of different organs which can be functional equivalents (e.g. leaves and stems). For example, the phyllids of mosses, the thalli of liverworts, the micro- and megaphylls of vascular plants, and the cladodes of many succulents are frequently dorsiventral and possess modifications that enable them to deal with the conflicting demands of gas exchange and water conservation. Hence, everyday

185

experience indicates that there are principles underlying the 'design' of photosynthetic organs. However, by emphasizing the similarities among these structures, the potential exists to trivialize significant differences among them.

2. There is a tremendous tolerance in how plants cope structurally with their physiological requirements. Provided an organ meets some minimum functional level of performance, its shape, size, and internal structure could vary widely. Hence, the relationship between form and function may be sufficiently relaxed to permit marked variation in form before function is impaired. This explanation stands in contrast to the next one and with it sets up a polarized view of adaptation.

3. The diversity seen in plant form could reflect extremely 'fine-tuned' adaptations. Thus, (2) and (3) argue for either a very loose or a very tight correlation between form and function, yet both predict the same outcome, high morphological/anatomical diversity.

4. The diversity of plant morphology may reflect phylogeny. Some plant lineages are more ancient than others. As such, their structural features can be viewed as legacies. Clearly, the shape, size, and anatomy of some plants differ because of their phylogeny. The bryophytes are limited in vertical growth because they lack vascular tissue. Many gymnosperms rely on abiotic seed dispersal because they do not produce fruits. Nonetheless, this explanation confuses the notion of 'primitiveness' with 'archaism'. The bryophytes are an ancient and hence archaic lineage of plants. However, they are extremely successful in their ecological niches, as the conifers are in theirs.

Clearly these four explanations are not mutually exclusive. By selecting them, I wish only to emphasize the differences among them and perhaps highlight some of their strengths and weaknesses. Regardless of philosophical inclination, however, it is obvious that each explanation attempts to translate form into the parlance of function, if for no other reason than to deny the purported relationship between the two. Accordingly, both the adaptationist's and the non-adaptationist's explanations for the diversity of plant form require a critical evaluation of how well plants perform the tasks thought to be essential to growth and reproduction. Without a quantitative basis for comparison, arguments concerning adaptative morphology are likely to degrade into opinion.

Fortunately, the relationship between form and function in plants is amenable to quantitative analysis. For example, each of the major physiological requirements for photosynthesis can be translated into the syntax of an appropriate discipline of engineering. Thus, gas exchange, light interception, mechanical stability and hydraulics are essential to photosynthesis and vegetative growth. And each in turn can be quantified according to principles of physics and engineering. An essential issue, however, is whether or not such analyses yield sufficient evidence to support one or more explanations for the morphological diversity of plants.

The present chapter outlines some of the structural requirements for photosynthesis in terrestrial plants. These requirements will be shown to result in conflicting structural solutions that must be reconciled in order to perform all the functions essential to growth. Because there is no single 'best' method to reconcile these conflicting design requirements, there is no single 'optimal' plant shape. Accordingly, morphological diversity is an expected outcome of the assumptions made in the analysis. This diversity in form is compounded by the simple and evident conclusion that a variety of morphological factors can be covaried to yield structures with comparable levels of performance. Therefore, underlying principles dictate plant shape. These principles can be used to deduce biologically significant generalizations concerning the relationship between form and function, but beyond them there exists a plethora of variation reflecting idiosyncracies of environment, development, and phyletic history.

8.2 BASIC REQUIREMENTS

All terrestrial plants must fulfill the basic requirements for survival, growth, and reproduction:

1. They must intercept sunlight which is the energy source for photosynthesis (Geller and Nobel, 1986; Nobel, 1977, 1980);
2. They must sustain the weight of portions elevated above ground (Archer and Wilson, 1973; Gordon, 1978; Wainwright *et al.*, 1976);
3. They must exchange gases with the external atmosphere (Bewley, 1979; Caemmerer and Farquhar, 1981);
4. They must absorb and conduct liquids (Carlquist, 1975; Giordano *et al.*, 1978; Rand, 1983); and,
5. They must reproduce (Grime, 1979).

Some of these functions have been the object of intensive theoretical and empirical examination. Gas exchange and water loss, for example, have occupied the attention of ecophysiologists for decades. The conflicting demands between the need to exchange CO_2 and O_2 with the atmosphere and the need to conserve water figure prominently in every textbook and discussion of stomata. Other design constraints are equally obvious though frequently ignored (such as the increased bending moment imposed on leaves and stems as their vertical descent angle increases from 0° to 90°). The issue raised by these examples is more germane than the particulars, however. Each of the five requirements for growth and survival has its own 'design specifications'. Some of these may stand in conflict with others. A large surface area to relative volume maximizes light interception and gas exchange but also maximizes water loss. Large deflection angles can maximize light interception but they also maximize bending stresses. The reconciliation of these antagonistic 'optimal' shapes or postures can set limits on the range of possible morphologies. In addition, some of the requirements are dependent upon others. A not so obvious example is the dependence of the flexural rigidity of primary plant tissues and organs on water content.

Can each of the four basic requirements for plant growth and survival be mathematically quantified and interrelated? The answer is a qualified yes and depends on the extent to which precision is required. If we neglect reproduction for the time being, the remaining four functions of a plant (light interception, mechanical stability, gas diffusion, and hydraulics) can be interrelated by means of geometric parameters relevant to plant size, shape, and internal structure.

The 'geometry' of light interception and gas diffusion

Most plants get 'bigger' because of indeterminate growth resulting from continued meristematic activity. Assuming that the ratio of surface area to volume must be maintained or kept above some minimum value during ontogeny, a reasonable question is 'What geometries are capable of increasing in absolute size while maintaining reasonably consistent surface area to volume ratios?' Spheres are essentially useless 'geometric solutions'. However, oblate and prolate spheroids, and cylinders are fairly good. For example, an increase from 0.1 to 10 mm in the major semiaxis of an oblate spheroid (with a minor semiaxis of 0.1 mm) results in a 50% reduction in surface area/volume, while a comparable geometric change in a prolate spheroid results in less than a 22% reduction (Fig. 8.1). Provided that both the major and minor semiaxes of a spheroid can be covaried, these geometries can maintain, reduce or increase S/V over any range of values. Cylindrical geometries can 'grow' by means of an increase in length or an increase in unit radius. Changes in S/V are easily calculated. Elongation is much the preferred route to minimize a reduction in S/V, since for any length, S/V decreases sharply as the circumference of the geometry increases (Fig. 8.2).

Accordingly, if gas diffusion is a function of S/V, plant morphology would converge on prolate or cylindrical geometries, since these shapes can maintain or minimize the reduction of S/V as absolute size increases. However, the ability of spheroids and cylinders to intercept light involves geometry as well. If the ability to intercept light is taken as a function of the projected surface area divided by the total surface area, Ap/A, then oblate spheroids and cylinders are the optimal solutions. Analyses of spheroids and cylinders of different sizes and aspect ratios indicate that the maximum Ap/A is achieved by the extreme flattening of an oblate spheroid (Fig. 8.3). Nor-

Major semiaxis of prolate spheroid (mm)

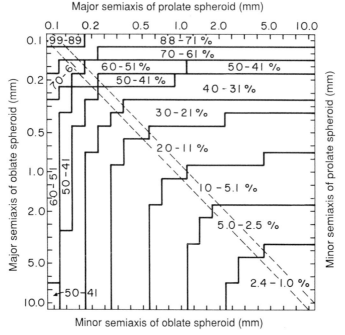

Fig. 8.1 Changes in the ratio of surface area to volume (S/V) as an initially spherical shape is altered into an oblate (lower half of diagram) or prolate spheroid (upper half of diagram). The diagonal dashed lines represent the 'trajectory' of a sphere as it increases in unit radius. All values for S/V are normalized as percentages of the initial sphere (99–98% values in upper left-hand corner). Because the absolute sizes of prolate and oblate spheroids increase, their S/V values decrease relative to that of their originally small spherical progenitor. A prolate spheroid with a minor axis equivalent to its spherical progenitor shows the least reduction in S/V as the major axis continues to increase (99–89% to 88–71%).

malizing all other Ap/A values for various geometries against this maximum indicates that as prolate spheroids 'elongate', the value of Ap/A decreases from 46% to 36% of $(Ap/A)_{max}$. Cylinders decrease from 64% to 79% of the $(Ap/A)_{max}$ as they elongate or thicken (Fig. 8.4).

Depending on their size and aspect ratio, oblate spheroids can modulate S/V and Ap/A in compensatory ways. Consequently, they make a highly versatile geometry with which to construct a leaf in which gas exchange, water loss, and light interception must be regulated. The quantity of light intercepted by a leaf influences its temperature and, through stomatal closure, CO_2 uptake. Large, nearly spherical leaves can conserve water due to their low S/V value and the reduced total direct solar radiation intercepted per day. However, such a leaf geometry has no preferred angle of

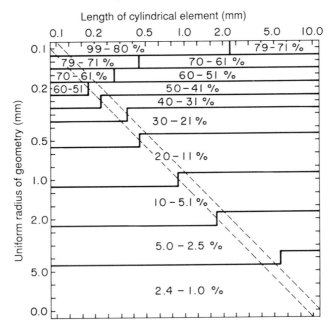

Fig. 8.2 Changes in the ratio of surface area to volume (S/V) as the dimensions of a cylinder capped by two hemispherical ends are varied. The diagonal dashed lines represent the trajectory of a sphere as its unit radius increases. All values of S/V are normalized as percentages against the highest S/V value given for the initial, small sphere. For any uniform radius, S/V decreases as the cylindrical element elongates. The smallest reduction in S/V (as a consequence of an increase in length) results from a very slender cylinder (diam = 0.1 mm).

orientation to the angle of solar radiation, i.e. Ap/A changes little as the solar angle varies diurnally. Therefore, this geometry can capitalize on any direction of sunlight at any time of day that water potential permits stomatal opening. Extremely flattened leaves have a high S/V and also have preferred angles of orientation to the sun (i.e., horizontal leaves intercept more sunlight at noon; vertical leaves intercept more sunlight in early morning or late afternoon) (Ehleringer and Werk, 1986). Provided that the leaf orientation angle is optimal for the times of day when water potential is high, these leaf geometries can maximize photosynthesis. It is probably not coincidental that most leaves are determinate in growth and reach an overall size and aspect ratio that remains constant during the period over which they function. Nor is it coincidental that solar tracking of leaves has evolved in many plant groups.

   In addition to leaf geometry, the geometry of leaf-arrangement on stems (phyllotaxy) can influence the capacity of leaf surfaces to intercept direct

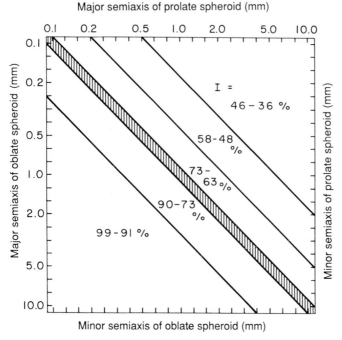

Fig. 8.3 Changes in the ability of spheroids to intercept light as their major and/or minor axes change. Light interception or I is defined as a function of the projected surface area/surface area as the solar angle varies from 0° to 180°. All values for I are normalized as percentages against that of an extremely 'flattened' oblate spheroid which maximizes I (lower left corner of diagram). Prolate spheroids are the least efficient geometries (upper right corner of diagram).

solar radiation. This relationship is potentially significant because leaf temperature and the rate of CO_2 uptake are influenced by the quantity of direct radiation received by leaves (Raschke, 1956; Gates, 1965, 1970; Mooney and Gulmon, 1979; Forseth and Ehleringer, 1980). Leaf orientation also influences light interception (Monsi and Saeki, 1953; Anderson, 1964; Shaver, 1978; Fisher and Honda, 1979; Herbert, 1983, 1984; Ehleringer and Werk, 1986), but may vary independently of the developmental pattern in which leaf primordia are initiated. Nonetheless, the geometry of leaf overlapping and hence the shadows cast by and upon neighboring leaves is influenced in part by phyllotactic pattern (Fig. 8.5). Historically, patterns of leaf arrangement have been dealt with in terms of selection pressures favoring light interception (Wright, 1873; Thompson, 1942; Richards, 1950) (see however, Sinnott, 1960, p. 153; Hardwick, 1986, p. 163).

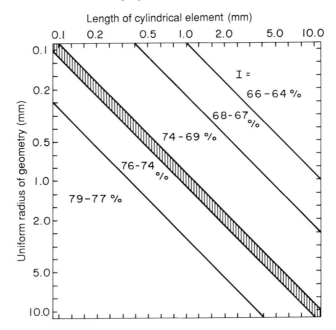

Fig. 8.4 Changes in value of I as the geometry of a cylinder is varied. All values of I are normalized as percentage against I_{max} which is produced by a flattened oblate spheroid (cf. Fig. 8.3).

The influence of leaf arrangement on light interception has not been quantitatively evaluated, because many factors can alter the direct influence of phyllotaxy on photobiology. The geometry of branching patterns and leaf fluttering can produce unpredictable canopy structures and complex light environments within individual plants (Monsi and Saeki, 1953; Fisher, 1986). Solar tracking leaf movements can reorient the laminae of leaves (Shackel and Hall, 1979; Forseth and Ehleringer, 1980; Werk *et al.*, 1983; Herbert, 1984). Similarly, leaf shape and the allometry of leaf expansion and petiole elongation can reduce the overlapping of photosynthetic surfaces. Hence, it is not too surprising that the role of phyllotactic patterns on light interception has been considered inconsequential (Hardwick, 1986). However, phyllotaxy can act as a potentially limiting factor in light interception, particularly for species with rosette growth habits in which the influence of complex branching patterns and leaf fluttering may be largely ignored as factors influencing light interception (Nobel, 1986).

Based on computer simulations, the quantity of light intercepted by single stems bearing equivalent numbers of leaves and total leaf area, is dependent upon leaf shape, phyllotactic pattern, and internodal distance. For any

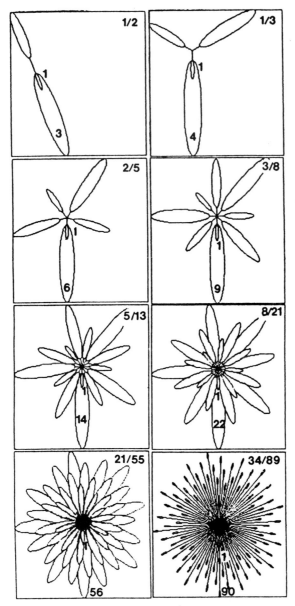

Fig. 8.5 Computer simulated 'rosettes' (as seen from above) corresponding to eight phyllotactic patterns. Each pattern corresponds to a phyllotactic fraction (shown in upper right hand corner) which conventionally represents the number of leaves found (denominator) between successively overlapping leaves along a spiral course around the stem (number of courses is given in the numerator). This fraction multiplied by 360° yields the leaf-divergence angle (cf. Sinnott, 1960). In all simulations, leaf-insertion is in the clockwise direction viewed from above (from Niklas, 1988).

elliptical or round leaf shape, light interception is maximized by a phyllotactic pattern producing a divergence angle near 137.5°, e.g. 8/21, 21/55 and 34/89 (Fig. 8.6). This angle represents the limit approached by the series of Fibonacci fractions which characterizes the various phyllotactic patterns commonly seen in plants (1/2, 1/3, 2/5, 3/8, 5/13, . . . 34/89, or 180°, 120°, 144°, 135°, 138.462°, . . . 137.528°, respectively) (cf. Fig. 8.5). However, as the internodal distance between successive leaves increases, the differences among the capacities of different phyllotactic patterns to intercept sunlight diminish. For example, if a leaf width to length ratio of 1 : 20 is used, the

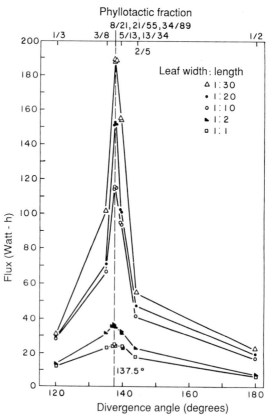

Fig. 8.6   Predicted effect of leaf shape and phyllotactic pattern on the quantity of light intercepted by simulated rosettes. Each datum represents the mean of three simulations in which rosettes with equivalent numbers of leaves (56) and total leaf area (350 cm²) were evaluated at 10.5 h daylengths for three different latitudes (0°, 20°N, 40°N). Each line represents the consequences of changing the leaf aspect ratio (width to length of leaf lamina; petiole lengths are constant at 1.0 cm). In all cases, leaf divergence angles (8/21, 21/55, 34/89) near 137.5° (vertical dashed line) yield the maximum values while fractions 1/3 and 1/2 yield the lowest values. Solar radiation intercepted by rosettes is given on the basis of total watt-h (from Niklas, 1988).

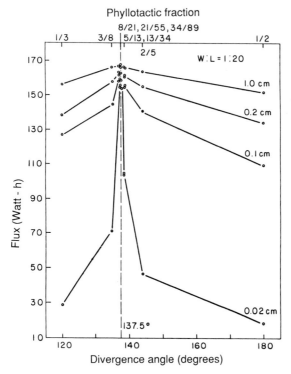

Fig. 8.7 Computer-predicted consequences of changing the internodal distances in rosettes with different phyllotactic patterns but equivalent leaf shape (width: length = 1:20), leaf number (56), and total leaf area (350 cm²). Although the 'ideal divergence angle' (137.5°) consistently yields the maximum quality of light intercepted by leaves, internodal elongation results in the convergence of computed values of flux. Each datum represents the mean of three simulations at 0°, 20°N and 40°N for 10.5 h daylengths (see Fig. 8.6) (from Niklas, 1988).

phyllotactic patterns of 3/8 (135°), 5/13 (138.462°), and 8/21 (137.1°), 13/34 (137.65°), 21/55 (137.455°), and 34/89 (137.528°) yield indistinguishable capacities to intercept sunlight for internodal distances equal to or greater than 1.0 cm (Fig. 8.7). The fractions 1/2 (180°), 1/3 (120°), 2/5 (144°) and 3/8 (135°) persistently yield light interception values that are significantly lower than the other phyllotactic arrangements examined, regardless of relatively large internodal distances. If a leaf width to length ratio of 1:10 is used, then simulations yield similar results, except that shorter internodal distances are required to yield insignificant differences among the 3/8, 5/13, 8/21, 13/34, 21/55 and 34/89 patterns (Fig. 8.8). Hence rounder leaves produce closer values of light-interception capabilities as internodal distances increase from 0.2 mm to 1.0 mm. Nonetheless, a variety of leaf

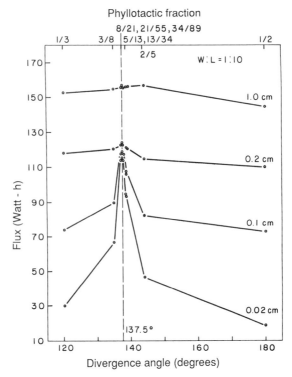

Fig. 8.8   Computer-predicted consequences of changing the internodal distances in rosettes with different phyllotactic patterns but otherwise equivalent morphologies. Leaf aspect ratio equals 1 : 10. See Fig. 8.7 for further explanation (from Niklas, 1988).

shapes yield a single maximum in a graph plotting light interception against phyllotactic pattern for short internodal distances ('rosette' morphologies) (cf. Fig. 8.6).

These simulations indicate that for rosettes with short internodal distances, phyllotactic patterns can be an important feature in ecophysiology. Significantly, most plants with a rosette growth habit or a rosette stage in their development have phyllotactic patterns with divergence angles between 135° and 138° (e.g. *Plantago*, *Lactuca* and *Solidago*), or if the herbaceous growth form has a decussate pattern of leaf arrangement, the rosette form is bijugate. However, for plants with an herbaceous growth habit and relatively large internodal distances (> 1.0 mm), the pattern of leaf arrangement may be far less significant, particularly since potentially 'inefficient' phyllotactic arrangement can be obviated by adjusting leaf size and shape.

Mechanical stability and hydraulics

Even a simple geometric analysis indicates that cylindrical plant axes and flattened leaves can maintain constant ratios of surface area to volume as size increases. The convergence seen in the shape of phyllids, dorsiventral thalli of many liverworts, and micro- and megaphylls is not incompatible with the insights given by analytical geometry. Similarly, it is not difficult to explain why cylindrical, vertically erect axes characterize the most primitive photosynthetic structures of early vascular land plants. Cylindrical axes which elongate apically, maintain or minimize S/V reduction and can increase in length by three orders of magnitude before their light gathering efficiencies are significantly reduced. With the evolution of leaves, oblate spheroid geometries could be elevated on cylindrical axes. Hence, nearby obstructions could be overtopped by a hybrid geometry. The advantages to overtopping, however, could only be achieved by plants whose tissues could accommodate the mechanical stresses of orthotropic growth. Hence, mechanical stability was an inherent problem for the earliest vascular land plants and all their descendents (McMahon and Kronaur, 1976).

Cylindrical axes are excellent load-bearing structures. They are radially symmetrical along their transverse plane and hence have no inherent bias in the ability to resist an axial load that is eccentrically applied. Depending upon their cross-sectional area and length, cylinders can withstand remarkably high vertical loadings before they deflect and eventually buckle. The extent to which a cylindrical plant axis can grow vertically depends on its flexural rigidity, R, which in turn depends upon two features: (1) the geometry of the cross section (e.g. whether it is solid or hollow); and (2) the physical properties of its constituent tissues. The second moment of inertia, I, mathematically expresses the geometric contribution to flexural rigidity. The bulk elastic modulus, E, describes the material properties of the tissues. $R = EI$. The critical load, P_{cr}, which a cylindrical plant axis can bear (i.e. the weight that will just produce large vertical distortions) is related to EI by $P_{cr} = \pi^2 \, EI/4\ell^2$, where ℓ is the length of the axis (Timoshenko and Gere, 1961; Timoshenko and Goodier, 1970). From this equation it is obvious that as axis length increases, P_{cr} dramatically decreases for any value of EI. This equation illustrates a significant biophysical constraint on plant form. The length of vertical plant axes cannot increase indefinitely unless the cross-sectional area of the axis, or I, can be increased and/or unless E can be increased (Fig. 8.9).

Many plants can increase the transverse area of older portions of their stems by secondary growth of the periderm and vascular tissue. The production of secondary wood also modifies the elastic modulus of stems since more liquified tissue is added. By contrast, many plants, both extinct and living, lack the capacity for secondary growth. In many of these taxa,

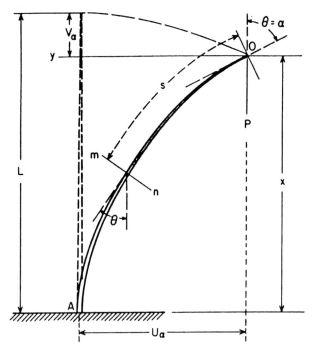

Fig. 8.9   The bending of a cylindrical plant organ (shown slightly tapered) can be related to the large deflection in an ideal column fixed at its base (A) and free at its upper end (O) with an applied apical load (P) larger than the critical load. The curvature of the column, $K = d\theta/ds$, produces a deflection angle ($\theta = \alpha$) at the column's tip. This angle and the vertical (V_α) and horizontal (Y_α) displacements provide descriptors for the geometry of deflection (redrawn from Niklas and O'Rourke, 1987, Fig. 1, which in turn was modified from Timoshenko and Gere, 1961, Fig. 2–28 p. 76).

stem or axial elongation due to primary growth ceases after some period of growth. Hence, ℓ reaches a maximum value which presumably is less than that which would buckle or deflect significantly under the stem's weight. Hence, many grasses, sedges and monocots with cylindrical axes, as well as *Equisetum* have determinate vertical growth. However, many other taxa lack secondary growth and possess indeterminate growth of vertical cylindrical stems or leaves. These structures eventually grow vertically until they can no longer support their own weight and deflect or mechanically fail.

The mechanical stability of a vertical cylindrical plant axis or leaf is much more complex than that of an abiotic structure because of the dependency of E on the extent to which tissues are hydrated (Robichaux, Holsinger and Morse, 1986). The equation $P_{cr} = \pi^2 EI/4\ell^2$ is based on the mechanical theory of elastic stability which in turn makes two assumptions: (1) the elastic modulus, E, is assumed not to change with time and to be uniform throughout the axis; and, (2) the second moment of inertia, I, is also assumed to be constant. Both of these assumptions are violated by plants (cf. Silk, Wang and Cleland, 1982). E varies as a function of the water content of tissues and as a function of tissue age and cellular composition. I can vary

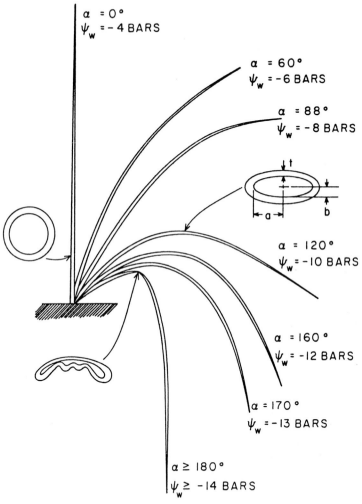

Fig. 8.10 Changes in the deflection angle (α) of the leaves of chive as water potential decreases. Full turgor ($\psi_\omega = -4$ bars) yields a vertical leaf; with a circular or near circular, hollow cross section. Progressive tissue dehydration produces greater deflection angles and regional ovaling in leaf cross section until 'crimping' and permanent mechanical failure of leaves occur (redrawn from Niklas and O'Rourke, 1987, Fig. 2).

even in stems or cylindrical leaves that lack secondary growth. This is particularly obvious if the stem or leaf is hollow. When such an organ bends, it experiences cross-sectional compression and tension and becomes oval.

The relationship between flexural rigidity, R, and hydraulics can be illustrated by examining the behavior of chive leaves as they dehydrate. Chive (*Allium schoenoprasnum* var. *schoenoprasnum*) has nearly cylindrical leaves which grow at their base due to an intercalary meristem. Young bulblets grown in water-saturated soil produce a first flush of leaves that grow vertically even when a small weight is added to the tip. However, if plants are water-stressed, the leaves begin to deflect from the vertical when a small weight is added. As the leaf water potential decreases from full turgor ($\Psi_\omega = -4$ bars) the extent to which leaves deflect increases until, at $\Psi_\omega = -13$ bars, leaves permanently buckle under their own weight and that of the applied weight (Fig. 8.10). Buckling occurs due to localized 'crimping' of the leaf near its base. From detailed measurements of leaf geometry, the changes in R due to tissue water loss can be calculated. Since $R = EI$ and I can be measured directly, the change in E as Ψ_ω varies can be determined. The data from chive leaves indicate that E decreases by over 52% just before leaves permanently buckle (Fig. 8.11). Similarly, I decreases by over 40%. Since R is the product of E and I, the weakening of both the material properties and the geometry of the leaf leads to mechanical failure.

Chive leaves illustrate a balance of conflicting functional requirements. It is economical to produce a hollow stem or leaf because this construction requires less material and hence a lower metabolic investment in comparison to a solid organ of comparable external dimensions. And because it has a lower weight, a hollow organ can grow taller than a solid one of equivalent diameter. (Assuming an identical unit weight of plant tissue, a hollow cylindrical stem can grow 26% taller than its solid cylindrical counterpart.) Yet hollow organs have a lower flexural rigidity than solid ones (assuming equal values for E) and can more easily buckle. Consequently, the mechanical design of a cylindrical organ treads a fine line between the principles of 'Economy in Design' and 'Design for Safety' – a line which is not rigidly fixed because of its dependence on the hydraulic status of the organ.

A useful dimensionless ratio to calculate the effect of 'thinning' a cylindrical organ is given by the equation

$$\frac{(EI)_a}{(EI)_u} = (1 - t/R)^4 + \frac{E_a}{E_u} [1 - (1 - t/R)^4]$$

Where $(EI)_a$ is the flexural rigidity of an annular (or tabular) plant cross section, $(EI)_u$ is that of a uniform plant cross section, R is the plant

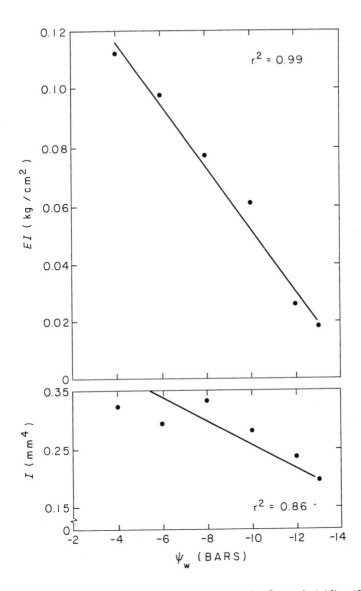

Fig. 8.11 Linear regression analyses of changes in the flexural rigidity (*EI*) and second moment of inertia (*I*) as water potential decreases from −4 bars to −13 bars. These data are based on measurements taken from chive leaves (cf. Fig. 8.10). *I* was computed according to the equation for a thick-walled tube, $I = \pi[ab^3 - (a - t) (b - t)^3]/4$ where t = thickness (0.24 mm) and a and b are the major and minor semiaxes of the elliptical cross section of leaves (see inserts in Fig. 8.10). *EI* was computed from the equation given in the text (modified from Niklas and O'Rourke, 1987, Fig. 3).

radius, t is the thickness of the annular walls, and E_a and E_u are the appropriate elastic moduli of the tissues composing the annular and uniform cross sections. This equation indicates that regardless of the value of E_a/E_u, a hollow stem has 50% of the flexural rigidity of a solid stem of comparable diameter when $t/R < 0.15$.

This prediction can be tested by examining species within a single genus that produce hollow and solid cylindrical plant organs of relatively comparable height. Similarly the equation predicts that no species had a t/R ratio less than 0.15, i.e. that stem wall thickness is not less than 15% of a stem radius.

Light interception, mechanical stability and fecundity

Vertical growth and the display of photosynthetic tissues above neighboring plants or abiotic obstructions to sunlight appears to be essential for the survival of terrestrial green plants. With few exceptions, the plant body of the major plant groups demonstrates orthotropic growth. Within the bryophytes, the gametophytic generation of mosses and a variety of liverworts produce vertical or nearly erect portions. And despite the rhizomatous growth habits of the majority of pteridophytes, all vascular non-seed plants elevate photosynthetic tissues above the ground. Clearly, the advantages conferred by vertical growth can be achieved only when the mechanical properties of tissues and organs are capable of supporting the bending moments imposed by vertical growth. Consequently there is a trade-off between mechanical stability and light interception that must be reconciled.

The complexity of reconciling the benefits of displaying vertical photosynthetic tissues with the mechanical stresses induced by vertical growth has been explored by a variety of authors. The difficulties in a numerical or analytical approach to this problem lie in mathematically defining the costs and profits involved in vertical growth and in the need to make simplifying assumptions, such as calculating static loads rather than complex dynamic loadings (induced for example by pressure induced drag due to wind). Nonetheless, attempts to deal with mechanical stability and light interception are heuristically useful since they can set boundary conditions on the problem of plant design.

For example, Niklas and Kerchner (1984) developed a computer model capable of assessing the extent to which different geometries of branching in early vascular leafless plants could intercept sunlight. In addition the model quantified the total bending moment induced by the three-dimensional, vertical axes of these simulated plants. This model was used to evaluate the morphological changes in branching that could maximize light interception and minimize the total bending moment (Figs. 8.12 and 8.13). The working hypothesis of this study, albeit a naive one, was that during the course of

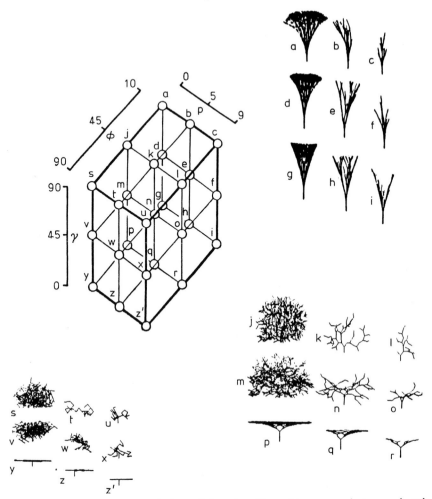

Fig. 8.12 A theoretical morphological domain of branching sporophytes produced by covarying three geometric features: (1) the branching angle (ϕ), which in these simulations is symmetrical; (2) the rotation angle, γ, between successive pairs of internodes; and (3) the probability of branching, p. Since the branching angle is symmetrical, all the sporophytes in this domain lack a main vertical 'trunk', other than the basalmost internode.

early land plant evolution there were selective pressures leading toward higher light-gathering efficiencies and lower mechanical instability. The model predicted a sequence of morphological alterations that coincided with the major changes in branching patterns seen in the evolution of Devonian vascular plants (Fig. 8.14). Consequently, remarkably few assumptions were required to produce an almost direct correspondence

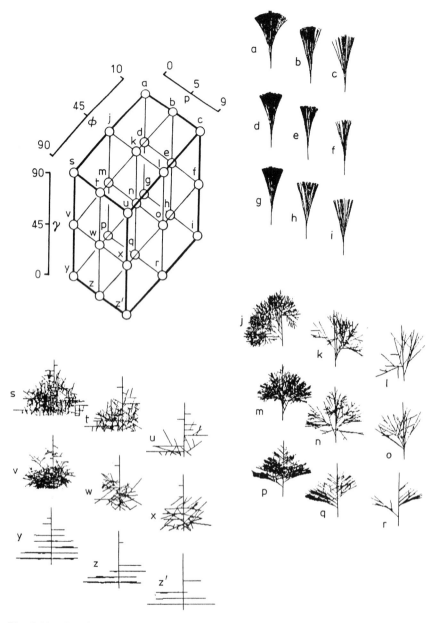

Fig. 8.13 Another portion of the theoretical morphological domain of branching sporophytes results when the branching angle (ϕ) is unequal. In this figure, each branching geometry has a main vertical 'trunk' produced by a branching angle, ϕ, which varies from 10° to 90° (upper right to lower left stick-figures). Reduction in the rotation angle (γ) produces planated, lateral branching figures (lower left stick-figures).

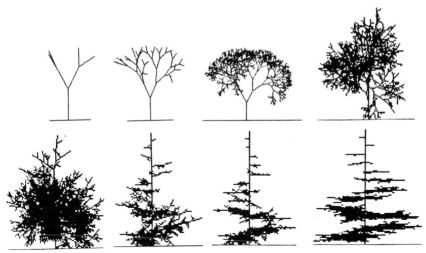

Fig. 8.14 A series of morphologies identified by computer analysis which search through the entire domain of branched sporophytes for geometries that progressively maximize light interception. The search started within the portion of the domain that coincided with the geometries of the most archaic vascular plants known from the fossil record (e.g. *Cooksonia*, an Upper Silurian plant). By progressively widening the search to encompass more diverse geometries a series of more efficient morphologies was identified. This series morphologically coincides with many of the geometric changes envisioned during the transition from a *Cooksonia*-like ancestor to more evolutionarily derived taxa, such as the trimerophytes (upper left to bottom right).

between paleobotanical observations and theoretical expectations about plant shape.

This approach can also be applied to extant plants. Niklas and Kerchner (1984) based their model on plants consisting solely of cylindrical photosynthetic axes. A variety of extant plants possess these features, e.g. *Psilotum, Ephedra, Salicornia*. Consequently, the model can be applied to these taxa without significant modifications in either computerized format or underlying assumptions. This can be illustrated for *Salicornia europaea* L. (Chenopodiaceae), a succulent, annual halophyte which is common to salt marshes and tidal flats in North America and Europe (Ellison and Niklas, unpublished). This species grows commonly in patches lacking other vegetation. Over time, however, it is over grown and excluded by perennial plants such as *Spartina patens, Distichlis spicata* and *Juncus gerardi. S. europaea* colonizes new patches at low density. However, seeds generally land within 20 cm of the parent plant and produce dense populations the following year. Computer simulations of *S. europaea* were based on 100 plants, each growing in early successional (low density monocultures), and late successional populations (beneath perennial canopies). The branching angle, Ø, internode ratios, probability of branching, and photo-tropism were determined for each of these three successional phases. The data for these observations are given in Table 8.1. These data were used to assess light-gathering capabilities of *S. europaea* plants in each of the three successional phases, and to calculate their load-bearing capabilities. It is then possible to predict which branching geometries maximize light inter-ception and minimize the total bending moment. Computer simulations indicate that the total bending moment is highest for plants with high probabilities of branching and branch continuation. Real *S. europaea* plants, however, never topple under their own weight or wind-induced drag. Consequently, light interception capabilities will be emphasized here. Changes in the total projected surface area of model plants (which heuristi-cally equates branching geometry to light-gathering capabilities) as a func-tion of the branching probability of the leader, the lateral branch continua-tion probability, leader internode ratio (LIR) and branching angle (Ø) are shown graphically as 'adaptive landscapes' in Figs. 8.15–8.17. Each land-scape represents a tensor-matrix composed of the probability of branching, the probability of lateral branch continuation and light-gathering efficiency. A single landscape is shown for each of the nine possible pairings of the three values of Ø (30°, 45°, 60°) and the three values of LIR (0.5, 0.75, 1.0).

For the branching angle (Ø = 30°) and low LIR ratio (0.5), model plants with the lowest light-gathering efficiencies are ones with a high branching probability and a high probability of branch continuation. The model predicts that increases in light-gathering efficiency will occur with decreased branching. Light-gathering capability increases with a simultaneous decrease in branching and branch continuation probabilities and an increase in LIR (or height) (Fig. 8.15). For Ø = 45° and Ø = 60° the pattern is identical (Figs 8.16–8.17). It should be noted that Ø = 60° is not found in

Table 8.1   Morphological parameters (means ± 1 standard deviation; $n = 100$) of *S. europaea* from field populations at different successional stages. For each parameter, values with different superscripted numbers are significantly different among the three successional stages ($P < 0.05$, Scheffe test of multiple com-parisons among means).

Suces-sional stage	Branch angle (°)	Internode ratio	Probability of leader branching	Probability of branch continuation	Phototropism
Early	33.0 ± 0.0004[1]	0.77 ± 0.221[1]	0.74 ± 0.057[1]	0.92 ± 0.158[1]	0.31 ± 0.024[1]
Middle	45.0 ± 0.0004[2]	0.85 ± 0.282[2]	0.11 ± 0.006[2]	0.10 ± 0.019[2]	0.29 ± 0.038[2]
Late	43.8 ± 0.0002[2]	0.84 ± 0.311[2]	0.13 ± 0.012[2]	0.09 ± 0.023[2]	0.35 ± 0.087[1]

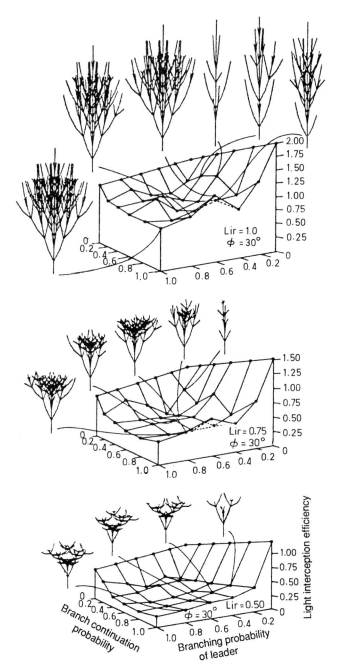

Light interception efficiency

Fig. 8.15 Adaptive landscapes of *Salicornia europaea* when $\phi = 30°$ and LIR = 0.5 (bottom); 0.75 (middle); 1.0 (top). Each landscape represents a tensor-matrix composed of the branch continuation probability, the branching probability of the leader and the light interception efficiency. Selected plants are shown which fall along the shortest path from least efficient to most efficient on each landscape. Simulations are based on field data given in Table 8.1.

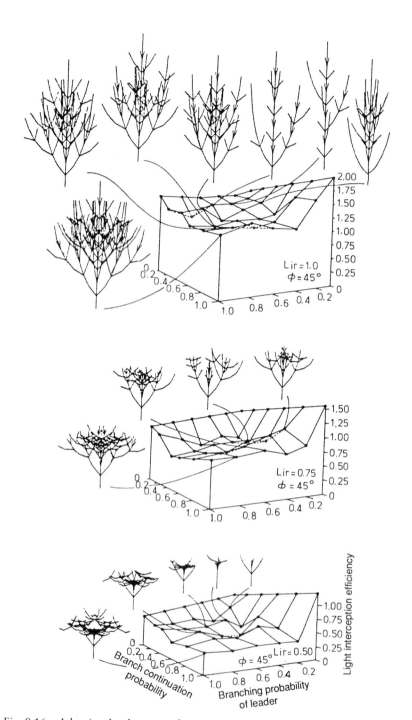

Fig. 8.16 Adaptive landscapes when $\phi = 45°$ and LIR = 0.5 (bottom); 0.75 (middle); 1.0 (top). Plants shown as in Fig. 8.15.

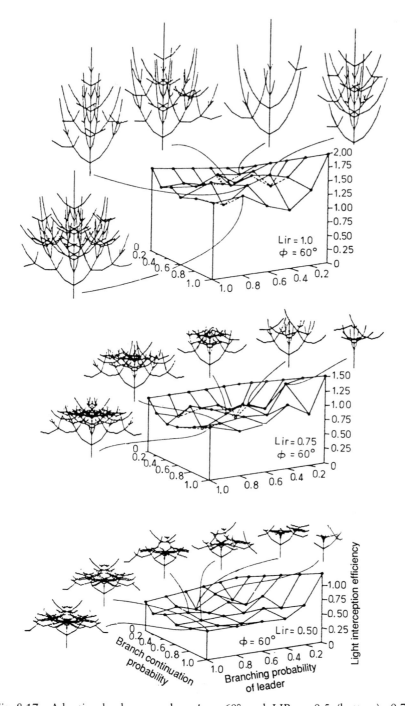

Fig. 8.17 Adaptive landscapes when $\phi = 60°$ and LIR $= 0.5$ (bottom); 0.75 (middle); 1.0 (top). Plants shown as in Fig. 8.15.

real *S. europaea* plants. This value was used to generate hypothetical *S. europaea* plants to see if the computer could design an even more efficient plant than occurs in nature. The computer was successful but the result, of course, was irrelevant!

The analyses indicate that for any pair of $\emptyset$ and LIR values, the plant form that minimizes physical load and maximizes light-gathering efficiency is one with few branches and a low probability of branch continuation. For a given value of $\emptyset$ and constant probability of branching and branch continuation, an increase in LIR results in an increase in light-gathering efficiency. Although the steepness of the adaptive landscape does not change with changes in LIR for any fixed value of $\emptyset$, the overall 'height' of the landscape increases. That is, as LIR increases, both the lower and higher ends of the landscape increase. Qualitatively, these results indicate that the least efficient model plants are those with numerous branches, small branching angles, and short internodes. Tall, spindly and unbranched geometries could exist, therefore, in light-limiting situations, while small bushy plants could exist in situations where light is not a limiting factor.

These computer-generated predictions agree with the morphological trends of *S. europaea* found in low and high density monocultures and in late successional situations (= beneath perennial canopies). Qualitatively, plants growing in early successional habitats are bushy and have distal internodes that are much smaller than older ones. Conversely, plants growing later in succession are unbranched, elongated and have equal-sized internodes. Ellison (1986) has shown that these morphological changes are a direct response to a reduction in light availability. *S. europaea* grown under perennial canopies or under artificially shaded conditions show identical reductions in branching and increased internodal ratios (etiolation). Accordingly, plants found in light-limiting situations are morphologically akin to computer-generated plants that maximize light interception and minimize bending moments.

Interestingly, branching in *S. europaea* is also correlated with fecundity. The number of flowers and hence seeds produced by a plant is highly correlated with branch number ($r^2 = 0.87$, $F_1,470 = 3266.7$, $P < 0.0001$ ANOVA). Computer-simulated plants which maximize light-gathering are sparsely branched and have few internodes. These plants would be predicted to produce fewer flowers and seeds and so reduce the seed bank in patches previously occupied by dense stands. In the natural populations examined, cylindrical plants often have fewer than 20 internodes and produce less than 120 seeds. Early successional plants produce more than 1000 times as many seeds as middle and late successional plants (Ellison, 1986).

When light, a primary resource, is limited, the 'optimal' geometric strategy (predicted by the computer) is a form that maximizes light-gathering efficiency, a nearly cylindrical form. When light is not limiting, a large fecund plant is predicted. These expectations are precisely what is seen in real populations of *Salicornia europaea*. This concordance between observations and expectations does not prove an ultimate cause–effect relationship, but it does suggest that light interception is the proximate cause of changes in the form of *S. europaea* during crowding.

8.3 LONG-TERM TRENDS IN PLANT EVOLUTION

How much of plant evolution can be understood or predicted in terms of the physical laws that influence the physiological requirements and the mechanics of growth in a terrestrial environment? This question has occupied the attention of numerous botanists, mathematicians and engineers for over a century and has been repeatedly answered in the context of biomechanical analyses (Schwendener, 1874; Thompson, 1942; Mark, 1967; Gordon, 1978). At least in theory, many of the physiological and structural requirements for the growth and survival of a terrestrial plant are amenable to mathematical analyses. From these studies the geometric compromises to reconcile the various design requirements can be predicted. There are, however, obvious limitations to this approach. The number of physiological, morphological, and developmental factors that can be handled in any single mathematical treatment must be limited to yield analytical, closed-form solutions. Thus, some variables must be grouped together or simply ignored for the sake of convenience. The use of simplifying assumptions immediately jeopardizes the biological integrity of the modelling procedure. Additionally, biomechanical or mathematical treatments of organic form must incorporate the developmental features that characterize the group of organisms being examined. These features are often imperfectly known or assumed to be invariable despite the knowledge that developmental patterns can be altered during the course of ontogeny or phylogeny. Consequently, many analyses of organic form and evolution are based on mathematical simplifications and must be considered crude approximations that at best shed light on broad patterns.

However, even from a cursory treatment it is evident that the morphological expression of terrestrial plants is limited by the physical constraints imposed upon the performance of metabolic and structural requirements for growth. Analyses of the various geometries capable of reconciling the various design requirements for a terrestrial plant indicate there is no single 'optimal' plant form. Rather, there is a limited domain in which various geometries are possible, each of which is as satisfactory as any other. A species may express more than one geometric 'solution', either as a consequence of (1) a diplobiontic life cycle, in which each generation has a different ecologic tolerance or reproductive requirement, or (2) as a result of a phenotypic plasticity that responds to environmental cues. Each of the morphological expressions of form may represent a geometry that maximizes the performance of one or more tasks but always at the expense of performing others in a less than optimal way.

Nonetheless, mathematical analyses of the geometries capable of reconciling the variety of tasks essential to vegetative growth identify two general morphological categories. These are the dorsiventral or oblate spheroid and the cylindrical axis. Each can be modified to yield more complex but reiterative morphologies. Lobing of the oblate spheroid results in a thalloid, dorsiventral form which mimics the general appearance of the gametophytes of many embryophytes. Similarly, dichotomization of a cylinder results in more complex branched geometries that mimic the morphology of the sporophytes of many archaic tracheophytes. A third category, which is modelled after the prolate spheroid, reflects a geometric solution that is virtually indistinguishable from that of cylindrical morphologies. However, neither the oblate nor prolate morphologies are capable of indeterminate orthotrophic growth. Consequently, the vertical display of photosynthetic tissues and elevation of reproductive organs require the use of cylindrical geometries.

The continued elongation of a vertical cylindrical axis results in a steady reduction in the performance level of many physiological tasks. In addition, it produces a mechanically unstable structure if the geometry of the cross-section of the axis is structurally limited, as for example in plants lacking secondary growth. 'Anatomical solutions', which can maximize the use of a limited quantity of tissues (primary growth) are possible. These involve the location of mechanically reinforcing tissues at either the center or the periphery of the cross-section of the axis. A centrally located 'mechanical strand' can be modified to transport fluids; a peripherally located 'rind' of tissue can accommodate gas exchange without a reduction in its mechanical reinforcement. Accordingly, the development of a hydrome (or vascular tissue) and of a hypodermal layer of thickened cells should evolve when cylindrical plant axes reach heights and girths that exceed the mechanical tolerances of parencyhma.

The evolutionary appearance of sporophytes with branched cylindrical axes must have involved the reconciliation of the mechanical loadings accompanying orthotropic growth with the vertical display of photosynthetic tissues. Analyses of the geometric solutions to vertical growth of branched plants indicate that morphological changes are contingent upon the geometry of disposition and quantity of mechanically reinforcing tissues. With the advent of secondary growth the geometric repertoire of branching sporophytes was unlimited. The maximization of light interception by branched axes most likely involved (1) an increase in the branching angles, (2) an asymmetry in the branching angle, producing a 'main' vertical axis with 'lateral' subordinate branches, (3) an acropetal decrease in the size of lateral branching systems, and (4) the planation of lateral branching (cf. Fig. 8.14).

Many of the geometric and anatomical adjustments predicted by theory occurred during the early evolution of tracheophytes (Banks, 1968; Chaloner and Sheerin, 1979; Raven, 1977; Stewart, 1984). However, it is not clear whether these adjustments represent the consequences of natural selection for more efficient morphologies. The sporophytes of geologically younger tracheophytes generally possess a higher proportion of primary xylem to ground tissue, which would lead to more efficient longitudinal transport and mechanical reinforcing systems. The transition from a rhyniophyte ancestor to a trimerophyte descendant involved numerous adjustments in the geometry of branching which coincide with those predicted to enhance light interception and mechanical stability (cf. Niklas and Kerchner, 1984; Zimmermann, 1952). In addition, the planation of lateral branching systems has occurred in a variety of plant lineages. However, both theory and the paleobotanical data indicate that there is no 'optimal' plant morphology. Within each of the many tracheophyte lineages, idiosyncracies in morphology and anatomy were 'evolutionarily tolerated', some for lesser and others for greater periods of geologic time. With the advent of each successive anatomical or morphological 'innovation', the domain of possible phenotypic expression was enlarged and taxa appear which occupy all or most of the available geometries within the domain. There is little or no evidence that an individual lineage underwent successive modifications such that its course of evolution gravitated toward an 'optimal' plant form. Rather, it appears more reasonable to consider some morphological trends in early tracheophyte evolution as the statistical consequences of rapid radiation within an initially uninhabited environment. Indeed there is evidence to suggest that some morphological changes decreased the efficiency with which certain tasks were performed. Increased branching of leafless sporophytes, regardless of the extent to which geometry can be adjusted, almost invariably leads to an increase in self-shading thereby reducing the amount of sunlight intercepted in proportion to the total bulk of the plant. Increased vertical growth and branching can have beneficial consequences, such as an increase in the dispersal distance and the quantity of spores or other propagules. But from the perspective of light interception, continued branching of a leafless plant can result in a physiological determinacy in light-gathering efficiency.

The use of biomechanical analyses leads to testable predictions about the course of plant evolution. In addition, it provides a phenomenological basis for the interpretation of morphological changes within the history of tracheophytes. This is in marked contrast to descriptive analyses which are characteristically immune to quantitative assessment.

A classic example of a 'descriptive' attempt to deal with plant evolution is the Telome Theory of W. Zimmermann (1952; see also Stewart, 1964, 1984; Stebbins, 1950, pp. 476–8). With the aid of a few 'processes', such as planation, overtopping, fusion, reduction and recurvature, a series of branches or 'telomes' can be conceptually molded into virtually any desired end-product. Yet, no evolutionary rationale is embedded within the 'theory' to explain the occurrence or the chronology of these processes. At the least, biomechanical analyses test hypotheses based on form–function relationships. Thus for example, the planation and determinate growth of lateral branching systems is not predicted 'to account for the evolutionary appearance of leaves' but rather as processes that increase light interception, regardless of subsequent morphological changes. Overtopping is not seen as a 'process' but as a statistical consequence of unequal branching that results in photosynthetic benefits upon which natural selection can act. Reduction is not necessarily the product of some 'need to economize' but rather a modification in the allometry of organs favored (or not) by its affect on function. The Telome Theory may provide a useful lexicon with which to describe morphological changes, but biomechanical analyses provide a conceptual basis for using the grammar of functional morphology.

8.4   ADAPTATION AND GANONG'S PRINCIPLE

If we are ignorant of the life history, development, and ecological relationships species we must maintain a completely open mind and an agnostic position concerning the adaptiveness or nonadaptiveness of its distinguishing characteristics. Even in the case of better-known species, neither the adaptive nor the nonadaptive quality of a particular character should be assumed unless definite evidence is available concerning that character. (Stebbins, 1950, p. 119)

The theme of this chapter has been that plant form has responded to the physical limitations imposed by the often conflicting demands made by a photosynthetic metabolism. Consequently, plant form has been viewed as an adaptation to the functional requisites of photosynthesis. However, the existence of physical limitations on form and the apparent concordance between morphologies that are predicted to deal with these constraints and those that actually occur do not provide proof for adaptation. If natural selection causes the evolution of an adaptation, then the adaptation must be shown to contribute to the fitness of the individual or its ancestor. In the case of green plants, photosynthetic efficiency is often cited as a general index of relative fitness. The net assimilation rate of a plant can be used as an index of the metabolic reserves available for vegetative and reproductive growth. However, the partitioning of photosynthetic resources among vegetative growth and reproduction is complex owing to a variety of life-history traits that modify relative fitness. Indeed, it is not proved nor generally accepted that plant fitness is directly related to photosynthetic efficiency. It may be argued, for instance, that a vast number of individual plants produce more metabolic reserves than are utilized during their lifetimes. Even the resolution of this debate will not in and of itself resolve the issue. The determination of how much of plant evolution can be understood in terms of the physical constraints on form by function requires both an analysis of the range in actual plant form and an analysis of the biophysical factors that influence form. This can be illustrated in the context of the 'case histories' presented here.

For example, in the discussion of phyllotactic patterns and the capacity to intercept sunlight, it was possible to show that the pattern of leaf initiation could influence the efficiency of light interception when internodal distances are small, as for example in species with rosette growth habits. Nonetheless, leaf shape and internodal distances were shown to represent potential genotypic or phenotypic variables that could be altered to circumvent the potential 'developmental constraint' of phyllotaxy. Leaf shape, size, and opacity, as well as internodal distances, are not necessarily linked to leaf-arrangement, but rather they can vary independently and in response to the environment. For example, of the approximately 260 species of *Plantago* currently recognized, most have the same phyllotactic patterns (13/34 or 21/55). However, there exists a diversity in leaf size and shape. Of the 16 North American species, seven have lanceolate or linear leaves (e.g. *P. maritima*, *P. pusilla*, *P. elongata*, *P. aristata*) whereas eight have ovate or broadly ovate leaves (e.g. *P. rugelii*, *P. major*, *P. cordata*, *P. eripoda*). The remaining species, *P. indica*, has opposite leaves and is herbaceous with long internodes. Among some species (*P. major* and *P. lanceolata*), the phenology of leaf senescence is associated with leaf overlapping which in turn is dictated by phyllotactic pattern. For example, species with a 13/34 phyllotaxy will have a leaf-overlap after 34 leaves, while those with a 21/55 pattern can produce 54 leaves before two leaves overlap. *P. lanceolata* has rosettes with significantly more leaves (23 ± 5.3, $n = 50$) than *P. major* (15 ± 4.7, $n = 50$), and neither species has rosettes with more leaves than would occur in a single phyllotactic sequence (34 or 55 leaves). Older leaves senesce before they are significantly overlapped by younger leaves. The width to length ratio of *P. lanceolata* and *P. major* leaves are $1:6$ and $1:3$, respectively. Computer simulations predict that *P. lanceolata* can maintain more leaves in a rosette than *P. major* because of the differences in their respective leaf geometry. Significantly, the leaves of *P. major* possess petioles which are often equal in length to the lamina. Consequently, the laminae of younger leaves overlap the petioles of older leaves more frequently than the laminae of older leaves. Significantly, the only North American species of *Plantago* with opposite leaves (an arrangement predicted to produce significant leaf overlapping in rosette growth habits) has lanceolate leaves (which reduce overlapping) and internodal distances equal to or greater than leaf length (which increases light interception).

Although computer simulations based on biophysical assumptions can often lead to predictions borne out by empirical studies, the extent to which these simulations predict biological reality is dictated by the scope of the assumptions. For example, based on field observations of *Salicornia europaea*, computer simulations predicted an 'adaptive landscape' of *Salicornia* morphology in terms of light-gathering efficiency. Comparisons between predicted morphological trends (maximizing light-gathering) with increased intraspecific density yield compatible results. These simulations were based on the assumption of a vertical, branching plant. However, among the five species of *Salicornia* known from North America, one species, *S. virginica* has a perennial, rhizomatous growth habit. This species produces vertical, erect shoots which rarely branch and a single individual can produce a 'population' of erect shoots that mimics the light environment of a high density, late successional community of *S. europaea*. In this regard, simulations based on *S. europaea* (which predict spindle-shaped, sparsely branched plants in late successional situations) are compatible with the behavior of a single individual of *S. virginica*, which shades itself because of its clonal growth. Therefore, a comparison between the anticipated behavior of annual, non-rhizomatous and perennial rhizomatous species within the genus *Salicornia* yields equivalent predictions about morphological responses to limiting light conditions.

The example to show the relationship between mechanical stability and hydraulics (deformation of chive leaves in response to changing tissue water content), illustrates an even greater morphological variation compatible with a biophysical limitation. The geometry and the physical properties of plant tissues dictate the flexural rigidity of plant organs. Both the geometry and the physical properties of organs can be changed as a consequence of the immediate environment (water potential of the soil) and by ontogenetic modification accompanying growth and development. Although the physical principles that dictate flexural rigidity remain constant from one plant group to another, the morphological versatility even within a single genus precludes predictions as to 'optimal' shape. The deposition of collenchyma, extraxylary fibers, and other potentially 'reinforcing tissues', as well as the potential to produce septate pith or nodal septa, provides a large domain of morphological variation that can equally well deal with the mechanical deflection of vertical organs during periods of water stress. Within the genus *Allium*, for example the ratio of thickness to diameter of cylindrical leaves can vary over an order of magnitude. And thicker walled organs reach approximately greater vertical heights.

Compounding the numerous ways in which biophysical constraints can be dealt with, is that serially homologous organs need not be formed in a fixed sequence. Nor is the ontogenetic development of homologous organs always similar. A variety of plant organs can assume the same function and a single plant organ can assume a variety of functions. Expressing this as the principle of 'metamorphosis along the lines of least resistance', Ganong (1901) argued that selective pressures will modify the organ whose ontogeny or shape or previous function can most easily respond. Selection due to persistent or repetitive physical demands placed on shape has resulted in dramatic convergence in the shape of plant organs. Thus, the cladodes of *Opuntia* and the leaves of many plants are geometrically and functionally similar, while leaves, stems, and peridermal outgrowths have accommodated the function of protection (spines, thorns and prickles). As reiterated by Stebbins (1950, pp. 496–9), Ganong's principle suggests that biophysical constraints on plant form can be dealt with in a diverse number of ways through the modification of cells, tissues and organs. Although 'metamorphosis along the path of least resistance' makes it difficult to predict which organs respond to selection pressures, biophysical analysis can be used to quantify the performance of an organ in relation to its function.

8.5 REFERENCES

Anderson, M. C. (1964) Studies of the woodland light climate. I. The photographic computation of light conditions. *J. Evol.*, **52**, 27–41.

Archer, R. R. and Wilson, B. F. (1973) Mechanics of the compression wood formation. *Plant Physiol.*, **51**, 777–82.

Banks, H. P. (1968) The early history of land plants. in *Evolution and Environment* (ed. E. T. Drake), Yale University Press, New Haven, pp. 73–107.

Bewley, J. D. (1979) Physiological aspects of desiccation resistance. *Ann. Rev. Pl. Physiol.*, **30**, 195–238.

Caemmerer, S. von and Farquhar, G. L. (1981) Some relations between the biochemistry of photosynthesis and the gas exchange of leaves. *Planta*, **153**, 376–87.

Carlquist, S. (1975) *Ecological Strategies of Xylem Evolution*. University of California Press, Berkeley, CA.

Chaloner, W. G. and Sheerin, A. (1979) Devonian macrofloras. *Palaeontology*, Special Papers No. **23**, 145–61.

Ehleringer, J. R. and Werk, K. S. (1986) Modifications of solar-radiation absorption patterns and implications for carbon gain at the leaf level. in *On the Economy of Plant Form and Function* (ed. T. J. Givnish), Cambridge University Press, Cambridge, pp. 57–82.

Ellison, A. M. (1986) The ecology of *Salicornia europaea* in a southern New England salt marsh. PhD dissertation, Brown University, Providence, Rhode Island.

Farquhar, G. L., von Caemmerer, S. and Berry, J. A. (1980) A biochemical model of photosynthetic CO_2 assimilation in the leaves of C_3 species. *Planta*, **149**, 78–90.

Fisher, J. B. (1986) Branching patterns and angles in trees. in *On the Economy of Plant Form and Function* (ed. T. J. Givnish), Cambridge University Press, Cambridge, pp. 493–524.

Fisher, J. B. and Honda, H. (1979) Branch geometry and effective leaf area: a study of *Terminalia*-branching pattern. I. Theoretical Trees. *Am. J. Bot.*, **66**, 633–44.

Forseth, I. and Ehleringer, J. (1980) Solar tracking response to drought in a desert annual. *Oecologia*, **44**, 159–63.

Ganong, W. F. (1901) The cardinal principles of morphology. *Bot. Gaz.*, **31**, 426–34.

Gates, D. M. (1965) Energy, plants and ecology. *Ecology*, **46**, 1–16.

Gates, D. M. (1970) Physical and physiological properties of plants. in *Remote Sensing*. National Academy of Sciences, Washington, DC, pp. 224–52.

Geller, G. N. and Nobel, P. S. (1986) Branching patterns of columnar cacti: influences on PAR interception and CO_2 uptake. *Am. J. Bot.*, **73**, 1193–200.

Giordano, R., Salleo, A., Salleo, S. and Wanderlingh, F. (1978) Flow in xylem vessels and Poiseuille's law. *Can. J. Bot.*, **56**, 333–8.

Gordon, J. E. (1978) *Structures – or Why Things Don't Fall Down*. Plenum Press, New York.

Grime, J. P. (1979) *Plant Strategies and Vegetation Processes*. Wiley, Chichester.

Hardwick, R. C. (1986) Physiological consequences of modular growth in plants. *Phil. Trans. R. Soc. Lond.*, **B313**, 161–73.

Herbert, T. J. (1983) The influence of axial rotation upon interception of solar radiation by plant leaves. *J. Theor. Biol.*, **105**, 603–18.

Herbert, T. J. (1984) Axial rotation of *Erythrina herbacea* leaflets. *Am. J. Bot.*, **71**, 76–9.

Mark, R. E. (1967) *Cell Wall Mechanics of Tracheids*. Yale University Press, New Haven.

McMahon, T. A. and Kronaur, R. E. (1976) Tree structures: deducing the principle of mechanical design. *J. Theor. Biol.*, **59**, 1–24.

Monsi, M. and Saeki, T. (1953) Uber den Lichtfaktor in den Pflanzengesell schaften und seine Bedeutung fur die Stoffproduction. *Jpn. J. Bot.*, **14**, 22–52.

Mooney, H. A. and Gulmon, S. L. (1979) Environmental and evolutionary constraints on the photosynthetic characteristics of higher plants. in *Topics in Plant Population Biology* (eds O. T. Solbrig, S. Jain, G. B. Johnson and P. H. Raven), Columbia University Press, New York, pp. 316–37.

Niklas, K. J. (1988) The role of phyllotactic pattern as a 'developmental constraint' on the interception of light by leaf surfaces. *Evolution*, **42**, 1–16.

Niklas, K. J. and Kerchner, V. (1984) Mechanical photosynthetic constraints on the evolution of plant shape. *Paleobiology*, **10**, 79–101.

Niklas, K. J. and O'Rourke, T. D. (1987). Flexural rigidity of chive and its response to water potential. *Am. J. Bot.*, **74**, 1033–44.

Nobel, P. S. (1977) Water relations and photosynthesis of a barrel cactus, *Ferocactus acanthodes*, in the Colorado Desert. *Oecologia*, **27**, 117–33.

Nobel, P. S. (1980) Interception of photosynthetically active radiation by cacti of different morphology. *Oecologia*, **45**, 160–6.

Nobel, P. S. (1986) Form and orientation in relation to PAR interception by cacti and agaves. in *On the Economy of Plant Form and Function* (ed. T. J. Givnish), Cambridge University Press, Cambridge, pp. 83–103.

Pearcy, R. W. and Ehleringer, J. (1984) Comparative ecophysiology of C_3 and C_4 plants. *Plant Cell Env.*, **7**, 1–13.

Rand, R. H. (1983) Fluid mechanics of green plants. *Ann. Rev. Fluid Mech.*, **15**, 29–45.

Raschke, K. (1956) The physical relationships between heat-transfer coefficients, radiation exchange, temperature, and transpiration of a leaf. *Planta*, **48**, 200–38.

Raven, J. A. (1977) The evolution of vascular land plants in relation to supracellular transport processes. *Adv. Bot. Res.*, **5**, 153–219.

Richards, F. J. (1950) Phyllotaxis: its quantitative expression and relation to growth in the apex. *Phil. Trans. Roy. Soc. London, B* **235**, 504–64.

Robichaux, R. H., Holsinger, K. E. and Morse, S. R. (1986) Turgor maintenance in Hawaiin *Dubautia* species. The role of variation in tissue osmotic and elastic properties. in *On the Economy of Plant Form and Function* (ed. T. J. Givnish), Cambridge University Press, Cambridge.

Schmitt, M. R. and Edwards, G. E. (1981) Photosynthetic capacity and nitrogen use efficiency of maize, wheat, and rice: a comparison of C_3 and C_4 photosynthesis. *J. Exp. Bot.*, **32**, 459–66.

Schwendener, S. (1874) Das Mechanissche Princip in anatomischen Bau der Monocotylen mit vergleichenden Ausblicken auf die ubrigen Pflanzenklassen. Verlag von Wilhelm Engelmann, Leipzig.

Shackel, K. A. and Hall, A. E. (1979) Reversible leaflet movements in relation to drought adaptation of cowpeas, *Vigna* unguiculata (L.) Walp. *Aust. J. Plant Physiol.*, **6**, 265–76.

Shaver, G. S. (1978) Leaf angle and light absorbance of *Arctostaphylos* species (Ericaceae) along elevational gradients. *Madrono*, **25**, 133–8.

Sinnott, E. D. (1960) *Plant Morphogenesis*. McGraw-Hill, New York.

Silk, W. K., Wang, L. L. and Cleland, R. E. (1982) Mechanical properties of the rice panicle. *Plant Physiol.*, **70**, 460–4.

Stebbins, G. L., Jr. (1950) *Variation and Evolution in Plants*. Columbia University Press, New York.

Stewart, K. D. and Mattox, K. R. (1975) Comparative cytology, evolution and classification of the green algae, with some consideration of the origin of other organisms with chlorophylls a and b. *Bot. Rev.*, **41**, 104–35.

Stewart, W. N. (1964) An upward outlook in plant morphology. *Phytomorphology*, **1**, 456–70.

Stewart, W. N. (1984) *Paleobotany and the Evolution of Plants*. Cambridge University Press, Cambridge.

Thompson, D'Arcy, W. (1942) *On Growth and Form*. 2nd edn, Cambridge University Press, Cambridge.

Timoshenko, S. P. and Gere, J. M. (1961) *Theory of Elastic Stability*. McGraw-Hill, New York.

Timoshenko, S. P. and Goodier, J. N. (1970) *Theory of Elasticity*. McGraw-Hill, New York.

Troughton, J. H., Card, K. A. and Hendy, C. H. (1974) Photosynthetic pathways and carbon isotope discrimination by plants. *Carnegie Inst. Washington Yearbook*, **73**, 768–80.

Wainwright, S. A., Biggs, W. D., Currey, J. D. and Gosline, J. M. (1976) *Mechanical design in organisms*. John Wiley and Sons, New York.

Werk, K. S., Ehleringer, J., Forseth, I. N. and Cook, C. S. (1983) Photosynthetic characteristics of Sonoren Desert winter annuals. *Oecologia*, 59, 101–5.

Wright, C. (1873) On the uses and origin of arrangements of leaves in plants. *Mem. Am. Acad. Arts Sci.*, 9, 379–415.

Zimmermann, W. (1952) Main results of the 'Telome Theory'. *Paleobotanist (Lucknow)*, 1, 456–70.

CHAPTER 9

Evolution and adaptation in Encelia (Asteraceae)

JAMES R. EHLERINGER and CURTIS CLARK

9.1 INTRODUCTION

In this chapter, we will focus on ecological and systematic studies of *Encelia* (Asteraceae) in order to gain insight into the evolution and adaptive radiation that has occurred within this genus of arid land shrubs. The ecological studies have emphasized the role of changes in leaf morphological and physiological characters in adapting to a range of thermal and precipitation regimes, while the systematic studies have placed an emphasis on cladistic approaches to understanding phylogenetic relationships within the genus. What emerges is that *Encelia* is a genus of closely related adaptive units, each particularly tuned to a specific physical environment. Some of these adaptive units have arisen as a result of divergent speciation, whereas others have arisen as a result of hybridization.

9.2 ADAPTATION TO ENVIRONMENT

While it is easily accepted that organisms are adapted to the environments in which they naturally occur, it is often much more difficult to assess the adaptive value of individual, specific features of an organism isolated from either the rest of the biology of that organism or the evolutionary circumstances under which the adaptation arose. For organisms that are long-lived (not annual) and where recruitment of new individuals into the population occurs at irregular intervals, it can also be equally difficult to provide an unequivocal measure of that organism's fitness in terms of either specific adaptive features or differential reproductive contributions to the next generation. These constraints, however, should not be deterrents to the study of adaptive features in plants. Below we present our approach for analyzing specific adaptive features in *Encelia*.

Reproductive success of an adult organism can be measured, and what is normally observed is that it is a positive, linear function of plant size (Fig. 9.1). For interspecific comparisons, though, interpretations based upon absolute differential reproductive outputs can at times provide misleading results as is the case for two *Encelia* species common to the Sonoran Desert. For instance, from Fig. 9.1 it might be argued that at equivalent sizes, *E. farinosa* outreproduces *E. frutescens* by a factor of three or four. Yet achene mass in *E. frutescens* is about five times greater than in *E. farinosa*, and as a consequence, the mass allocated to reproduction as a function of plant size is very similar. Selective forces relating to seedling establishment in the different habitats where *E. farinosa* and *E. frutescens* each occur are the likely explanation for differences in achene mass. Within each species, though, a clear pattern emerges – the likelihood of an established plant contributing to future generations is a function of its size. If longevity is

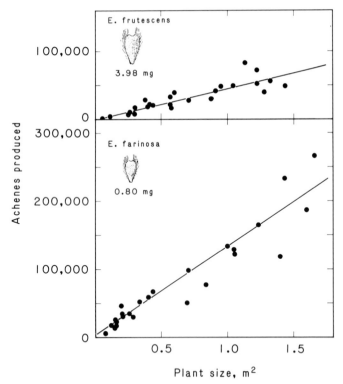

Fig. 9.1 The relationship between the number of achenes produced and plant size (m²) for *Encelia farinosa* and *Encelia frutescens*, based on sampling of natural plants in the Ibex Mountains near Death Valley in spring 1980. Achene mass is the average of 100 full, mature achenes.

similar among genotypes, this suggests that for perennial plants such as *Encelia*, the adaptive significance of variations in specific features might be evaluated in terms of how they affect the ability of that plant to achieve a larger size.

This is the approach that has been used in our studies of *Encelia*, where the interest has focused on the adaptive significance of variations in leaf energy balance parameters. Leaves interact with their surrounding environment in certain definable ways and so it is possible to describe and quantify the interactions between leaf morphological and physiological characters and the abiotic and biotic environments (Fig. 9.2). Once the impact of the environment on leaf energy balance (i.e. leaf temperature and transpiration) and physiological activity (i.e. photosynthesis) is understood, it is possible to scale up from the single leaf to the entire plant to determine the consequences on whole plant net carbon gain and to the potential for affecting growth and plant size.

The energy balance analysis provides a mechanism whereby the significance of certain, specific leaf characters (such as size, shape, angle and pubescence) can be evaluated largely independent of other aspects of the organism's biology. However, it is possible that a specific morphological or physiological character such as pubescence, for example, may directly or indirectly serve multiple functions for a plant (such as a reflecting surface, a diffusion barrier to retard water loss, or dehydration layer for insects which

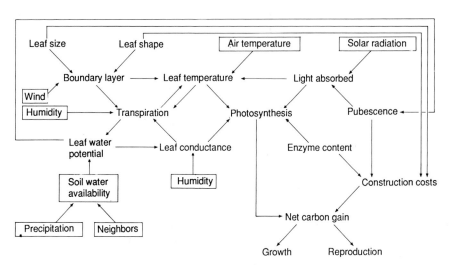

Fig. 9.2 A flow diagram of the interactions between the abiotic and biotic environmental factors (boxed), leaf morphological parameters and physiological process as they affect net carbon gain, growth and reproduction.

might lay their eggs on the leaf surface). These possible functions, some of which we consider less likely in *Encelia*, must be considered as potential limitations in interpreting the adaptive patterns described over the next several sections.

For the discussion which follows it is also important to distinguish between two different levels of adaptation, both of which have a genetic basis and could be selected for independently. The first involves selection for the presence or absence of a feature which is thought to have adaptive value to an organism. If present, the adaptation can be thought of as being constitutive. The second level is selection for plasticity in that feature. For a quantitatively variable character, this can be viewed as the phenotypic plasticity of a single genotype for the feature in question.

9.3 *ENCELIA*: A MODEL SYSTEM FOR THE STUDY OF ADAPTATION

Encelia is a genus of 15 species of frutescent or suffrutescent shrubs common to the arid lands of southwestern North America and western South America. Different species were initially recognized largely on the basis of differences in floral (ray and disk flower colors) and leaf morphology characters (size, shape and pubescence) (Blake, 1913). Species are clearly distinguishable in the vegetative phase; with only one major exception the species distributions are parapatric (Fig. 9.3). *Encelia* species (except *E. canescens* of South America) are obligate outcrossers and will freely hybridize with each other given the opportunity. Where species distributions overlap, hybridization is common (fertile flowers but no introgression), and these plants appear to occupy intermediate habitats (Kyhos, Clark and Thompson, 1981).

The distributions of *Encelia* species fall into generally distinct climatic zones at the macroenvironment level (Fig. 9.4). Some species occupy a broad environmental range, whereas others are very limited in their distributions. For those with a broad distribution, there is often a large variation in certain morphological characters along environmental gradients. There is a strong tendency for members of the *californica* clade to occupy regions of predictable seasonal precipitation and a more oceanic climate (*E. farinosa* is the exception). In contrast, members of the *frutescens* clade occur in habitats with less-predictable precipitation and a strongly continental climate.

Leaf pubescence is highly variable within the genus. Species from the wetter habitats are glabrous or only lightly pubescent, whereas those from drier habitats tend to be progressively more pubescent. Within a single species, the extent of leaf pubescence development is population-dependent, again with those populations from drier locations being more pubescent

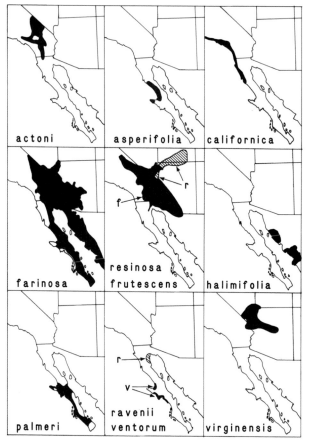

Fig. 9.3 Distributions of *Encelia* species in North America.

(Ehleringer, 1980; Ehleringer *et al.*, 1981). In contrast, pubescence development on reproductive structures is much more conservative (Charest-Clark, 1984).

It is clear from casual observation that each species of *Encelia* possesses a different suite of features evidently adaptive to its environment. It is more difficult to determine which of these features are in fact adaptations of a given species rather than attributes of a group of related species, even when there is clear experimental evidence of their adaptive nature. One may address the presence of adaptation through experiment; one must address the origins of adaptation through evolutionary history. These complementary approaches together provide a powerful research program for the study of adaptation.

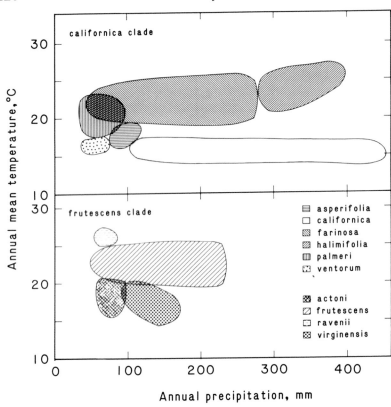

Fig. 9.4 Climatic envelopes for *Encelia* species ordinated by annual mean temperature and annual precipitation to show how the species fall into distinct climatic regions.

The origin of an adaptation unique to a single species can be studied in light of that species' ecology, but if the adaptation is shared by a group of species, its origin may not be reflected by its current adaptive value in any of the species possessing it. It is not unusual for adaptive features in such situations to be exaptations (Gould and Vrba, 1982), features that originally served a different adaptive function than they do now.

In order to distinguish such situations, one must have a hypothesis of the phylogenetic relationships of the species studied. Perhaps the most powerful means for generating such hypotheses are those provided by cladistic (phylogenetic) systematics (Wiley, 1981; Hennig, 1966). The basic premise of cladistics is that clades (groups consisting of all the descendants of a common ancestor) can be diagnosed by shared unique homologies (synapomorphy). The methods of phylogenetic analysis allow the construc-

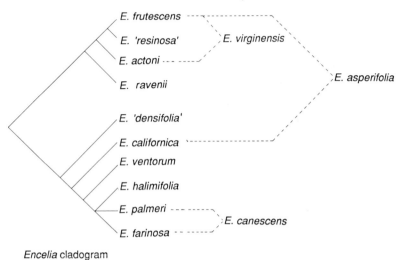

Encelia cladogram

Fig. 9.5 Cladogram for the genus *Encelia*, based on Clark (1986).

tion of a cladogram, a graphic hypothesis of relationships among species. This hypothesis is tested by the discovery of additional homologies, which may or may not support the hypothesized relationships.

Phylogenetic analysis of *Encelia* demonstrates that the genus can be divided into two major clades, named the *californica* clade and the *frutescens* clade after constituent species (Clark, 1982, 1986; Charest-Clark and Clark, 1984; Proksch and Clark, 1984, 1986; Clark and Sanders, 1986). The existence of each of these clades is supported by several synapomorphies, and their membership has remained unchanged with the addition of new characters to the analysis over a five-year period. Relationships within each clade are not as well substantiated, although they, too, have not changed greatly with the addition of new characters. The most recent cladogram of the genus (Clark, 1986) is presented in Fig. 9.5.

One unusual feature of the genus, confirmed both by phylogenetic analysis and traditional methods, is the origin of diploid species by hybridization (Clark and Kyhos, 1979; Clark, Kyhos and Thompson, 1980). Three species, *E. asperifolia*, *E. virginensis* and *E. canescens*, appear to have originated by this means. The first of these derives from ancestral species representing each of the two major clades.

Structure and function: leaf hairs and reflectance

Within the genus *Encelia*, four distinctive kinds of hairs appear on leaf surfaces (Fig. 9.6) (Clark, Thompson, and Kyhos, 1980). Not all hair types

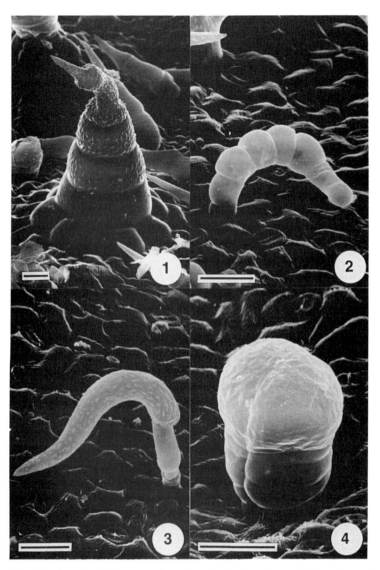

Fig. 9.6 The four leaf hair types found on the leaf surfaces of *Encelia* species. 1: broad, multicellular-based uniseriate hair. 2: moniliform hair. 3: narrow, unicellular-based uniseriate hair. 4: biseriate glandular hair. Bar indicates 25 mm. (From Ehleringer and Cook, 1987.)

appear on any single species, but there are groupings of different hair combinations associated with species groupings as described in the previous section. Of these hairs, only one hair, the narrow unicellular-based uniseriate hair, is associated with increased leaf reflectance (Ehleringer and Cook, 1987). This uniseriate hair consists of 3–4 basal cells and an elongated terminal cell. In some species, this distal cell is 1–5 times the length of the basal cells; in other species the distal cell expands up to several hundred times the length of the basal cells.

One consequence of the extension of the distal cell of the unicellular-based uniseriate hair is a change in leaf spectral properties. This is best illustrated by contrasting the absorptance characteristics of *E. californica* and *E. farinosa* (Fig. 9.7). In *E. californica*, the uniseriate hairs are relatively short, whereas in *E. farinosa*, the distal cell is at least two orders of magnitude longer than the basal cells.

The *E. californica* leaf in Fig. 9.7 is absorbing approximately 85% of the solar radiation between 400–700 nm (visible waveband). Light transmittance is approximately 5%, indicating that the cuticle is responsible for

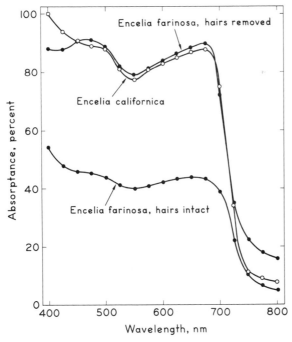

Fig. 9.7 Leaf absorptance spectra between 400–700 nm for intact leavse of *Encelia californica* and *Encelia farinosa* and for an *Encelia farinosa* leaf in which the reflective pubescence coating has been removed. (From Ehleringer and Björkman, 1978a.)

reflecting about 10% of the incident visible sunlight. These values are in agreement with previous measurements on many other glabrous-leaved species (Gates, 1980; Lin and Ehleringer, 1983). In contrast, the *E. farinosa* leaf in Fig. 9.7 is absorbing only 42% of the incident solar radiation between 400–700 nm, and absorptance values as low as 29% have been reported for this species (Ehleringer, Björkman and Mooney, 1976).

The decreased absorptance by *E. farinosa* leaves is the result of an increased leaf reflectance, and is not due to any changes in chlorophyll absorptance or concentration or leaf transmittance (Ehleringer and Björkman, 1978b). This can be seen by removing the hairs with either forceps or a razor. The result of the *E. farinosa* leaf hair removal is to produce an absorptance spectrum almost identical to that of *E. californica* (Fig. 9.7). This figure illustrates an additional point – the hairs appear to be a blanket reflector in the visible wavelengths, reflecting all wavelengths equally well. Although this is true for the visible wavelengths, which contain about half the sun's energy, it does not hold for the near infrared band (700–3000 nm), which contains the other half of the sun's energy. Wong and Blevins (1967) demonstrated that dry cellulose has a very high reflectance in the near infrared and that water was the primary factor responsible for absorptance by leaves in this waveband. Since the distal cells in mature leaves are dead (in contrast to the basal cells which may remain alive), it is not surprising to observe that the near infrared absorptance of pubescent *E. farinosa* leaves is close to zero.

The percentage leaf absorptance to visible wavelengths (400–700 nm) is not the same as the percentage absorptance to the entire solar waveband (400–3000 nm), because of this change in near infrared absorptance. The 400–700 nm waveband is of primary importance to photosynthesis, whereas the 400–3000 nm waveband is of importance to energy balance considerations. This should be kept in mind when considering the influences of solar radiation on plant performance as illustrated in Fig. 9.2. Ehleringer (1981) showed that there was a linear relationship between the two absorptance coefficients, with the 400–700 nm absorptances varying from 87 down to 29%, while the 400–3000 nm absorptance varied from 51 down to 10%.

The extent of leaf pubescence development in *E. farinosa* is strongly influenced by environmental factors (Ehleringer, 1982). Of these, leaf water potential (varying with soil water availability) appears to be the most important parameter. The developmental control over the development of leaf pubescence resides in the apical meristem. The control exerted seems to be that the amount of pubescence developed in new leaves will depend on the driest environmental conditions seen during that growing season. Thus, leaves produced during the season always have lower leaf absorptances even if they were produced during a wetter period toward the end of an otherwise relatively dry growing season. During the long, semi-dormant periods between growing seasons, the apical meristem controls are reset through some as yet unknown mechanism. Interestingly, though, if the apical meristem is destroyed during the growing season, the newly activated axillary meristems of well-watered plants will produce high-absorptance leaves whereas the new leaves produced from existing apical meristems will continue to produce low-absorptance leaves (Ehleringer, 1982).

Small topographic changes in soil depth and rockiness influence soil water storage and availability and thus affect leaf water potential. The presence of neighboring plants also utilizing this resource will exacerbate these effects. These strong interactions between environment and development of leaf pubescence make it difficult to evaluate precisely the magnitude of genetically based variations in leaf absorptance within a single population under natural conditions. However, given this constraint it is reasonable to estimate that within the Death Valley *E. farinosa* population surveyed by Ehleringer (1983), genetic differences may contribute as much as 5–6% of the observed absolute absorptance differences. From much more limited sample sizes under controlled experimental garden conditions, Ehleringer (1986a) observed only a 1–2% maximum difference in leaf absorptances among individuals within a single population.

Among *E. farinosa* populations, there can be large differences in the mean minimum leaf absorptances observed during drought periods. These observations provide an approximate indication of the extent of interpopulation variation in this character. There is a positive correlation between the mean minimum observed leaf absorptance and mean precipitation. Thus, *E. farinosa* populations derived from wetter habitats such as Riverside, California (278 mm precipitation primarily in winter–spring) and Hermosillo, Mexico (244 mm precipitation primarily in the summer) have minimum leaf absorptances (mean ± 95% confidence interval) of 63.4 ± 1.8% and 61.7 ± 1.5%, respectively, whereas populations from Palm Desert, California (109 mm precipitation) and Death Valley, California (42 mm precipitation) have minimum leaf absorptances of 44.9 ± 2.0% and 35.3 ± 1.9%, respectively.

Energy balance considerations: transpiration and temperature

Leaves interact with their surrounding physical environment through the processes of energy and gas exchange as presented in Fig. 9.2. The tight coupling between leaf and environmental parameters influences the rates of exchange and determines leaf temperatures, which in turn affects other metabolic processes. It is here where the effects of changes in leaf pubescence on function and in particular on energy balance can first be seen. Energy balance equations were first developed by Raschke (1960) and Gates

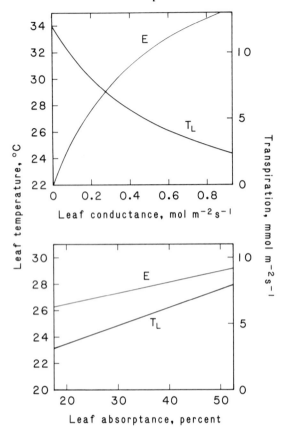

Fig. 9.8 Leaf energy budget calculations of the effects of changes in leaf absorptance (400–3000 nm) and in leaf conductance to water vapor on leaf temperature and transpiration rate. For these simulations, the air temperature was 25°C, relative humidity was 30%, total solar radiation was 1000 W m^{-2}, leaf width was 2 cm, and wind speed was 1 m s^{-1}.

(1962) and have been used to predict the consequences of changes in either the physical environment or of leaf characters on both leaf temperature and transpiration rates.

Fig. 9.8 (lower) shows the effect that a change in leaf pubescence (via its changing leaf absorptance) from a glabrous appearance (absorptance of 50% such as might appear in *E. californica* leaves) to heavily pubescent (absorptance of 20% such as might appear in *E. farinosa* leaves) will result in a nearly 4°C change in leaf temperature. The transpiration rate also changes as a function of leaf temperature, independent of any changes in leaf conductance. This is because the driving force for transpiration rate is

Table 9.1 Calculated values of photosynthesis, transpiration, and leaf temperature for *E. farinosa* under midday summer conditions. The energy budget calculations used the following values: wind speed 1 m s^{-1}, soil temperature 50 °C, air temperature 40 °C, total solar radiation 1000 W m^{-2}, 10% diffuse solar radiation, sky infrared radiation 350 W m^{-2}, air vapor density 10 g m^{-3}, leaf angle 25°, leaf width 4 cm, and leaf conductance 0.09 mol m^{-2} s^{-1}. Photosynthetic rate based on response curves from Ehleringer and Mooney (1978). (From Ehleringer and Werk, 1986.)

	Green leaf	White leaf
Absorptance (%)		
400–700 nm	85	40
400–3000 nm	50	17
Leaf temperature (°C)	43.5	37.5
Transpiration (mmol m^{-2} s^{-1})	6.1	4.1
Photosynthesis (% of maximum)	36	82

the difference in water vapor pressures between leaf and air, and the leaf water vapor pressure is an exponential function of temperature. Thus, the changes in pubescence between *Encelia* species can have a considerable impact on both water loss and metabolic rates.

As a specific example, consider an *E. farinosa* leaf under typical midday summer conditions with an air temperature of 40°C (Table 9.1). The expected leaf absorbtance (400–3000 nm) will be approximately 17%. Using a conservative estimate of 0.09 mol m^{-2} s^{-1} for the leaf conductance, the calculated leaf temperature is 37.5°C, which is 2.5°C below air temperature. However, if the same leaf had a 400–3000 nm leaf absorptance of 50% (typical of green leaves), the calculated leaf temperature becomes 43.5°C, a full 3.5°C above air temperature. As a consequence of this 6°C difference in leaf temperatures, the expected transpiration rate decreases from 6.1 to 4.1 mmol m^{-2} s^{-1}, resulting in a 33% decrease in the water loss rate.

Large changes in leaf temperature can also be achieved without changing leaf absorptance if the leaf conductance is allowed to increase (Fig. 9.8, top). That is, it is possible to alter leaf energy balance in two different ways to achieve equal leaf temperatures in the same environment. One means is through reducing the amount of energy absorbed by the leaf, such as reducing leaf absorptance. The other is to maintain the energy absorbed at a high level, but to increase rates of energy loss through transpiration (latent heat cooling). While both mechanisms are feasible under arid land conditions, the latter has associated with it very high rates of water loss which may be difficult to achieve in water-limited habitats.

There has been speculation in the literature that changes in leaf hairiness

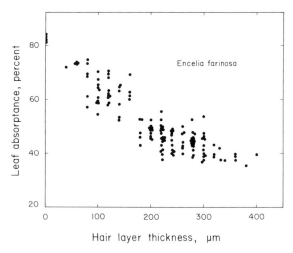

Fig. 9.9   The relationship between leaf absorptance in *Encelia farinosa* (a measure of pubescence abundance) and the thickness of the pubescence layer.

will have their primary effect on changing leaf boundary layer characteristics (Wooley, 1964; Wuenscher, 1970). The significance of leaf hairs could then become as a mechanism for retarding rates of water loss. In *E. farinosa*, the leaf boundary layer does increase slightly as leaf pubescence levels increase (Fig. 9.9), but these changes in boundary layer thickness are much too small to have much of an impact on either leaf temperature or rates of water loss.

Metabolic considerations: photosynthesis and temperature

In response to the seasonally changing thermal conditions in warm deserts, many plant species have the ability to make metabolic adjustments which allow these plants to function well under the new temperature regime (Berry and Björkman, 1980). Such adjustments are common in photosynthetic and respiratory processes. This ability to acclimate metabolically represents an alteration of both membrane- and protein-level subcellular features. The consequence of such changes in metabolic characteristics with environmental change are that metabolic activities occur at near-optimal temperatures and that thermally damaging or lethal temperature extremes are avoided.

Members of the genus *Encelia* are unusual in that photosynthetic activities do not acclimate to changes in growth temperature regimes. Whether grown under cool or hot temperatures and either in the field or under controlled environment conditions, the temperature dependence of photosynthesis does not seem to exhibit any acclimation (Fig. 9.10). There

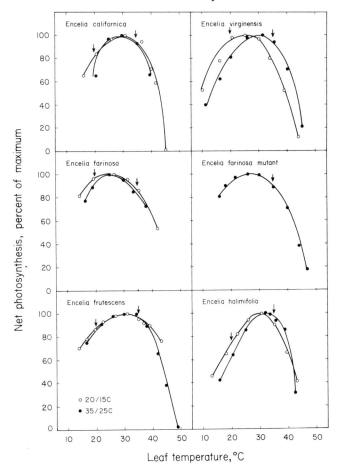

Fig. 9.10 The relationship between photosynthetic rate, expressed as a percentage of the maximum value, and leaf temperature for different species of *Encelia*. (Based in part on data from Ehleringer and Björkman (1978b), Ehleringer (1983), and Comstock and Ehleringer, 1984.)

appears to be little or no change in either the optimal temperature for photosynthesis or in the upper, thermal-damaging temperature extreme. Of particular interest is that the basic photosynthesis–temperature response is very similar for all *Encelia* species be they from relatively cool coastal or hot interior habitats (Fig. 9.10). The optimal temperature appears to be 26–28°C and the upper thermal-damaging temperature approximately 45–47°C.

This lack of photosynthetic thermal acclimation takes on greater importance when we consider the actual air temperature regimes experienced in the hot interior desert regions. During the late spring and summer, it is common for air temperatures to be 40–42°C during midday and frequently as high as 45–46°C. From Table 9.1, it appears that *E. farinosa* leaves will achieve a leaf temperature that is 2.5°C below air temperature if they are heavily pubescent (low absorptance), but will have a leaf temperature 3.5°C above air temperature if they are not (high absorptance). As such, it is quite possible that *E. farinosa* leaves would occasionally experience thermally damaging and lethal leaf temperatures if they were lightly pubescent or glabrous. Thus, through its role in altering leaf energy balance and leaf temperature, leaf pubescence in *Encelia* species appears to be an essential feature for maintaining physiological activities during periods of high air temperatures.

Ecological studies

Earlier in Fig. 9.4 it was shown that *Encelia* species are differentially distributed in distinct climatic zones. Within each of the two major clades,

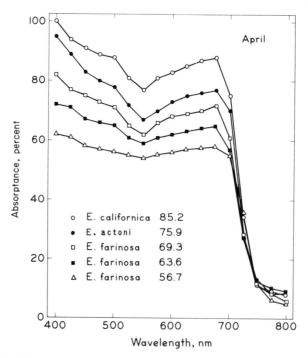

Fig. 9.11 Leaf absorptance spectra of intact leaves of *E. californica*, *E. actoni* and *E. farinosa* along an aridity gradient during April. The value adjacent to each species represents the leaf absorptance to solar radiation between 400–700 nm. (From Ehleringer, 1980.)

there is a general tendency for the leaves of the species occupying the drier sites to be more pubescent. Along a precipitation cline from the southern California coast inland, this progression from glabrous to pubescent-leaved species is associated with a clear difference in the leaf absorptances of the species along the transect (Fig. 9.11). The integrated leaf absorptance of *E. californica* leaves along the coast is 85.2%. Proceeding inland, *E. californica* is replaced by *E. actoni* with an absorptance of 75.9%, and then by *E. farinosa*, whose leaf absorptances range from 69.3% down to 56.7% with habitat aridity. Ehleringer, Björkman and Mooney (1976) demonstrated that the extent of leaf pubescence development (leaf absorptance) among plants at different locations was strongly negatively correlated with the amount of precipitation received during the current growing season.

Leaf absorptance values vary seasonally within a single population of all pubescent-leaved species in addition to the geographical variation along a precipitation cline. Leaves of *E. farinosa* produced after the first winter rains are lightly pubescent and have leaf absorptances near 80%. However,

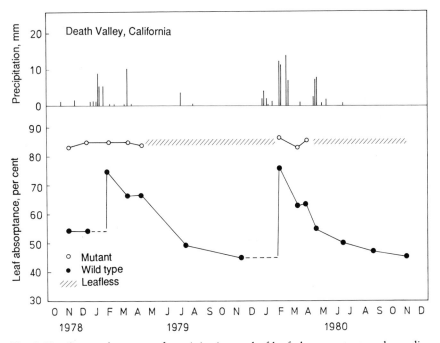

Fig. 9.12 Seasonal courses of precipitation and of leaf absorptance to solar radiation (400–700 nm) in leaves of *E. farinosa* in Death Valley, California. Also plotted are the *in situ* seasonal leaf absorptances for a mutant form of this species that lacks the reflective leaf hairs. (From Ehleringer, 1983.)

successive leaves produced during the growing season as soil moisture reserves become depleted are more pubescent and have lower leaf absorptances (Fig. 9.12). This seasonal change in leaf absorptance is not a time-dependent feature, but instead appears to be a direct response to soil moisture availability (Ehleringer, 1982). There appear to be no changes in leaf absorptances of the high-absorptance, glabrous-leaved species (Ehleringer, Björkman and Mooney, 1976; Ehleringer, 1980).

The strong relationship between pubescence development and soil moisture availability suggests that leaf pubescence might serve as a mechanism for modifying leaf energy balance and regulating leaf temperatures. The rationale behind this is that as air temperatures increase through the growing season, only two possibilities exist for reducing leaf temperatures: either an increase in transpiration rate is necessary or a decrease in the amount of energy absorbed is required. In going from wetter to drier habitats, transpirational cooling becomes less of a possibility and therefore a decreased leaf absorptance is a viable alternative. Given that the temperature optimum for photosynthesis in leaves of *Encelia* species occurs at 26–28°C (Fig. 9.10), leaf temperature regulation becomes important for much of the growing season.

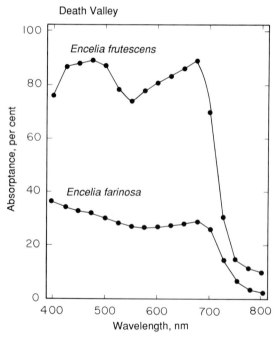

Fig. 9.13 Leaf absorptance spectra of intact leaves of *E. farinosa* and *E. frutescens* from Death Valley, California. (From Ehleringer, 1988a.)

There is one extremely important exception to the trend of increased leaf pubescence in more arid habitats. This is *E. frutescens*, a green, scabrous-leaved shrub which is widespread throughout much of the Sonoran Desert, and sympatric with *E. farinosa* over much of its range (Fig. 9.3). Its leaf absorptance spectrum is similar to that measured from leaves of the coastal *E. californica* and very much different from that of *E. farinosa* leaves (Fig. 9.13). The temperature dependence of photosynthesis by *E. frutescens* leaves is similar to that measured on *E. farinosa* leaves (Fig. 9.10) and so the same metabolic constraints on plant performance indicated for *E. farinosa* should apply to *E. frutescens*.

Despite the clear leaf absorptance differences, leaf temperatures of adjacent *E. frutescens* and *E. farinosa* shrubs are nearly identical. This occurs because *E. frutescens* leaves operate at a significantly higher leaf conductance than do *E. farinosa* leaves (Fig. 9.14). The greater leaf conductance in *E. frutescens* leaves results in a higher rate of water loss and therefore in a greater degree of transpirational cooling than occurs in *E. farinosa* leaves. In other words, *E. frutescens* leaves maintain a high leaf energy input, but

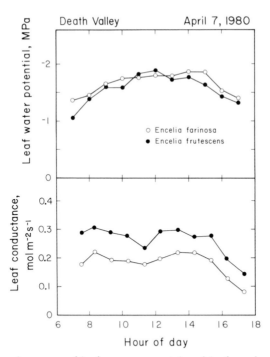

Fig. 9.14 Diurnal courses of leaf water potential and leaf conductance to water vapor of *E. farinosa* and *E. frutescens* from Death Valley, California. (From Ehleringer, 1988a.)

offset this with much greater latent heat cooling, whereas *E. farinosa* leaves reduce the leaf energy input via a reduced absorptance to compensate for the reduced transpirational cooling rate.

Leaves of *E. frutescens* shrubs are able to maintain higher transpiration rates than occur in *E. farinosa* because of a key difference in their micro-habitat distributions (Fig. 9.15). *E. frutescens* occurs almost exclusively in wash habitats, which characteristically have greater soil water availability because of the deeper soil profile and the greater runoff received. This microhabitat provides the resource necessary to maintain higher transpiration rates. In contrast, *E. farinosa* is common on the shallow-soil slopes and hillsides (Fig. 9.15). It is rarely found in active washes, and then only young individuals are observed. The most likely explanation for this is that *E. farinosa* shrubs are unable to tolerate the flashfloods which occur frequently in wash habitats. These flashfloods carry a torrent of rocks and debris which wash over the shrubs. *E. farinosa* stems are thick, brittle and break easily, whereas as *E. frutescens* stems are thinner and more pliable. Thus, a flashflood will do more damage to *E. farinosa* shrubs than to *E. frutescens* shrubs. Additionally, when stem breakage occurs, the stem loss has a greater impact on *E. farinosa* than *E. frutescens*, because the carbohydrate and nitrogen reserves for regrowth are contained within the thick stems of *E. farinosa*, but within the roots of *E. frutescens* (Ehleringer, unpublished observations).

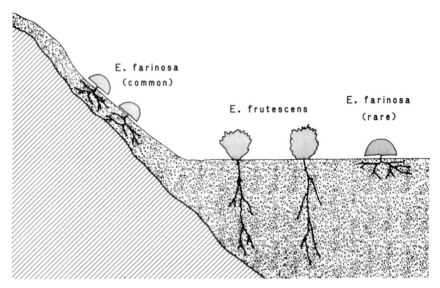

Fig. 9.15 Distributions of *E. farinosa* and *E. frutescens* in the wash and slope habitats in the Sonoran Desert. (From Ehleringer, 1988a.)

Thus far most of the discussion has focused on leaf-level properties and has inferred adaptive value to these characteristics at whole-plant levels. The link between leaf and physiological characters and whole plant performance was established in Fig. 9.2, but until now we have described no experimental results which show that the differences in energy balance characteristics translate into differences in whole-plant performance or fitness (as measured by reproductive output).

We can evaluate the significance of the pubescence to plant fitness through the use of common-garden experiments where plants are grown together under similar environmental regimes. To do this, *Encelia* species with contrasting leaf pubescence and transpiration characteristics were grown in common gardens under both natural and irrigated conditions at both Phoenix, Arizona (habitat for *E. farinosa* and *E. frutescens*) and at Irvine, California (habitat for *E. californica*). In addition, a naturally occurring mutant *E. farinosa* which lacks developed pubescence (Ehleringer, 1983) was planted.

The results of these common-garden experiments demonstrates that under desert conditions in Phoenix, the pubescent-leaved *E. farinosa* outperformed the other three (Table 9.2). Its growth rate was more than 85% higher than that of the glabrous-leaved *E. californica* and *E. farinosa* mutant. The growth rate of *E. frutescens* was 35% lower than that of *E. farinosa*, but was still higher than that of the other two green-leaved shrubs. When additional water was supplied through irrigation, *E. frutescens* shrubs in the Phoenix garden responded (increased transpirational

Table 9.2 Sizes of *Encelia* species in transplant gardens at Irvine, California and Phoenix, Arizona in June, 1980. Size was measured as the ground surface area occupied by plants. All shrubs were planted in October, 1978 and at that time were of equal size (0.01 m² area). Plants were grown under both natural precipitation and irrigated conditions. Irrigated plants received an additional 20 l water weekly. Data are means ± 1 SD of five shrubs. Ratio is the ratio of sizes of shrubs in Phoenix to those in Irvine under natural conditions. (From Ehleringer, 1988b.)

| | Phoenix | | Irvine | | |
	Natural	Irrigated	Natural	Irrigated	Ratio
Area (m²)					
E. californica	0.39 ± 0.07	0.49 ± 0.07	2.59 ± 0.55	3.05 ± 0.29	0.15
E. farinosa	0.72 ± 0.15	0.97 ± 0.13	0.89 ± 0.33	0.93 ± 0.52	0.80
E. farinosa mutant	0.37 ± 0.10	0.44 ± 0.06	0.79 ± 0.15	1.09 ± 0.06	0.38
E. frutescens	0.52 ± 0.10	0.72 ± 0.18	0.64 ± 0.28	0.77 ± 0.19	0.81

cooling) and its growth improved significantly while the growth of
E. californica and the *E. farinosa* mutant changed little.

Both the leaf-level and common-garden observations indicate that soil
water availability plays a key role in affecting leaf energy balance, metabolic
rate and therefore whole plant performance. It appears that in *E. farinosa*
leaves operate so as to reduce leaf absorptance only as soil moisture becomes
limiting and temperatures are above the thermal optimum for photosynthesis. One reason for this is that increased reflectance can directly reduce
net photosynthetic gains by reflecting photons that might otherwise be used
to carry out photosynthetic reactions (Ehleringer and Björkman, 1978b;
Ehleringer and Cook, 1984). In evaluating these tradeoffs, modeling efforts
indicate that leaves will become more pubescent only when the net carbon
gain through reducing leaf temperatures exceeds the net carbon loss associated with reflecting photons (Ehleringer and Mooney, 1978; Ehleringer,
1980). Coupled with this is a decrease in total plant water loss that may
allow plants to maintain metabolic activities longer into the drought periods
for an additional increase in plant net carbon gain.

However, implicit in this last statement is the assumption that water not
used by the plant at one time is then available for use at some later period.

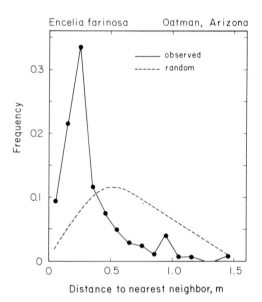

Fig. 9.16 The observed relationship between plant frequency and nearest-neighbor
distance for *E. farinosa* shrubs at Oatman, Arizona. Also plotted for comparison is
the expected frequency distribution for randomly distributed shrubs. (From
Ehleringer, 1984.)

This is not likely to be the case if plants are competing for water resources under natural conditions. If competition for water is occurring, then total plant performance is constrained by both absolute precipitation inputs as well as by rapid utilization by neighboring plants. This matter can be evaluated under natural field conditions as both *E. farinosa* and *E. frutescens* often form large monospecific stands.

If we look at the distributions of *E. farinosa* and *E. frutescens* shrubs, we find that their nearest-neighbor distributions show that plants are highly contiguous in their distributions (Fig. 9.16). For *E. frutescens*, this is easily explained since frequent flashfloods result in a myriad of erosion channels through the wash and plants are most common at those locations where recent erosion has not occurred. For *E. farinosa*, it appears that plants are most common along small microrivulets on the hillside where sheet flow might occasionally occur.

The consequence of the contiguous plant distribution in *E. farinosa*, though, is that neighbors strongly compete for water and that the rapid utilization of this limiting resource is largely responsible for the observed variation in plant sizes within a population (Fig. 9.17). When these competing neighbors are experimentally removed, there is significantly more soil water available resulting in higher leaf water potentials (Fig. 9.18); greater

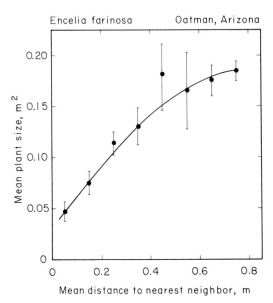

Fig. 9.17 The relationship between mean plant size and the mean distance to the nearest neighbor for *E. farinosa* shrubs at Oatman, Arizona. (From Ehleringer, 1984.)

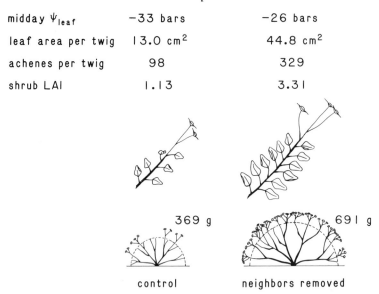

midday Ψ_{leaf}	−33 bars	−26 bars
leaf area per twig	13.0 cm²	44.8 cm²
achenes per twig	98	329
shrub LAI	1.13	3.31

369 g   691 g

control   neighbors removed

Fig. 9.18 The effects of nearest-neighbor removals on the water relations, biomass and reproduction in *E. farinosa* shrubs at Oatman, Arizona. (Based on data from Ehleringer, 1984.)

water availability results in greater rates of carbon gain, which translates into a larger plant canopy and leaf area (Fig. 9.18). The culmination of these changes is that reproductive activity is enhanced by more than a factor of three.

9.4 EVOLUTION OF ADAPTATIONS WITHIN THE GENUS

Both *E. farinosa* and *E. frutescens*, the dominant desert species, show clear adaptations to their arid habitats. *E. farinosa* generally inhabits hillslopes with low soil water availability. Decreased leaf absorptance resulting from leaf pubescence reduces leaf temperatures, hence transpirational water loss. The reduction of photon flux available for photosynthesis that results from this is to some degree offset by both seasonal and interpopulational changes in pubescence density. Nitrogen and carbohydrate reserves for regrowth are stored primarily in above-ground tissues. In contrast, *E. frutescens* occurs in washes and other areas of higher soil water availability. Leaf absorptance is not appreciably decreased by pubescence; instead, leaf temperatures are reduced by increased transpirational water loss. Seasonal and/or interpopulational differences in leaf pubescence have little or no effect on absorptance. Nutrient and carbohydrate reserves are stored below ground.

Are these contrasts the result of adaptive evolution specific to each of the species, or are they characteristic of divergent patterns within the two major clades? We can begin to address this question by examining the hypothesized closest relatives of each of these species.

E. farinosa is related to *E. palmeri* of the central and southern Baja California peninsula and *E. canescens* of South America is probably derived from hybrids between them (Clark, 1986 and unpublished). Although *E. palmeri* and *E. canescens* have not been studied as intensively as *E. farinosa*, observations of these species allow us to infer similarities and differences between the two (Ehleringer, Björkman and Mooney, 1976). Like *E. farinosa*, both species have a dense pubescence on their leaf surfaces that varies seasonally and between populations, although absorptances are generally greater than for *E. farinosa*. The regions inhabited by *E. palmeri* and *E. canescens* are generally as arid as those in which *E. farinosa* occurs, but both species are more often found in areas of low topography and sandy soils, suggesting a need for more available soil moisture. Additionally, both *E. palmeri* and *E. canescens* occupy habitats with a strong oceanic influence in contrast to the drier continental habitat of *E. farinosa*. All this suggests that, while *E. farinosa* is not unique in reducing leaf temperature through decreased leaf absorptance, its lower absorptance, perhaps coupled with other factors, enables it to exist in drier sites than those inhabited by *E. palmeri* or *E. canescens*.

More distantly related are *E. californica* and *E. ventorum* of wetter, coastal environments in southern California and northern Baja California. Unlike either *E. palmeri* or *E. farinosa*, both species have a sparse leaf pubescence with no reduction in absorptance. The cooler and moister coastal environments that they inhabit, coupled with drought-deciduousness in *E. californica*, enable *E. californica* and *E. ventorum* to complete seasonal carbon gain without the necessity of an altered leaf energy balance.

From members of the *californica* clade one might infer a general evolutionary trend toward increased leaf pubescence. In contrast, examination of the *frutescens* clade suggests an opposite pattern. *E. frutescens* is most closely related to its variety *resinosa*, which will be elevated to specific status on morphological grounds (Clark and Kyhos, in preparation). Variety *resinosa* occurs in the deserts surrounding the Colorado River and its tributaries in northern Arizona and southeastern Utah. Like *E. frutescens*, it has glabrous, highly absorptive leaves with sparse pubescence. The region it inhabits, however, is not as dry as that occupied by *E. frutescens*; both higher precipitation and lower temperatures contribute to a generally greater soil water availability. This, plus the greater water-holding capabilities of the sandstone-derived rock common to the region, may explain the occurrence of variety *resinosa* in many other sites in addition to watercourses.

Two less closely related species in the *frutescens* clade, *E. actoni* and *E. ravenii*, are both more pubescent than either *E. frutescens* or variety *resinosa*. Reductions in leaf absorptance are greatest in *E. ravenii* (in the range of those achieved by *E. palmeri*) and intermediate in *E. actoni* (see Fig. 9.11). Both species occur most often on hillslopes. *E. actoni* inhabits wetter, semi-desert environments than *E. ravenii*, which is found in only a few sites in the harsh desert of northeastern Baja California. Thus, the pattern emerging in the *frutescens* clade is one of decreasing emphasis on leaf pubescence as a reflective structure. The trend is also from hillside to wash habitats, which allows the plants to utilize enhanced transpirational cooling as the alternative mechanism for regulating leaf temperatures.

At present we can not definitively comment on the original habitat of the ancestral *Encelia*. Our impression is that the genus originated in a temperate semi-arid climate, perhaps northern Mexico, and then invaded drier habitats. There is likely to have been substantial migration during recent glacial periods. Whether or not the two clades originally represented radiation into different desert regions is unknown. Presently, there are two centers of diversity. The first is in central Baja California, which probably reflects a new isolated refugium from past glacial periods. The second is in the northern Mohave Desert where both the *californica* and *frutescens* clades come together. This contact may have occurred only recently in evolutionary terms. From this standpoint, it is interesting to note that two species of these regions (*E. asperifolia* and *E. virginensis*) are thought to have arisen from hybridization and one of these (*E. asperifolia*) represents a hybrid origin derived from parents of different clades.

The objective of this chapter has been to show that a combination of systematic and ecological approaches to the study of adaptation in *Encelia* provides more insights into the evolution of the genus than available from either methodology alone. Clearly, this integrated approach will continue to bear fruit in the study of *Encelia*, and we are convinced of its value for similar studies of other genera as well.

9.5 REFERENCES

Berry, J. A. and Björkman, O. (1980) Photosynthetic response and adaptation to temperature in higher plants. *Ann. Rev. Plant Physiol.*, **31**, 491–543.

Blake, S. F. (1913) Contributions from the Gray Herbarium of Harvard University, n.s. No. 41. II. A revision of *Encelia* and some related genera. *Proc. Amer. Acad. Arts.* 49, 346–96.

Charest-Clark, N. (1984) Preliminary scanning electron microscopic study of the peduncle, phyllary, and pale trichomes of *Encelia* (Asteraceae: Heliantheae). *Crossosoma*, **10**, 1–7.

Charest-Clark, N. and Clark, C. (1984) Comparison of trichomes of the capitulum to leaf trichomes of the *Encelia californica* clade (Asteraceae: Heliantheae). *Am. J. Bot.*, **81**(5), Part 2, p. 152 (Abstract).

Clark, C. (1982) Hybridization in *Encelia* (Compositae: Heliantheae) and its effect on phylogenetic analysis. *Bot. Soc. Am.*, Publ. 162 (Abstract).

Clark, C. (1986) The phylogeny of *Encelia* (Asteraceae: Heliantheae). *Bot. Soc. Am.*, **73**, 757 (Abstract).

Clark, C. and Kyhos, D. W. (1979) Origin of species by hybridization in *Encelia* (Compositae: Heliantheae). *Bot. Soc. Am.*, Publ. 157 (Abstract).

Clark, C., Kyhos, D. W. and Thompson, W. C. (1980) Evidence for the origin of diploid species in *Encelia* (Compositae: Heliantheae) by hybridization. *2nd Int. Congr. Syst. Evol. Biol.*, Abstracts.

Clark, C. and Sanders, D. L. (1986) Floral ultraviolet in the *Encelia* alliance (Asteraceae: Heliantheae). *Madrono*, **33**, 130–5.

Clark, C., Thompson, W. C. and Kyhos, D. W. (1980) Comparative morphology of the leaf trichomes of *Encelia* (Compositae: Heliantheae). *Bot. Soc. Am.*, Publ. 158 (Abstract).

Comstock, J. and Ehleringer, J. R. (1984) Photosynthetic responses to slowly decreasing leaf water potentials in *Encelia frutescens*. *Oecologia*, **61**, 241–8.

Ehleringer, J. R. (1980) Leaf morphology and reflectance in relation to water and temperature stress. in *Adaptation of Plants to Water and High Temperature Stress* (eds N. C. Turner and P. J. Kramer), Wiley, New York.

Ehleringer, J. R. (1981) Leaf absorptance of Mohave and Sonoran Desert plants. *Oecologia*, **49**, 366–70.

Ehleringer, J. R. (1982) The influence of water stress and temperature on leaf pubescence development in *Encelia farinosa*. *Am. J. Bot.*, **69**, 670–5.

Ehleringer, J. R. (1983) Characterization of a glabrate *Encelia farinosa* mutant: morphology, ecophysiology and field observations. *Oecologia*, **57**, 303–10.

Ehleringer, J. R. (1984) Intraspecific competitive effects on water relations, growth and reproduction in *Encelia farinosa*. *Oecologia*, **63**, 153–8.

Ehleringer, J. R. (1988a) Comparative ecophysiology of *Encelia farinosa* and *Encelia frutescens*. *Oecologia*, (in press).

Ehleringer, J. R. (1988b) Comparative performance of *Encelia* species differing in leaf reflectance and transpiration rate under common garden conditions. *Oecologia*, (in press).

Ehleringer, J. R. and Björkman, O. (1978a) Pubescence and leaf spectral characteristics in a desert shrub, *Encelia farinosa*. *Oecologia*, **36**, 151–62.

Ehleringer, J. R. and Björkman (1978b) A comparison of photosynthetic characteristics of *Encelia* species possessing glabrous and pubescent leaves. *Plant Physiol.*, **62**, 185–90.

Ehleringer, J. R., Björkman, O. and Mooney, H. A. (1976) Leaf pubescence: effects on absorptance and photosynthesis in a desert shrub. *Science*, **192**, 376–7.

Ehleringer, J. R. and Cook, C. S. (1984) Photosynthesis in *Encelia farinosa* Gray in response to decreasing leaf water potential. *Plant Physiol.*, **75**, 688–93.

Ehleringer, J. R. and Cook, C. S. (1986) Leaf hairs in *Encelia* (Asteraceae). *Am. J. Bot.*, **74**, 1532–1540.

Ehleringer, J. R. and Mooney, H. A. (1978) Leaf hairs: effects on physiological activity and adaptive value to a desert shrub. *Oecologia*, **37**, 183–200.

Ehleringer, J. R., Mooney, H. A., Gulmon, S. L. and Rundel, P. W. (1981) Parallel evolution of leaf pubescence in *Encelia* in coastal deserts of North and South America. *Oecologia*, **49**, 38–41.

References

Ehleringer, J. R. and Werk, K. S. (1986) Modifications of solar-radiation absorption patterns and implications for carbon gain at the leaf level. in *On the Economy of Plant Form and Function* (ed. T. J. Givnish), Cambridge University Press, London.

Gates, D. M. (1962) *Energy Exchange in the Biosphere*. Harper and Row, New York.

Gates, D. M. (1980) *Biophysical Ecology*. Springer-Verlag, New York.

Gould, S. J. and Vrba, E. S. (1982) Exaptation – a missing term in the science of form. *Paleobiology*, 8, 4–15.

Hennig, W. (1966) *Phylogenetic Systematics*. University of Illinois Press.

Kyhos, D. W. (1971) Evidence of different adaptations of flower color variants of *Encelia farinosa* (Compositae). *Madrono*, 21, 49–61.

Kyhos, D. W., Clark, C. and Thompson, W. C. (1981) The hybrid nature of *Encelia laciniata* (Compositae: Heliantheae) and control of population composition by post-dispersal selection. *Syst. Bot.*, 6, 399–411.

Lin, Z. F. and Ehleringer, J. R. (1983) Epidermal effects on spectral properties of leaves of four herbaceous species. *Physiol. Plant.*, 59, 91–4.

Proksch, P. and Clark, C. (1984) New chromenes and benzofurans from the *Encelia* alliance (Asteraceae: Heliantheae) and their systematic significance. *Am. J. Bot.*, 71(5), Part 2, p. 183 (Abstract).

Proksch, P. and Clark, C. (1987) Systematic implications of chromenes and benzo-furans from *Encelia* (Asteraceae). *Phytochemistry*, 26. 171–74.

Raschke, K. (1960) Heat transfer between the plant and the environment. *Ann. Rev. Plant Physiol.*, 11, 111–26.

Sanders, D. L. and Clark, C. (1987) Comparative morphology of the capitulum of *Enceliopsis*. *Am. J. Bot.*, 74, 1072–86.

Wiley, E. O. (1981) *Phylogenetics*. John Wiley & Sons, New York.

Wong, C. L. and Blevins, W. F. (1967) Infrared reflectances of plant leaves. *Aust. J. Biol. Sci.*, 20, 501–8.

Wooley, J. T. (1964) Water relations of soybean leaf hairs. *Agron. J.*, 56, 569–71.

Wuenscher, J. E. (1970) The effect of leaf hairs of *Verbascum thapsus* on leaf energy exchange. *New Phytol.*, 69, 65–73.

Editors' commentary on Part 4

The study of adaptation in plants has primarily emphasized questions of function: How does it work? In addition, interactions between plant and environment have been examined: How does the adaptation contribute to the plant's ability to grow and reproduce in its habitat? And, from a comparative standpoint, how do populations of the same species adapt to contrasting environments, and how do different species adapt to the same environment?

Adaptation is intrinsically a complex subject because it is rarely clear how to separate the responsible characters (and character complexes) from non-contributing characters. And it is frequently difficult to demonstrate that a particular adaptation evolved as a response to a particular environment or whether its presence reflects phylogenetic legacy.

At what level should we answer the question 'How does it work?' Should we require a description that includes information about the transcription and translation of relevant genes or is knowledge of physiological function sufficient? Must we include information about changes in character expression under different environments, so-called reaction norms? How much information is appropriate before we switch the question to one having to do with fitness functions, before we ask about the phenotypic variance of the relevant characters and the relationship between this variance and contribution of genes to succeeding generations? As in much of biology, the multiplication of appropriate questions is easy. The questions must be posed, but they must not be multiplied so that they stifle inquiry.

The two chapters in this section approach the study of adaptation from different standpoints. The section provides a transition from the consideration of genetics, morphogenesis and phylogenetic reconstruction emphasized in the previous two sections to the population and community orientation of the next two sections. The two chapters present broadly complementary treatments of how different factors govern photosynthetic performance.

In Chapter 8 Karl Niklas focuses on biomechanical and geometric requirements to evaluate how well plants perform certain 'tasks' related to photo-

249

synthesis. He demonstrates that each of the physiological requirements such as gas exchange, light interception, mechanical support, and hydraulics can be quantified via an appropriate engineering analysis into geometric parameters of size, shape, orientation, and internal structure. He finds that although changes in geometry can optimize the performance of single tasks, this necessarily reduces the performance of other tasks so that 'there is no single "optimal" plant form'.

His analysis of the influence of leaf size and shape, phyllotaxy, and internode length on the efficiency of light interception in *Plantago* species provides an interesting example of diverse design solutions. Thus, in some species, he finds a correlation between the phenology of leaf senescence and the degree of leaf overlapping dictated by phyllotactic pattern; whereas, in a different species, he finds one between leaf shape, internode distance and phyllotaxy. Overall, it appears that numerous morphological factors can be covaried to produce structures and arrangements with comparable levels of performance.

In contrast to Niklas' approach utilizing functional morphology, Ehleringer and Clark emphasize energy balance parameters (leaf temperature and transpiration), physiological activity (photosynthesis) and the plant microenvironment to account for net carbon gain and, thereby, increased growth and size. The information is then used to explain how related species are able to inhabit contrasting environments. In the case of *Encelia farinosa*, it turns out that increased leaf pubescence appears to provide a major adaptation that permits populations to live in hot dry desert sites. The increased pubescence results from vastly elongated distal trichome cells. The cells form a 'blanket-like' covering over much of the plant body, reflecting a substantial proportion of the solar radiation. This reduces leaf temperature some 6°C below air temperature (relative to a glabrous leaf) and reduces transpiration by about one-third, thereby, permitting continued photosynthesis and other physiological activities. What is remarkable is how much physiological consequence can be assigned to the change in this single character. The authors note the potential role of reproductive rates and patchy dispersion of plants in 'wetter' microhabitats as additional components of adaptation. Demographic and population genetic studies are now appropriate.

A coordinated study of other *Encelia* species revealed that in the group related to *E. farinosa*, several species also utilize increased pubescence to reduce leaf temperature, but *E. farinosa* displays an extreme form of the character. In contrast, a different clade of Encelias utilizes epidermal resins rather than pubescence to reduce water loss. The information from related species suggests that the origin of pubescence was an adaptation that took place in an ancestor of *E. farinosa* and was not a unique adaptation to this species. The two clades inherited different adaptations to similar environmental problems. Associated with the different phylogenetic legacies are differences in other morphology and in microhabitat.

Although these two chapters provide significant information about characters that serve adaptive functions, they tell us little about the genetic and developmental modifications that led to their evolution. Such evidence is relevant because we wish to determine if the acquisition of pubescence, for example, should be regarded as simple in Sachs' sense. The genetic basis of the presence/absence of pubescence and its density and distribution on the plant body has been studied in cotton (Lee, 1985) and, at least in this plant, it seems to be an easily manipulated character. In *Encelia*, it may be that a simple genetic input has had a complex physiological consequence, but this remains to be determined.

REFERENCES

Lee, J. A. (1985) Revision of the genetics of the hairiness–smoothness system in *Gossypium. J. Hered.*, 76, 123–6.

PART 5

Genetics and ecology of populations

CHAPTER 10

Natural selection of flower color polymorphisms in morning glory populations

MICHAEL T. CLEGG and BRYAN K. EPPERSON

10.1 INTRODUCTION

A mechanistic analysis of natural selection requires two kinds of information. First, the causal relationship between an environment and natural selection on a phenotype must be established. In other words, why is one phenotype better adapted to a particular environment while a second phenotype is less well adapted? What does this differential adaptation mean in terms of probabilities of survival and reproductive success, and how are these differentials manifested during the life history of the organism? Second, it is necessary to establish the genetic basis of the phenotypic differences that confer differential adaptation. Genetic analysis is simplest for major gene polymorphisms that are determined by alleles at a single locus. In addition, the formal theory of population genetics can be readily used to make predictions about rates of evolutionary change for single locus polymorphisms.

These two problems of relating phenotypes to genotypes and of determining the mechanisms of environmental adaption are easy to state but are frequently difficult to resolve experimentally. The difficulties of detecting selection are well illustrated by the neutrality controversy that dominated evolutionary genetics throughout the 1970s, and which claimed that much of electrophoretic variation is irrelevant to natural selection (Kimura, 1983; Kimura and Ohta, 1971). Enzyme polymorphism was found to be abundant and the genetic analysis of this class of variation was relatively simple; however, it proved very difficult to establish that measurable fitness differences accompany electrophoretic differences (Lewontin, 1985).

A classical tenet of ecological genetics has been that major gene polymorphisms (where the phenotypes represent discrete morphological alternatives) are indicators of some form of balancing selection (Ford, 1965). The relationship between phenotype and environment is sometimes apparent for major gene polymorphisms so that it is possible to identify potential mechanisms of selection. These potential mechanisms can then be investigated systematically in experimental and field studies. This program of ecological genetics has been successful in analyses of industrial melanism (Kettlewell, 1961; Lees, 1981), mimicry (Sheppard, 1967) and in many other classical studies of natural selection. On the other hand, there are a great number of similar cases where selection has been detected through the analysis of genotypic frequency change, but where the environmental mechanisms of selection remain obscure.

It is not surprising that analyses of the mechanisms of natural selection have been most successful where environments have been altered in specific ways by the introduction of insecticides, antibiotics, air pollutants, heavy metal contamination, etc. (Bishop and Cook, 1981). These man-driven environmental changes can be isolated from the vast complex of other factors that impinge on living organisms. Moreover, the bulk of these agents act through viability selection by reducing growth and survival rates of the susceptible genotype. This stage of the life cycle presents fewer problems because the complex genotypic interactions associated with mating, and the processes of gamete production and differential fertility are absent.

There are thus four elements to consider in the analysis of natural selection:

1. The genetic determination of the relevant phenotypic differences;
2. The detection of selection in terms of gene or genotypic frequency change;
3. The demographic stages at which selection operates;
4. The identification of the environmental mechanisms of natural selection.

During the past eight years we have attempted to apply these four points in studies of the genetic transmission of a set of flower color genes in the common morning glory (*Ipomoea purpurea* L.). Our goal has been to develop a complete account of the ways in which natural selection acts upon a system of major gene polymorphisms in plant populations. Initially our work has focused on the effect of flower color variation on mating systems. There are genotypic differences in rates of self-fertilization between flower color morphs of morning glory, owing to differential pollinator service. Accordingly, the so-called mating system modifier genes, such as the flower color polymorphisms in morning glory populations, are expected to be selected because such genes bias their own transmission in populations (discussed below). In this chapter we will use experimental results from our morning glory work to illustrate some of the complexities associated with the study of natural selection in plant populations.

10.2 THE EVOLUTION OF MATING SYSTEM MODIFIER GENES

Elementary theoretical considerations

There are two categories of explanations for the evolution of self-fertilization in plant species. One category is ecological and Stebbins (1950), in his classic book *Variation and Evolution in Plants*, provides a penetrating review of the ecological factors that promote the evolution of self-fertilization that is still timely. Stebbins (1950) noted that self-fertilization is frequently a feature of annual plants that occur in extreme or harsh environments and he argued that self-fertilization may be advantageous because it insures pollination in very low density populations, or in the absence of pollinators. He also argued that self-fertilization can be advantageous because it prevents genetic recombination and thereby preserves favorable gene combinations. Allard (1975) has confirmed this latter hypothesis in a major series of experimental studies on the genetics of inbreeding plant populations.

There is another potential reason for the evolution of self-fertilization that is purely genetic and arises from a transmission bias that occurs in populations when separate genotypes have different probabilities of self-fertilization. Fisher (1941) was the first to study a mathematical model that assumed a single genetic locus with genotypic differences in probabilities of self-fertilization and he showed that a gene that causes increased selfing would increase to fixation, assuming that the selfing gene was transmitted through outcrossing at a rate proportional to its frequency in the population. The intuitive explanation for this result is simple and rests on the fact that two copies of the selfing gene are transmitted to progeny that arise through self-fertilization. If a particular genotype has a higher probability of self-fertilization than other genotypes in the population, the selfing gene will be disproportionately transmitted to its progeny. Furthermore, if the genotype with an enhanced frequency of selfing also contributes pollen to outcross matings in proportion to its frequency in the population – that is the selfing genotype suffers no reduction in male (pollen) fertility, then the selfing gene has a net advantage in transmission.

There has been much recent interest in this explanation for the evolution of self-fertilization, because it focuses on the role of asymmetries in genetic transmission as an evolutionary force (Charlesworth and Charlesworth, 1979; Ross, 1984). At the present time theoretical investigations of genetic models of mating system evolution have advanced well beyond empirical work (e.g. Gregorius, 1986). There is substantial experimental evidence for genetic variation in outcrossing rates in self-compatible plant species (e.g. Harding, 1970; Horovitz and Harding, 1972; Rick, Fobes and Tanksley, 1979; Schoen, 1982), however, it has been difficult to find discrete genetic variants that are polymorphic within populations and that act as mating system modifier genes. This difficulty has contributed to the paucity of experimental work in an area of active theoretical research. The morning glory system appeared to offer a particularly good experimental model for the study of the evolution of mating system modifier genes.

The experimental system

The common morning glory may have been introduced into the southeastern United States during the eighteenth century as a horticultural plant. Following its presumed introduction into the southeast, the morning glory escaped cultivation and became established as a weedy species in cultivated fields and in disturbed environments. The horticultural forms of morning glory are characterized by a series of showy flower color polymorphisms and naturalized and weedy populations also exhibit a wide range of flower color morphs. *I. purpurea* is believed to be a native of the highlands of central Mexico where the domestication of the plant and selection of the flower color variants are assumed to have occurred.

Our work has been directed towards understanding the role of the flower color morphs in modifying the mating system of this predominantly outcrossed, but self-compatible species. To establish a genetic basis for this work, we have shown that ten of the flower color phenotypes are determined by at least four genetic loci (Ennos and Clegg, 1983; Epperson and Clegg, 1988). Three of these loci, which are relevant to the mating system modification work, are:

1. The W/w locus that determines three codominant phenotypes (white, light and dark);
2. The P/p locus that determines a dominant blue versus a recessive pink phenotype;
3. A third locus (denoted A/a) that is epistatic to the other flower color loci and which yields pure white flowers in the recessive aa homozygote, independent of the genotypic constitution at the other loci (Epperson and Clegg, 1988).

We also surveyed southeastern US populations of *I. purpurea* for isozyme variation and found low levels of enzyme polymorphism. One esterase locus and one phosphoglucomutase locus showed moderate levels of polymorphism out of more than 20 enzyme systems surveyed. The low level of enzyme variation may be a consequence of selection of the horticultural types from a restricted foundation population, although we have scant information on the history of domestication of *I. purpurea*. Despite low levels of enzyme polymorphism, *I. purpurea* exhibits substantial morphological variation in US populations and striking mutant phenotypes are common. We now believe that some of the mutant phenotypes observed may result from high levels of mutator activity (discussed on page 269).

The *I. purpurea* flower color morphs modify the mating system

In the southeastern US, *I. purpurea* is predominantly pollinated by bumblebees (*Bombus pennsylvanicus* and *B. impatiens*). Our initial goal was to ask whether pollinator behavior varied among maternal flower color morphs and then to ask whether the transmission of marker genes was correlated with differential pollinator behavior. Because *I. purpurea* is a self-compatible species, we reasoned that undervisited morphs should exhibit a higher level of self-pollination. We also hypothesized that the white phenotypes would be undervisited because experimental work on bumblebee behavior had shown that bumblebees prefer dark flowers (Heinrich, Mudge and Deringis, 1977).

A plot in a natural population of *I. purpurea* near Athens, Georgia was observed over two consecutive years. The frequency of the white morph in the observation plot averaged about 10% over-years and over-days of observation within years. The frequency of pollinator visits to individual flowers was recorded during a two hour period (8:00 am−10:00 am) of each morning of observation as was the frequency of paired successive visits (e.g. white−white, white−blue, blue−pink, etc.). The resulting data (Table 10.1) showed that white flowers received approximately half the visits expected based on a random choice model (Brown and Clegg, 1984).

Flowers in the observation plot were tagged by color phenotype and seed were harvested at maturity. The resulting seed progeny were analyzed for their genotype at a polymorphic esterase locus and the proportion of self-fertilization was estimated based on the mixed mating model. This

Table 10.1 Observed and expected numbers of visits of pollinators to white, pink and blue flower color morphs together with the correlation in paired successive visits partitioned into nearest neighbor and non-nearest neighbor visits. The estimated frequency of self-fertilization by color morph (*s*) is also reported. (After Brown and Clegg, 1984.)

Morph	*Total flights*		*Flights between nearest neighbors*		*Flights between non-nearest neighbors*		*s*
	Obs	*Exp*	*Obs*	*Exp*	*Obs*	*Exp*	
white	54	95	12	25	42	71	0.52
pink	432	438	121	130	197	201	0.27
blue	691	643	205	183	486	460	0.31

model assumes that self-fertilization occurs with probability s and that random outcrossing occurs with probability $t(= 1 - s)$. The model further assumes that male gametes are drawn randomly from the total population and that successive outcross events within families are independent. (See Clegg (1980) for further discussion of the mixed mating model.) The estimates by morph phenotype are given in Table 10.1 for white, pink and blue flowers, respectively. The estimate of s for the blue and pink flowers do not differ significantly, whereas the white flower estimate is significantly different from both the pink and blue flower estimates. These results establish that white flowers are undervisited and that the transmission of genes is correlated with the pattern of pollinator behavior. We therefore conclude that the flower color polymorphisms constitute a system of mating system modifier genes.

One feature of the analysis of the pollinator flight data was a marked tendency of the bumblebees to make nearest-neighbor visits. The correlation among paired successive visits (which measures flower color constancy) indicated strong positive assortative mating for all color phenotypes when partitioned into the subset of nearest-neighbor visits. This result suggests that the spatial patterning of the color phenotypes in the field can influence the potential for mating system modification (Brown and Clegg, 1984). The problem of spatial patterning as a confounding influence on studies of genetic transmission will be considered further below.

If we assume that the genes determining flower color are mating system modifier genes, then based on the theory of Fisher, we expect the white phenotype to be the predominant phenotype in natural populations. This is not the case. Surveys of natural and weedy populations in the southeastern US indicate that the frequency of the w allele is about 10 per cent (Epperson and Clegg, 1986). There are several potential explanations for this discrepancy.

One category of explanation is historical and assumes that local populations have been recently established and that there has been insufficient time for the white gene to increase in frequency. It is difficult to reject historical explanations because they posit past scenarios that can not be observed. Nevertheless, the available evidence provides little support for an historical (or non-equilibrium) explanation. Numerical iteration of the equations of the Fisher model, using parameter values similar to those estimated in the *I. purpurea* case, show that the expected gene frequency change would be rapid. Moreover, anecdotal evidence obtained from farmers dealing with morning glory infestations indicates that some local populations are more than 75 years old. This is consistent with reports of morning glory populations in the southeastern US by the early 1800s (Pursh, 1814; Eaton, 1833). Finally, there is no particular reason to suppose that the white flower morph would have been the minority phenotype initially, because this morph is also a common horticultural type.

A more compelling explanation for the low frequency of the white phenotype is that natural selection opposes the transmission advantage of the white gene. One point where natural selection is likely to act is through pollen fertility because white flowers may not contribute pollen to the outcross pool in proportion to the frequency of white plants in natural populations. A reduced frequency of pollinator visitation is likely to be correlated with a reduced contribution of pollen from white flowers to the outcross pool. This is known as the 'pollen discounting hypothesis' and theoretical investigations of the Fisher model, where pollen discounting is included, show that conditions for the spread of a selfing gene are more stringent (Nagylaki, 1976; Lloyd, 1979; Schoen and Lloyd, 1984; Holsinger, Feldman and Christiansen, 1984). To test the pollen discounting hypothesis, it is necessary to devise methods for estimating the male contribution of different flower color phenotypes to fertilization events. Because male gametophytes are small and ephemeral, the genotypes of particular male gametophytes contributing to successful fertilization events can not be observed. We therefore use marker genes together with statistical models of the mating process to estimate male reproductive success.

Statistical models impose certain assumptions about homogeneity of the population units sampled. These models also make assumptions about the reproductive biology of the plant species under investigation. It is frequently the case that these simplifying assumptions are violated by natural plant populations. To illustrate these problems, we now discuss some of the methodological difficulties of applying statistical models to the study of natural populations of morning glory.

10.3 OPERATIONAL DEFINITION OF A POPULATION

The first and most basic problem concerns the operational definition of a population. The abstract concept of a population is fundamental to evolutionary theory, because the transmission of genes through evolutionary time occurs in populations. To paraphrase Stebbins (1950), individuals belong to a population because they share a set of spatial and temporal relationships that allow interbreeding among the population members. The notion of a population is also fundamental for a second reason: natural selection is most effective when it operates between individuals within a population (individual selection). Therefore the goal of theoretical population genetics has been to investigate the statistical rules that define breeding relationships among members of a population and then to combine breeding rules with models of selection to study the dynamics of genetic change.

The empirical study of genetic change faces a serious obstacle in defining populations in an operational or sampling sense. Although it is easy to discuss the abstract concept of population, operational definitions are arbitrary and do not represent actual breeding relationships. Consider the morning glory populations: how can we identify population units in an interbreeding sense and how is interbreeding shaped by the spatial connections among a collection of individuals? The usual empirical solution is to adopt a sampling strategy that accords with the investigator's subjective impression of potential breeding relationships. Because flower color is readily observed in morning glory populations, it is relatively easy to investigate the spatial patterns of phenotypic variation. Indeed casual observations suggest that some flower color phenotypes occur in patchy distributions and are non-randomly distributed in space.

To assess the extent to which local populations are composed of a non-random mosaic of phenotypic patches, a detailed survey of 12 local populations of *I. purpurea* was conducted. Each population was sampled on a regular lattice and flower color type was recorded for each position on the lattice. The sampling of color type by spatial location within a population permits the use of autocorrelation statistics to analyze spatial structure (Sokal and Oden, 1978). The results of the spatial autocorrelation analyses for the P/p locus revealed a significant positive autocorrelation of color phenotype over short to moderate distances, indicating a non-random spatial distribution and a tendency for the color phenotypes to be aggregated together in patches. The average patch size for pink or blue phenotypes is relatively constant across populations with approximately 120 plants in flower on any given day per patch. (The total number of flowering and non-flowering plants per pink or blue patch ranges from 240 to 720 (Epperson and Clegg, 1986).) Interestingly, this is not the case for the white locus. The distribution of white phenotypes is either random in space (no significant spatial autocorrelation), or when patches can be detected, they range from one-half to one-sixth the size of patches for the P/p locus (Epperson and Clegg, 1986).

Plant populations with limited pollen dispersal are expected to exhibit a patchy genotypic distribution (Turner, Stephens and Anderson, 1982) and the expected patch dimensions for isolation by distance models have been determined through Monte Carlo simulation (Sokal and Wartenberg, 1983; Epperson, unpublished data). Comparisons of these theoretical results with the morning glory data indicate that patch dimensions for the P/p locus are in accord with expectations based on isolation by distance for a selectively neutral locus. In contrast, the difference in patch dimensions for the W/w locus can not be accounted for by genotypic differences in breeding system, because simulation studies indicate that average patch size is relatively insensitive to variation in neighborhood size in the range observed for the morning glory populations (see Epperson and Clegg (1986) for detailed arguments). The selective elimination of white phenotypes appears to be the most plausible explanation for the absence of patch structure. (Potential mechanisms of selection against white phenotypes are discussed further below.)

There are three facts about spatial structure that must be considered in drawing inferences from population samples. First, pollinators tend to visit nearest-neighbors and consequently breeding relationships are defined by spatial proximity. (Pollen carry-over in morning glory (Ennos and Clegg, 1982) and other insect pollinated plants blurs spatial relationships to some degree.) The second fact is that the pink/blue phenotypes are spatially aggregated and the interaction of nearest-neighbor pollination with spatial aggregation causes population substructure. Finally, the mating system of the white phenotype is the result of an interaction between increased self-fertilization and reduced spatial aggregation; the latter should tend to promote a higher effective level of outcrossing at the W/w locus. In the face of these complicated spatial and breeding relationships, inferences about genetic transmission based on samples from natural populations are seriously compromised. To understand why complicated spatial relationships are such a serious problem it is useful to discuss mating system estimation models.

10.4 MATING SYSTEM ESTIMATION

The primary statistical model for mating system estimation, as applied to self-compatible plants, is the mixed mating model. The mixed mating model assumes that male gametes are drawn independently from an outcross pool, so that the probability of drawing a particular pollen type is independent of the location of the female parent in the population. This assumption is clearly violated in morning glory populations by the tendency of pollinators to make nearest-neighbor visits and by the spatial autocorrelation of flower color morphs. Spatial substructuring causes s to be overestimated by as much as twice its true population value (Ennos and Clegg, 1982).

There is a second more subtle violation of the mixed mating model associated with insect pollination. This second problem arises because family structured data are used to estimate plant mating systems and certain assumptions must be made about the way in which male genes are drawn for transmission within maternal families. To make the problem clear, consider the mixed mating model formulation of the conditional probability of drawing n_1 and n_2 progeny of genotypes A_1A_1 and A_1A_2, given a maternal genotype of A_1A_1. This probability is

$$P(n_1,n_2 \mid s, A_1A_1) = C(s + tp)^{n_1}[t(1 - p)]^{n_2}$$

where C is a combinatorial coefficient and where p denotes the frequency of the A_1 allele in the pollen pool. This probability expression is violated because successive outcross events within a family are unlikely to be independent in insect-pollinated plants. Insect pollinators visit a plant and carry a load of pollen to the next plant visited, so that pollen is drawn from one or perhaps two male parents, rather than independently from the entire population of male gametes.

To facilitate mating system estimation for insect pollinated plants, an alternative estimation model, called the 'one pollen parent model', that assumes that all outcross events within a family derive from a single male parent, has been developed (Schoen and Clegg, 1984, 1986). Nevertheless, the one-pollen parent model still assumes that the male parent is chosen randomly from the population, without regard to the spatial location of the female parent. Unfortunately, when complex non-random spatial patterns exist, statistical models fail because it becomes difficult or impossible to specify the true sampling distribution.

An alternative and very promising approach for the characterization of plant mating structures is based on the estimation of male parentage using mulitlocus isozyme markers. The use of a large number of marker genes makes it increasingly probable that only one, or a small number of plants in a population could contribute a particular multilocus male gamete to a given seed progeny. Two approaches to the estimation of male parentage have been introduced. The first approach is based on paternity exclusion and requires a complete genotypic census of the population (Ellstrand, 1984). The second approach is based on maximum likelihood estimation of male parentage (Meagher, 1986). These methods offer a powerful empirical approach to a wide variety of fundamental questions such as determining multiple parentage within fruits (Ellstrand and Marshall, 1986), measuring interpopulational gene flow (Ellstrand and Marshall, 1985), characterizing the distribution of distances between mating pairs and determining the fertility distribution of male parents (Meagher, 1986). Nevertheless, these methods do not allow us to follow the transmission of particular genes, such as the alleles of the W/w locus. To follow the transmission of a particular gene, the patterns of linkage disequilibria between the locus of interest and the marker loci must be known. For these reasons, and because of the difficulties associated with sampling natural populations, we have turned to the experimental manipulation of artificial populations to investigate the transmission of the flower color genes during reproductive phases of the life cycle.

Experiments to estimate male reproductive contribution by flower color morph

To estimate pollen discounting for the white and pigmented phenotypes controlled by the W/w locus, an experimental design was developed utilizing a polymorphic esterase locus in linkage disequilibrium with the W/w locus (Schoen and Clegg, 1985). The transmission of particular alleles at the esterase locus could then be used to provide information about the transmission of associated alleles at the W/w locus. Implementation of this design required extensive breeding experiments to construct lines with the appropriate combination of marker and floral pigment genes.

The experimental plant materials were grown in a greenhouse and placed into field plots in fixed spatial designs. Twelve plants with white flowers and twelve with pigmented flowers were placed in a 4 × 6 grid in a random spatial arrangement. A design with equal frequencies of the two morphs was chosen to maximize the number of outcross events observed and consequently to maximize statistical power for the estimation of pollen pool gene frequencies. Pollinator behavior was observed and seed progeny were collected from maternal plants of known esterase genotype and floral pigment phenotype and the genotypic distribution of the seed progeny was determined. Joint maximum likelihood estimation of morph specific outcrossing rates and pollen pool gene frequencies was based on the observed family distributions. Two experimental designs that differed in the pattern of association between esterase alleles and W/w alleles were used to control partially for genetic background effects.

The results of the pollen discounting experiments are consistent over designs (Table 10.2) and indicate an advantage for the w gene in male transmission. In addition, the estimated outcrossing rates give no indication of a difference between color morphs in selfing rates. The pollen discounting hypothesis predicted, to the contrary, a reduction in w gene transmission. These results raise two separate questions. First, why did the experiments fail to replicate the field observations on differential selfing rates, and second, why does the w allele have an apparent advantage in male transmission?

Rates of self-fertilization depend on flower color morph frequencies

Frequency-dependent pollinator behavior is a potential explanation for the equal selfing rates estimated for pigmented and white flower morphs in the

Table 10.2 Estimates of morph specific outcrossing rates and male gamete contribution from pollen discounting experiments. Standard errors are given in parentheses. (After Schoen and Clegg, 1985.)

	Outcrossing rate		Pollen frequencies	
Design	Colored morph	White morph	Colored morph	White morph
1	0.617	0.723	0.347	0.653
	(0.200)	(0.592)	(0.272)	(0.272)
2	0.777	0.696	0.333	0.667
	(0.259)	(0.245)	(0.204)	(0.204)

pollen discounting experiments. It is well established that pollinators discriminate against rare species and Levin (1972) has found evidence for intraspecific discrimination against rare corolla shape morphs in phlox. The observation plots in the natural population studied contained about 10% white flowers, whereas the experiments of Schoen and Clegg (1985) utilized 50% white flowers. If pollinator behavior is frequency dependent, the discrepancy between the field observations and the experiments can be accounted for.

To test for frequency-dependent pollinator behavior, a series of artificial populations was established using greenhouse-grown plants (Epperson and Clegg, 1987b). Prior to each day of observation, 20 plants were arranged in a 4×5 grid and the frequency of the white morph was set at 15 or 20%, 50% and 80 or 85% in designs 1, 2 and 3, respectively. Pollinator behavior was observed in 24 separate experiments over two consecutive years. In addition, the selfing rate for each flower color morph was estimated during the second year's experiments using the transmission of marker alleles to seed progeny from the esterase and the W/w loci.

The data on pollinator behavior show a strong discrimination against the white morph when white is in the minority (Table 10.3). Interestingly, the frequency-dependent behavior is asymmetrical with respect to flower color morph, because there is no evidence for discrimination against the pigmented morphs when rare. Moreover, the estimates of selfing rates are consistent with the pollinator observations in showing a significant increase in s for the white morphs when rare ($s = 0.42$ and 0.11 for white and pigmented morphs, respectively), and no significant difference between morphs when frequencies are equal or when the white is the majority type (Epperson and Clegg, 1987b).

A frequency-dependent transmission bias, like that observed, is expected to lead to a neutral polymorphism for the white gene. Assuming no pollen discounting, the white gene should be favored through differential self-

Table 10.3   Observed and expected frequencies of pollinator visits to white flowers in experimental plots, averaged over days of observation within years, together with Chi-square tests of the null hypothesis of equal visitation rates. (After Epperson and Clegg, 1987b.)

Frequency of w allele	Visits to white flowers 1983			Visits to white flowers 1984		
	Obs	Exp	χ^2	Obs	Exp	χ^2
0.15–0.20	6	19.2	12.49	86	143.2	28.56
0.50	117	118.5	0.04	223	228.5	0.26
0.80–0.85	132	135.2	0.51	267	260.0	1.03

fertilization when rare. On the other hand, all phenotypes self-fertilize at the same rate when white is common, eliminating any transmission differential among genotypes after the white morph reaches a certain threshold frequency. This may partly account for the observed frequency of the white gene in natural populations. However, two contradictory observations must be reconciled. First, the absence of a patchy distribution for white flowers seemed to indicate selection against the white phenotype. And, second the pollen discounting experiments showed a transmission advantage in outcrosses involving w allele bearing pollen.

The w allele may be favoured during gametophytic stages of the life cycle

Three kinds of experimental evidence suggest a gametophytic advantage for w bearing pollen. First, some backcross and F_2 families show distorted Mendelian ratios, where the w allele is transmitted more frequently than expected based on equal segregation (unpublished data). However, the segregation tests are ambiguous because they are heterogeneous over families; some families fit Mendelian expectations while in other families the w allele is transmitted preferentially. The second source of evidence for a gametophytic advantage for w allele bearing pollen comes from the pollen discounting experiments described above. In these experiments w was transmitted in outcross pollen significantly more often than expected (Schoen and Clegg, 1985). The third source of evidence, derives from a series of experiments to measure the effect of sequential pollination on male gene transmission (Epperson and Clegg, 1987a).

Equal mixtures of w and W allele bearing pollen were applied to receptive stigmas on a series of maternal test plants. Several different plants were used as pollen donors in these experiments, allowing heterogeneity over pollen donors to be detected. The test materials were manipulated so that the transmission of the w allele could be monitored by following the transmission of an esterase allele in complete linkage disequilibrium with w. The results of these experiments are homogeneous over w pollen donors and indicate an average advantage in male transmission of w over W, however, the results are heterogeneous over W pollen donors (Epperson and Clegg, 1987a). Of five W pollen donors, one is equivalent to the w donors in male gene transmission, two are slightly less effective and two are substantially less effective.

The statistical heterogeneity of two of the experiments on male gene transmission is troubling. It is possible that genetic background effects are confounded with w gene transmission. For example, if a gametophytic factor is linked to, and in linkage disequilibrium with the W/w locus, then an apparent advantage of the w allele would be observed along with heterogeneity over pollen donors. This apparent advantage would, of course, not result from any differential effect of the W/w locus during gametophytic stages of the life cycle. The fact that three different lines of experimental evidence all point to a differential transmission of the w allele is comforting. Nevertheless, larger experiments with a wider diversity of genetic backgrounds are necessary to completely resolve the possibility of gametophytic selection.

The timing of pollination events has a strong influence on male gene transmission

Observations of pollinator behavior show that the typical flower may receive many visits during mornings when pollinators are very abundant (as many as 20 visits). (The morning glory flower is only receptive for a single morning, after which it closes and abscises from the plant.) White flowers that are undervisited relative to colored flowers may still receive several pollinator visits over the course of a morning, when pollinators are abundant. Multiple pollinator visitation poses several questions regarding the transmission of the flower color genes. First, why does the white morph exhibit a higher rate of self-fertilization, if the probability of at least one pollinator visit during a morning is high? And second, how is the probability of fertilization determined by sequential order in a temporal series of pollinations?

To address these questions a series of artificial pollination experiments were performed. The experiments involved applying pollen of known genotype to maternal plants of known genotype (with respect to the W/w locus and an esterase marker locus in complete linkage disequilibrium with the W/w locus). Several treatments were established. First, pollen of two genotypes was applied in an equal mixture to stigmas (treatment T_1). Second, pollen of one genotype was applied to a stigma immediately followed by application of pollen of a second genotype (treatment T_2). Third, pollen of one genotype was applied to the stigma followed by application of pollen of a second genotype at 30 (T_3), 60 (T_4) and 120 (T_5) minutes. The second pollen parent in treatment T_2 was only half as effective in fertilization, despite the zero time delay. The second pollination in T_3–T_5 was almost completely ineffective (Epperson and Clegg, 1987a). This result suggests that most successful fertilizations are due to the first pollination of the day. Even though white flowers are visited in nature, they tend to be visited later in the morning; however, any self pollen will have an advantage in fertilization, if it arrives at the stigma first.

Maternal fertility differences associated with the white flower phenotype

Data on the fertility of maternal plants are much more limited, however, reduced seed set is expected when flowers are undervisited by pollinators.

Depending on the abundance of pollinators, white-flowered plants must rely on autopollination more often than plants with colored flowers. In one experiment, seed set for flowers receiving one or more visits was near the maximum possible value. (The average seed set per capsule was 4.98 (unpublished data) with a maximum of six seeds per capsule.) Seed set among autopollinated flowers was significantly reduced in the same experiments ($\chi^2 = 16.9$ with one degree of freedom), averaging 3.73 seed per capsule. In addition, Ennos (1981) showed that under conditions of auto-pollination seed set was much reduced (average about 1.6 seed per capsule). Finally in field studies, Stucky (1985) found seed set values of 2.8 under autopollination and 3.7 under open pollination (note that open pollination does not imply that all flowers received a pollinator visit).

The data on reduced seed set under autopollination can be used to obtain crude estimates on the expected maternal fertility disadvantage of under-visited flowers. At least three major factors must be taken into account in estimating maternal fertility loss in white flowers. These factors are (a) total pollinator visitation rate, (b) differential pollinator visitation to white flowers, and (c) seed set for autopollinated flowers relative to insect polli-nated flowers. Calculations based on these considerations suggest that a 50% reduction in visitation to white flowers can result in up to a 25% reduction in seed fertility. Despite these calculations, direct observational and experimental data have yet to be gathered to test the hypothesis that white flowers do have a reduced maternal fertility. Such data are essential to a complete description of the opportunities for selection during reproductive phases of the life cycle.

Loci determining white phenotypes may be affected by mutator activity

A surprising result of our genetic studies has been the discovery of an unstable allele at the A/a locus (denoted a*). Plants of genotype WW a*a*, that are expected to be phenotypically pure white due to the epistatic interaction of the A/a locus with the W/w and P/p loci, are characterized by flowers with pigmented sectors on white corollas (color determined by the P/p genotype). These results suggest that sectoring occurs when one or both of the unstable a* alleles has independently mutated to A in the somatic corolla tissue. (The pigment expressed by the P/p locus is evident in Aa heterozygotes.) We have analyzed F_2 progeny distributions from eight families and the results are completely in accord with Mendelian transmis-sion of the a* allele. Furthermore, reciprocal crosses showed no differences, indicating an absence of any detectable cytoplasmic effect on the sectoring phenomenon (Epperson and Clegg, 1987c).

Germ line mutations also occur at a high frequency. In one experiment, 42 out of 6770 a* alleles mutated to stable A alleles (yielding an average 0.6% germ line mutation rate). In addition, two of 417 a* alleles mutated to stable a and four out of 672 a* alleles showed a spontaneous threefold increase in mutation rate (as measured by somatic sectoring frequencies).

Our working hypothesis is that the unstable a* allele is the integration site of an active transposable element. Excision of the element results in mutation to a or to A depending on the nature of the excision event. The discovery of mutator activity associated with the white phenotypes was quite unexpected and adds yet another complication to our efforts to study the evolutionary dynamics of the flower color polymorphisms. Our current working hypothesis is that white phenotypes arise relatively frequently owing to mutator activity. New mutants would be expected to arise at random spatial locations in natural populations, which would in turn account for the random spatial distribution of white phenotypes observed in natural populations. We further hypothesize that white phenotypes are selected against through reduced seed set under conditions of autopollination, and therefore, that new white 'patches' are extinguished owing to a fertility disadvantage. These hypotheses are provisional and must be tested in natural and experimental populations.

10.5 CONCLUSIONS

Our work with morning glory populations illustrates some of the complications associated with the study of genetic transmission during reproductive phases of the life cycle. These complications arise for several reasons. First, plant life histories are complex. Practical considerations require that relatively homogeneous aspects of the life cycle, such as the mating cycle, gametophytic stages, pollen fertility, maternal fertility and viability, be investigated in separate experiments. This dissection of the life cycle into homogeneous episodes is likely to reveal unexpected sources of selection as illustrated by our studies with morning glory. A complete understanding of the various influences on the transmission of genes in plant populations can only be achieved through an experimental reduction of the life cycle into a series of elementary operations.

The second complication arises from the application of mathematical and statistical models to the study of natural populations. It is unlikely that simple assumptions regarding the random sampling of genes or genotypes are ever satisfied. This means that inferences about genetic transmission, based on natural population samples, are probably subject to biases of unknown magnitude. Perhaps the best that can be achieved is to sketch a picture of genetic transmission based on a number of separate and artificial experiments.

A third complication is presented by the complexities of the genetic system. We indicated that genetic background effects are a potential confounding influence in interpreting the evidence for gametophytic selection.

Problems of associated selection are notoriously difficult to control and are a potential trap for the unwary investigator. The discovery of mutator activity associated with the floral pigment loci adds yet another dimension to the complexity of the genetic system.

Finally, the utilization of insect pollinators means that the plant genetic system is influenced by the behavioral characteristics of the particular insect species utilized by the plant. The differential self-fertilization of the white phenotype is a consequence of the pollinator's behavior. Moreover, the asymmetry in the frequency dependent preferences of the pollinator produces quite different genetic consequences than would obtain if the pollinator exhibited an unconditional preference for one flower color. These complications can only be revealed through experimentation.

10.6 ACKNOWLEDGEMENTS

The work described in this chapter was supported in part by NSF grants BSR-8418381 and BSR-8614608.

10.7 REFERENCES

Allard, R. W. (1975) The mating system and microevolution. *Genetics*, **79**, s115–s126.

Bishop, J. A. and Cook, L. M. (1981) *Genetic Consequences of Man Made Change*. Academic Press, New York.

Brown, B. A. and Clegg, M. T. (1984) The influence of flower color polymorphisms on genetic transmission in a natural population of the common morning glory, *Ipomoea purpurea*. *Evolution*, **38**, 796–803.

Charlesworth, B. and Charlesworth, D. (1979) The evolutionary genetics of sexual systems in flowering plants. *Proc. R. Soc. London B*, **205**, 513–30.

Clegg, M. T. (1980) Measuring plant mating systems. *Bioscience*, **30**, 814–18.

Eaton, A. (1833) *Manual of Botany for North America*. Oliver Steele, Albany, NY.

Ellstrand, N. C. (1984) Multiple paternity within the fruits of the wild radish. *Am. Natur.*, **123**, 819–28.

Ellstrand, N. C. and Marshall, D. L. (1985) Interpopulation gene flow by pollen in wild radish, *Raphanus sativus*. *Am. Natur.*, **126**, 606–16.

Ellstrand, N. C. and Marshall, D. L. (1986) Patterns of multiple paternity in populations of *Raphanus sativus*. *Evolution*, **40**, 837–42.

Ennos, R. A. (1981) Quantitative studies of the mating system in two sympatric species of *Ipomoea* (Convolvulaceae). *Genetica*, **57**, 93–8.

Ennos, R. A. and Clegg, M. T. (1982) Effect of population substructuring on estimates of outcrossing rate in plant populations. *Heredity*, **48**, 282–92.

Ennos, R. A. and Clegg, M. T. (1983) Flower color variation in the morning glory, *Ipomoea purpurea*. *J. Hered.*, **74**, 247–50.

Epperson, B. K. and Clegg, M. T. (1986) Spatial autocorrelation analysis of flower color polymorphisms within substructured populations of morning glory (*Ipomoea purpurea*). *Am. Natur.*, **128**, 840–58.

Epperson, B. K. and Clegg, M. T. (1987a) First-pollination primacy and pollen selection in the morning glory, *Ipomoea purpurea*. *Heredity*, 58, 5–14.

Epperson, B. K. and Clegg, M. T. (1987b) Frequency-dependent variation for outcrossing rate among color morphs of *Ipomoea purpurea*. *Evolution*, 41, 1302–11.

Epperson, B. K. and Clegg, M. T. (1987c) Instability at a flower color locus in the morning glory. *J. Hered.*, 78, 346–52.

Epperson, B. K. and Clegg, M. T. (1988) Genetics of flower color polymorphisms in the common morning glory (*Ipomoea purpurea*). *J. Hered.* (in press).

Fisher, R. A. (1941) Average excess and average effect of a gene substitution. *Ann. Eugen.*, 11, 53–63.

Ford, E. B. (1965) *Genetic Polymorphism*. MIT Press, Cambridge, MA.

Gregorius, H.-R. (1986) Polymorphisms for purely cytoplasmically inherited traits in bisexual plants. *Genetics*, 112, 385–92.

Harding, J. (1970) Genetics of *Lupinus*. II. The selective disadvantage of the pink flower color mutant in *Lupinus nanus*. *Evolution*, 24, 120–7.

Heinrich, B., Mudge, P. R. and Deringis, P. G. (1977) Laboratory analysis of flower constancy in foraging bumblebees: *Bombus ternarius* and *B. terricola*. *Behav. Ecol. Sociobiol.*, 2, 247–65.

Holsinger, K. E., Feldman, M. W. and Christiansen, F. B. (1984) The evolution of self-fertilization in plants: a population genetic model. *Am. Natur.*, 124, 446–53.

Horovitz, A. and Harding, J. (1972) Genetics of *Lupinus*. V. Intraspecific variability for reproductive traits in *Lupinus nanus*. *Bot. Gaz.*, 133, 155–65.

Kettlewell, H. B. D. (1961) The phenomenon of industrial melanism in Lepidoptera. *Ann. Rev. Ent.*, 6, 245–62.

Kimura, M. (1983) *The Neutral Theory of Molecular Evolution*. Cambridge University Press, London.

Kimura, M. and Ohta, T. (1971) *Theoretical Aspects of Population Genetics*. Princeton University Press, Princeton, NJ.

Lees (1981) Industrial melanism: genetic adaptation of animals to air pollution. in *Genetic Consequences of Man Made Change* (eds J. A. Bishop and L. M. Cook), Academic Press, New York.

Levin, D. A. (1972) Low frequency disadvantage in the exploitation of pollinators by corolla variants in *Phlox*. *Am. Natur.*, 106, 453–60.

Lewontin, R. C. (1985) Population genetics. *Ann. Rev. Genet.*, 19, 81–102.

Lloyd, D. G. (1979) Some reproductive factors affecting the selection of self-fertilization in plants. *Am. Natur.*, 113, 67–79.

Meagher, T. R. (1986) Analysis of paternity within a natural population of *Chamaelirium luteum*. I. Identification of most likely male parents. *Am. Natur.*, 128, 199–215.

Nagylaki, T. (1976) A model for the evolution of self-fertilization and vegetative reproduction. *J. Theor. Biol.*, 58, 55–8.

Pursh, F. (1814) *Flora Americae Septentrionalis*. White, Cochrane, London.

Rick, C. M., Fobes, J. F. and Tanksley, S. D. (1979) Evolution of mating system in *Lycopersicon hirsutum* as deduced from genetic variation in electrophoretic characters. *Plant Syst. Evol.*, 132, 279–98.

Ross, M. D. (1984) Frequency-dependent selection in hermaphrodites: The rule rather than the exception. *Biol. J. Linn. Soc.*, **23**, 145–55.

Schoen, D. J. (1982) The breeding system of *Gilia achilleifolia*: variation in floral characteristics and outcrossing rate. *Evolution*, **36**, 352–60.

Schoen, D. J. and Clegg, M. T. (1984) Estimation of mating system parameters when outcrossing events are correlated. *Proc. Natl. Acad. Sci. USA*, **81**, 5258–62.

Schoen, D. J. and Clegg, M. T. (1985) The influence of flower color on outcrossing rate and male reproductive success in *Ipomoea purpurea*. *Evolution*, **39**, 1242–9.

Schoen, D. J. and Clegg, M. T. (1986) Monte Carlo studies of plant mating system estimation models: the one pollen parent and mixed mating models. *Genetics*, **112**, 927–45.

Schoen, D. J. and Lloyd, D. G. (1984) The selection of cleistogamy and heteromorphic diaspores. *Biol. J. Linn. Soc.*, **23**, 303–22.

Sheppard, P. M. (1967) *Natural Selection and Heredity*. Hutchinson University Library, London.

Sokal, R. R. and Oden, N. L. (1978) Spatial autocorrelation in biology. 1. Methodology. *Biol. J. Linn. Soc.*, **10**, 199–228.

Sokal, R. R. and Wartenberg, D. E. (1983) A test of spatial autocorrelation analysis using an isolation-by-distance model. *Genetics*, **105**, 219–37.

Stebbins, G. L. (1950) *Variation and Evolution in Plants*. Columbia University Press, New York.

Stucky, J. M. (1985) Pollination systems of sympatric *Ipomoea hederacea* and *I. purpurea* and the significance of interspecific pollen flow. *Am. J. Bot.*, **72**, 32–43.

Turner, M. E., Stephens, J. C. and Anderson, W. W. (1982) Homozygosity and patch structure in plant populations as a result of nearest-neighbor pollination. *Proc. Natl. Acad. Sci. USA*, **79**, 203–7.

CHAPTER 11

Genetic variation and environmental variation: expectations and experiments

JANIS ANTONOVICS, NORMAN C. ELLSTRAND and ROBERT N. BRANDON

We may think of the evolution of genetic systems as a course of evolution which, although running parallel to and closely integrated with the evolution of form and function, is nevertheless separate enough to be studied by itself.

(Stebbins, 1950)

11.1 INTRODUCTION

Because the diversity of genetic systems (defined as those characteristics of the organism influencing the rate of genetic recombination) is much greater in plants than in animals, such systems have been of particular interest to students of plant evolution. Ever since the work of Darlington (1939), it has been realized that the evolutionary forces acting on genetic systems are likely to be different from those acting on more conventional morphological and physiological traits. For example, a particular chromosome number or recombination frequency may have very little direct impact on the physiological functioning of the organism, on its survival and fecundity, but may have a marked effect on the evolutionary potential of the descendants of that individual. As Stebbins (1950) states, 'Hence in discussing the selective value of genetic systems we must consider primarily the advantages a particular system gives to the progeny of those who have it ... the immediate advantages or disadvantages of the system are of secondary importance.' In the early works of Darlington, Huxley, Mather and Stebbins, the hypothesis was proposed that particular genetic systems result from a compromise between the need for constancy so as to preserve adaptation to the immediate contemporary environment and the need for flexibility in the face of changing environments to which the species will

275

become exposed in the future. This idea was so forceful in its elegance and in its explanatory power that it became engrained in evolutionary biology more as a paradigm of how genetic systems actually do evolve, rather than as a hypothesis requiring rigorous formulation and testing. As a result the idea that genetic variation is in some sense an adaptation for coping with environmental variation and change has become an idea that pervades not only introductory texts, but also our evolutionary consciousness. It has also received support from a homomorphism of ideas between ecology and genetics. Environmental variation and diversity are seen as essential (and desirable) ingredients of ecological systems: surely then the genetic variation we see in organisms is there to adapt the organism to such environmental variation.

What we want to do in this chapter is to continue in the Stebbinsian tradition of examining closely the relationship between genetic variation and environmental variation. However, we wish to do so less from a comparative standpoint, but more from an experimental perspective. This we believe is in keeping with the rapid change in evolutionary biology that is bringing the science closer in methodology to other traditionally more experimental sciences such as developmental biology and physiology. Plants, given their experimental convenience, have a large role to play in this change.

First, we present a straw-man, the intent of which is to defuse any glib assumptions that genetic variation is necessarily an adaptation to environmental variation. Secondly, we outline some specific hypotheses that have been proposed for 'short-term' advantages to sexual reproduction in varying environments: this is because we feel such hypotheses are amenable to direct experimental test, whereas 'long-term' hypotheses are much less so. We emphasize however that such hypotheses are still crude, unquantified word-models, and that indeed we have few quantitative models that can serve as a focus for precise experimental quantification of the selective forces acting on genetic systems. Third, we emphasize that there has been considerable confusion about the concept of environment: this has served both to divert our thoughts into inappropriate directions as well as to distort our experimentation into focusing on inappropriate (and often overly complex) measurements. Finally, we present the results of an experiment which not only illustrates some of these pitfalls, but which illustrates the necessary, but not necessarily sufficient, conditions for genetic variation to be favored in a variable environment. In brief, the hypotheses generated by Ledyard Stebbins and other plant evolutionary biologists many years ago still demand our attention today: we are perhaps only now developing the conceptual and experimental tools that can examine these ideas both critically and creatively.

A-only   a-only   A and a mix

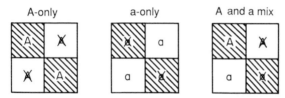

Fig. 11.1 Schematic illustration of fate of a genetically uniform progeny array (*A*–only, or *a*–only) on a heterogeneous environment vs. fate of a variable progeny array (*A* and *a* mix). Note: *A* survives in grey, shaded micro-habitats, and *a* in white unshaded habitats; progeny are dispersed at random over the micro-habitats.

11.2   A STRAW-MAN

Consider a habitat which is a mosaic patchwork of two soil types (uncontaminated vs. metal contaminated soil?) which we will call white and grey (Fig. 11.1). Let this mosaic be occupied by a haploid organism, such that genotype *A* survives in the grey microsites and genotype *a* in the white microsites. Then we can ask: will an individual be favored if it produces only *A* type or only *a* type progeny (asexually) or will it be favored if it produces a 50 : 50 proportion of *A* and *a*. Let us assume that the progeny are dispersed at random over the microsites. It is evident that the progeny of an individual that produces only *A* or *a* will have 50% survival (Fig. 11.1). However, it is also evident (Fig. 11.1) that the individual that produces two types of offspring will have no advantage, since *A* and *a* will each fall in the wrong microsite 50% of the time. There is thus no overall advantage to an individual with a genetic system that produces variable as opposed to uniform progeny.

The point of this straw-man is twofold. First, the facile presumption that genetic variance is 'adaptive' for environmental variance is clearly erroneous. We need to have precise hypotheses, where assumptions are explicitly stated, before the more generalized idea can be tested. (Most biologists could easily modify the above model, such that the variable progeny would be favored, for example, by assuming competition in microsites, such that the best genotype for that site wins.) Secondly, given that the numbers of individuals that occupy each microsite are limited, the straw-man model would maintain genetic variance in a population but would still not provide an individual advantage to variable progeny. Thus the observation of a correlation between genetic variation and environmental variation is also insufficient proof that environmental variation provides the selective forces for genetic systems that promote variation.

11.3 TOWARDS WORD-MODELS

The past decade has seen a re-structuring of our ideas on the evolution of genetic systems. Although it is generally agreed that many such systems may not have a large physiological cost (and hence 'direct' fitness effect) on the individual, it has been pointed out that there are major costs to outcrossing and sexual reproduction in terms of gene-transmission (Williams, 1975; Lloyd, 1980a,b; Uyenoyama, 1984). Concomitantly, physiological costs associated with outcrossing and sexual reproduction (such as costs of mate attraction) have also been re-emphasized. Because these transmission costs are large, yet sexual reproduction is a widespread phenomenon, commonly occurring throughout nearly every major group of eukaryotes, theoreticians have offered a large variety of models to account for the advantages of sex and recombination (summarized in reviews by Ghiselin, 1974; Williams, 1975; Maynard Smith, 1978; Lloyd, 1980a; Bell, 1982). These models all seek to find 'short-term' (i.e. single-generation) advantages in terms of the number of progeny that reproduce again in the following generation. Such advantages are often termed 'individual advantages' to contrast them with advantages that may accrue to the population (or 'group') as a whole. Use of the term 'individual advantage' is somewhat misleading, since it is most often assumed that the individual possessing a particular genetic system still has no advantage in terms of survival or offspring number, but that the advantage is accrued by the immediate progeny of that individual. Most of such 'short-term' selection models fall into two classes, those involving frequency-dependent selection and those involving changing or unpredictable environments.

The first class of models argue that sex will be favored if there is an advantage to being genetically different from the majority genotype (Levin, 1975; Jaenike, 1978; Glesener, 1979; Lloyd, 1980a; Hamilton, 1980; Price and Waser, 1982; Tooby, 1982). Such minority types are more likely to escape pathogens and predators. We have obtained experimental evidence in the grass *Anthoxanthum* that sexual individuals indeed have a substantial fitness advantage as a result of frequency-dependent selection (Antonovics and Ellstrand, 1984; Schmitt and Antonovics, 1986b). Under high sibling densities, genetically variable sexual progeny will also be favored because of reduced competition as a result of resource partitioning among diverse genotypes (Maynard Smith, 1978; Young, 1981; Price and Waser, 1982). Minority genotypes will compete less with each other if they use differing resources. However, we found no evidence for this in *Anthoxanthum*; fitness differences for sexual and asexual progeny did not increase with increasing density (Ellstrand and Antonovics, 1985).

The second class of models argues that the production of genetically variable progeny will be advantageous when environments are variable in time and space. In the case of temporal variation, individuals that produce variable progeny will be favored if the environment encountered by these progeny is likely to be different from that of the parents. Environmental states may change from generation to generation so as to be negatively autocorrelated (Maynard Smith, 1978), or environments may change over time in some directional manner (Treisman, 1976). Genetic variation may also be favored because it provides a mechanism of 'bet-hedging' in the face of environmental unpredictability. Different genotypes may be favored at different times, so resulting in a lower variance in fitness over time. Since such fitness values over time are integrated as the geometric mean rather than as the arithmetic mean (Gillespie, 1977; Lacey *et al.*, 1983), those genetic systems reducing variance (even at the expense of mean performance) would be favored.

In the case of spatial variation, one can envisage a continuum between one extreme where environments are so highly heterogeneous that all possible habitats are encountered with about equal probability in every generation (fine-grained environments *sensu* Levins, 1968), and the other extreme where environments change gradually over space so that new environments are only encountered in successive generations of dispersal (coarse-grained environments). In the former case, variable progeny will be favored if the appropriate genotype can either choose (behaviorally or by differential growth) the appropriate habitat (Maynard Smith, 1978; Bell, 1982), or if sib-competition or differential mortality results in the appropriate genotype displacing other genotypes in a given microsite (Maynard Smith, 1978; Williams and Mitton, 1973; Bulmer, 1980; Taylor, 1979; Barton and Post, 1986). In the case of gradually changing spatial environments, individuals that produce variable progeny will be favored if the environment encountered by these progeny is likely to be different from that of the parents. The gradually changing spatial environment models converge on those relating to temporal variation, in the sense that successive generations encounter different environments in space, rather than in time.

The difficulty with all these heterogeneous environment hypotheses is one of degree. Do environments in nature in fact change sufficiently enough and consistently enough to sustain sexual reproduction in this way? What do we mean by 'enough'? What environmental parameters should we measure? Indeed how can these models be operationalized so as to be amenable to experimental test? How can the selective forces acting on genetic systems be measured?

11.4 THE CONCEPT OF THE ENVIRONMENT

Whereas the concept of fitness has received extensive attention from both philosophers and empiricists interested in evolution, the concept of the environment has received only minimal or passing interest (Brandon, 1986). Yet for the theory of natural selection to have explanatory power with regard to how adaptations originate, the concept of environment is as important as that of fitness. The reason is simple and can be easily illustrated by an example. If we grew one plant on good soil, and another on poor soil, the one on good soil would probably survive better, grow larger, and have more seed. Although we might be tempted to say one plant had a greater 'fitness' than the other, we are in this case referring to properties of the environment rather than to properties of the phenotypes of those plants which would explain their differential success. (We assume phenotypes are not 'choosing' different soils.) In other words for the theory of natural selection to have explanatory power, we must compare the fitness of different phenotypes in identical environments. Conversely, two environments can be thought of as homogeneous (with regard to selection) if their effect on the relative fitness of phenotypes is the same. It is within such selectively homogeneous environments that differential fitness is the result of properties of the organism and within which the theory of natural selection therefore has explanatory power.

It is almost irresistibly tempting, in view of its popular usage, to equate environment with some measure of conditions external to the organism. Indeed whole fields of endeavour, the environmental sciences, are concerned with this. In biology however, we should recognize that there are really three quite disparate ways in which we can measure the environment, and these measures may produce quite different scales of heterogeneity (Fig. 11.2). The 'external environment' reflects properties of the environment that are measurable externally, without any necessary involvement of the organism itself. This type of environment is equivalent to the popular usage alluded to above.

The 'ecological environment' reflects properties of the external environment that influence the organism's contribution to population growth, or its reproductive value *sensu* Fisher (1930). This environment is measurable in terms of the demographic performance of individuals. It therefore follows that the scale of heterogeneity that is present will depend on the organisms (whether they be individuals, populations or species) used as the 'measuring instruments' (Antonovics, Clay and Schmitt, 1987). The use of organisms as environmental measuring instruments was first pioneered by Clements and Goldsmith (1924), although in these studies the focus was often the external environment, with the organism used as an inexpensive instrument which could be calibrated against more complex measuring instruments indicating the external environment.

The 'selective environment' reflects properties of the external environment that influence the differential contribution of genotypes to subsequent generations. This environment is measurable in terms of the fitness (in a

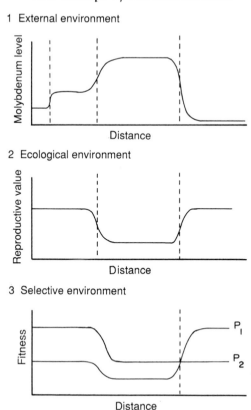

Fig. 11.2 Schematic illustration of how external, ecological and selective environments may show different scales of heterogeneity. We assume that molybdenum levels only reduce the performance of the organism when they are above a threshhold, and that two phenotypes, P_1 and P_a within the species react differentially only to extremely low molybdenum levels.

relative sense) of genotypes, and environmental heterogeneity is indicated by differential performance of genotypes in different regions or at different times (i.e. by genotype × environment interactions). The scale of environmental heterogeneity will therefore depend on the genotypes used to measure it. Curiously, but obviously, if two genotypes used to measure an environment are identical, then the selective environment for these genotypes will be uniform!

Two points should be noted. The first point is that the selective environment although measurable by the use of genotypes as 'phytometers', will also have correlates (often causal) with the ecological and external environments. The second point is that if ideas can be expressed and quantified in terms of the selective environment, perhaps the experimentalist need only take measurements of this one type of environment, and can safely ignore the other types so perhaps simplifying his task.

When we speak of the evolution of genetic systems, in heterogeneous environments, with which type of environment should we be primarily concerned? We will argue that our concern should be with none of the above (in their entirety), but with a subclass of the selective environment.

11.5   AWAY FROM WORD-MODELS

In this section, we return to two of the models of short-term advantages of sex in heterogeneous environments, and use them to illustrate how, at least in principle, these word-models can be operationalized into experimentally testable hypotheses. We have chosen the two models because they relate to large-scale variation in time or space and therefore represent our simplest, or perhaps most naive, expectations as to the type of environmental variation that may favor genetic variation.

Maynard Smith's 'hot genes, wet genes'

Maynard Smith (1978) developed a scenario whereby increased recombination between two gene loci would be favored in a temporally varying environment. At one locus, the A allele confers higher fitness in hot environments while the a allele confers higher fitness in cold environments. At the other locus, the B allele has a higher fitness when the environment is wet whereas the b allele has the higher fitness when it is dry. If hot/wet or cold/dry environments alternate with hot/dry and cold/wet environments, then direct translation of gene effects as above would favor the cis combinations (AB, ab) alternatively with the trans combinations (Ab, aB) thus favoring greater recombination among these loci. While this model is very explicit, as Maynard Smith admits, it is also very contrived. As a consequence it is hard to 'imagine' how it would work in nature. More to the point, it is difficult to see how it could be tested, not so much in principle, as in practice. Specific genes with specific environmental effects would have to be identified, their fitness effects established under a range of environments, and those environments then measured over successive time intervals.

Williams' 'cod–starfish model'

Williams (1975) pointed out that the genetic variance of sexually produced progeny is likely to be greater than the variance of asexually produced progeny. If the environment now changes such that only extreme progeny types are favored, it is more likely that a sexual progeny array would contain these extreme phenotypes. Sexual reproduction would thus be favored. This model, in contrast to that of Maynard Smith is extremely general, and in practice difficult to test because of this. One would need knowledge of what factors in the environment are changing (one may measure many, but miss the important one), one would need to know which phenotypes give a high performance in those environments, and one would need to know something about the heritability as well as within progeny genetic variance of those phenotypes.

Both these models are phrased explicitly or implicitly in terms of 'external environments'. However, an essential ingredient for these two models, indeed a necessary condition for them to work, is that there is genotype – environment interaction in fitness, i.e. that the environment be 'selectively heterogeneous'. Both models can be translated to fitness effects of genotypes in different environments. By way of illustration (Fig. 11.3), we have done this

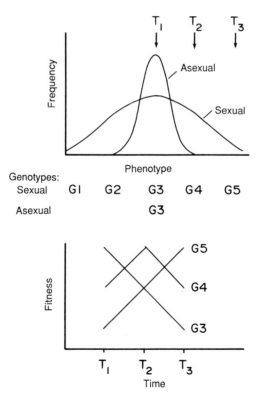

Fig. 11.3 Diagram showing how the greater fitness of a sexually produced off-spring array in the face of a temporally changing environment subsumes an underlying genotype–environment interaction in fitness over time, i.e. a selective environment that is variable in time.

explicitly for the 'cod–starfish model'. When this translation is carried out (as well as from intuitive reasoning) it is evident that we need a particular extreme form of genotype–environment interaction for the models to work.

We can recognize two types of genotype–environment interaction, namely those of the 'crossing type' and those of the 'non-crossing type' (Fig. 11.4). In the non-crossing type, the genetic correlation of phenotypes between environments will be positive and genetic variance will be high relative to genotype–environment interaction. In the crossing type, the genetic correlation of phenotypes between environments will be negative and genetic variance will be low relative to genotype–environment interaction. It can be shown (Shaw, 1986; after Tachida and Mukai, 1985) that a genotype–environment interaction variance component can be expressed as a function of the variances in the individual environments and the genetic correlation among environments. For the two-environment case:

$$2V_{GE} = (V_1 - V_2) + V_1 V_2 (1 - R)$$

where V_{GE} = genotype–environment interaction variance;
V_1, V_2 = genetic variance in environments 1 and 2, respectively; and
R = genetic correlation among environments.
Then,

$$V_{GE} - V_G = V_1 V_2 (1 - R) - V_1 V_2,$$

where $V_G = (V_1 + V_2)/2$, or total genetic variance.

From this it follows that if $V_{GE} < V_G$, then $R > O$ and the genotype–environment interaction is of the non-crossing type. Alternatively if $V_{GE} > V_G$ then $R < O$ and the genotype–environment interaction is of the crossing type.

For there to be an advantage to genetic variation in heterogeneous environments it is necessary that there be a crossing genotype–environment interaction in fitness. This is indicated empirically either by a negative genetic correlation in fitness across environments (see Via and Lande, 1985, for further discussion of this concept) or by a genotype–environment interaction effect that is greater than the genotype main effect. In other words, the environment not only has to be selectively heterogeneous but it has to be heterogeneous in a rather extreme way. While a crossing type of genotype–environment interaction in fitness is necessary for there to be selection for a 'more open genetic system' in a heterogeneous environment, it is clearly not a sufficient condition, particularly given that there may be transmission costs associated with such genetic systems. It is unclear from such generalized models exactly how large a particular genotype–environment interaction in fitness would have to be to favor a more open breeding system. Nor is it clear that this is even the correct way to phrase the issue. Perhaps we should instead be assessing the degree to which the

Genotype - environment interaction

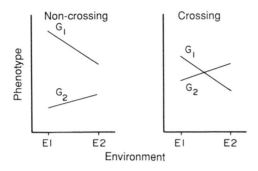

Genetic correlation across environments

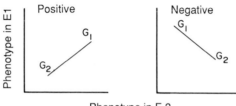

Fig. 11.4 Diagram showing difference between non-crossing and crossing genotype–environment interactions, and how this difference results in a positive and negative genetic correlation across environments.

descendants of a particular individual (with a particular genetic system) encounter a selective environment that is heterogeneous in space and time, and what the expected fitness is from the observed responses of the range of genotypes produced by that individual versus the expected fitness from a range of genotypes produced by an alternative genetic system (which could be imposed experimentally). It is probably true that we do not know for any genetic system whether the naturally occurring one generates a greater 'short-term' fitness than say a more open or more closed system.

We next describe an experiment carried out very early in our research program on the success of genetically variable vs. genetically uniform progeny. The experiment was naive in its goals: we simply wished to know whether different genotypes of *Anthoxanthum* responded differently to spatial environmental heterogeneity. We chose vegetation composition as our main measure of such heterogeneity, since not only was this an environmental variable that in all likelihood affected the performance of *Anthoxanthum* but it could also be documented relatively easily, without extensive instrumentation. We use this experiment both to illustrate the existence of genotype–environment interactions as well as to illustrate, after the event, the pitfalls of failing to distinguish external from selective environments.

11.6 AN EXPERIMENT

In 1978, we simulated dispersal of asexual progeny across a spectrum of environments normally encountered by a population of the perennial, self-incompatible grass, *Anthoxanthum odoratum* L., and asked if the environments were heterogeneous, whether the fitness of *Anthoxanthum* was affected by these environments, and whether different genotypes of *Anthoxanthum* reacted differently to different environments. We thus made separate measures of heterogeneity of the 'external', 'ecological' and 'selective' environments in this field. We further asked whether and at what scale genotype × environment interactions were greater than genotype main-effects, since such effects would have to be present for there to be an advantage to genetic variation in a heterogeneous environment.

Plants of *A. odoratum* were collected in May 1978 from a mown field on Duke Campus, Durham, North Carolina (see Fowler and Antonovics, 1981; Antonovics and Ellstrand, 1984, for further details). Collections were made from the area studied by Fowler and Antonovics (1981) and Fowler (1981); 22 genotypes of *Anthoxanthum* were randomly sampled as single tillers, grown in the greenhouse and cloned. In November 1978, 50 tillers of each genotype at the 3–4 leaf stage were individually weighed, and planted into paper tubes filled with soil from the field site. The tubes were 1 cm diameter and 5 cm long, and were made by sewing together folded strips of filter paper (see also Antonovics and Primack, 1982). The planted tillers were kept in the greenhouse for one week during which time they started to root. They were then transplanted into the field; a hole was made in the ground using an apple corer, the paper tube inserted, and the soil pushed back around the tube. The surrounding vegetation was left completely undisturbed. Each plant (hereafter referred to as a 'transplant') was marked with a plastic coated wire ring and a plastic toothpick.

The 22 cloned genotypes were planted 20 cm apart in random order in a linear array within each of 50 blocks, there being one transplant per genotype per block. These blocks were themselves arranged contiguously in a linear fashion along three transects spanning the field and approximately centered around the original collection site (Fig. 11.5). Over the next three years, the transplants were scored in May for tiller number and inflorescence number, and in November for tiller number. The 'environment' around each transplant was measured in May 1979 in terms of the surrounding vegetation, using a hexagonal 37-point cover grid (Fig. 11.6), with a spacing of 2 cm between points. The grid consisted of two layers of cross wires, with

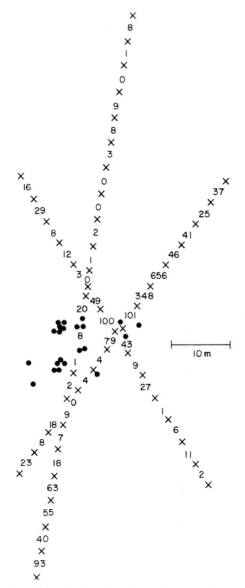

Fig. 11.5 Map of the field site showing the positions of the parental plants (large solid circles), the transect positions (bounded by '×' marks), and the average lifetime reproductive output (inflorescence number, × 10) of all 22 transplants within each transect position.

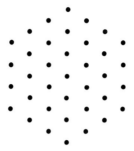

Fig. 11.6 Diagram of the hexagonal 37-point cover grid used in vegetation survey. The central point on the grid was placed so as to be directly above the position of a transplant.

the intersection of the wires corresponding to each point. The identity of the species below each point was noted, and per cent cover per grid calculated. Where the transplant itself was encountered, this was excluded from the analysis.

Fitness of an individual was estimated by its total reproductive output (in inflorescence number) over the duration of the experiment. Inflorescence number was highly correlated with the number of spikelets (Antonovics and Ellstrand, 1984).

Vegetation composition (the 'external environment')

There were 22 species (including the categories 'space' and 'mosses') with a vegetation cover greater than 0.1% (Table 11.1*a*). Principal components analysis using the covariance matrix of percentage cover of each species within each vegetation grid, gave one major component which accounted for 32% of the variance and weighted strongly on the most abundant species *Panicum sphaerocarpon*. The remaining nine axes accounted for another 62% of the variance, with a representation of 2–12% each (Table 11.2*a*).

The first ten principal component scores were subjected to a sequential cluster analysis, into 50 clusters using the FASTCLUS procedure of SAS (Ray, 1982a), and then into five using the CLUSTER procedure with Ward's criterion for clustering. The first cluster division (cluster D) accounted for 23.5% of the variance as opposed to 4–6% accounted for by each of the other divisions. Grids belonging to this cluster were generally found in the upper (northwest) part of the field and were associated with a high frequency of *Panicum sphaerocarpon* and above average frequencies of *Danthonia* and *Hypochoeris* (Table 11.3a). The other large cluster, A, had a very low frequency of *Panicum sphaerocarpon* but otherwise a greater

Table 11.1 (*a*) Overall percentage cover scores for species surrounding transplanted tillers. (*b*) Mean percentage cover (arcsin square root transformed, ×100 for 12 most abundant species, ×1000 for others) of each species for a given reproductive output class of *Anthoxanthum*. Significant differences ($P < 0.05$) are indicated by different letters according to Duncan's multiple range test. (*c*) F-values and significance levels for 'effect of' lifetime reproductive output class (R), and interaction of reproductive output class and genotype (G × R) on frequency of each species.

Species	(a) Percentage cover	(b) Percentage cover for each reproductive output class				(c) F-value for effect	
		0	1	2–9	>9	R	G × R
Panicum sphaerocarpon	20.16	37a	38a	14b	8b	20.96***	0.69
Space	14.20	35	31	32	30	1.85	1.03
Plantago lanceolata	13.95	29	28	32	31	0.46	0.93
Cynodon dactylon	9.54	20a	21a	26a	39b	17.39***	1.83***
Anthoxanthum odoratum	9.38	23	21	26	17	2.21+	1.05
Trifolium dubium	6.28	14	15	17	19	1.50	0.95
Andropogon virginicus	5.41	11ab	8a	15b	17b	3.20*	1.26
Salvia lyrata	4.81	10	12	14	9	1.16	1.19
Rumex acetosella	3.70	10a	8a	16b	19b	11.54***	1.35
Danthonia spicata	3.24	7ab	9a	3b	2b	2.79*	0.45
Poa pratensis	2.58	7	8	11	11	2.50+	1.19
Hypochoeris radicata	2.03	5	6	4	7	0.54	1.32+
Aira praecox	1.00	22a	25a	61b	35ab	3.17*	1.39*
Carex sp.	0.78	24a	16a	39ab	51b	1.88	0.75
Paspalum dilatatum	0.55	13	17	26	28	1.13	2.28***
Oxalis sp.	0.49	13a	19a	21a	42b	2.47+	1.48*
Trifolium repens	0.35	11	17	13	10	0.92	0.81
Trifolium arvense	0.28	8	3	6	16	0.91	0.92
Solidago sp.	0.24	4ab	17a	18a	0b	5.97***	4.77***
Moss spp.	0.19	7	5	15	0	1.15	0.90
Panicum anceps	0.18	6	11	7	2	1.14	1.13
Tridens flavus	0.33	8	6	10	2	0.32	0.31

Significance levels, *** = $P < 0.001$, ** = $P < 0.01$, * = $P < 0.05$, + = $P < 0.1$.

diversity, with the highest levels of 'space', *Anthoxanthum* and *Hypochoeris*, and above average levels of *Plantago*, *Salvia*, *Rumex* and *Poa*. Clusters B, C and E were smaller and were characterized by high frequencies of *Cynodon* and *Trifolium dubium*, *Andropogon* and *Plantago*, respectively.

To assess the scale of heterogeneity of the vegetation, the principal component scores were subjected to an autocorrelation analysis using the procedure AUTOREG of SAS (Allen, 1982) with lags of 1 to 100 transect

Table 11.2 (a) Percentage variation accounted for by first ten principal components. (b) The three species showing greatest correlation between their frequency and principal component score for each component (correlation coefficients, × 100 are indicated). (c) Average score for each component by reproductive output class, with significant differences indicated by different letters according to Duncan's multiple range test. (d) F-values and significance levels for effect of lifetime reproductive output class (R) and interaction of reproductive output class and genotype (G × R) on each principal component score.

Principal component	(a) Variation %	(b) Correlated species	(c) Scores for each reproductive output class				(d) F-value for effect	
			0	1	2–9	>9	R	R × G
1	32.2	Panicum sphaerocarpon (−94), Plantago (+18), Cynodon (+18)	−2.48a	−3.39a	13.56a	17.06a	17.88***	0.69
2	12.0	Plantago (−87), Space (+43), Andropogon (+14)	−0.01	−0.71	0.03	1.00	0.15	1.15
3	10.9	Space (−66), Cynodon (+63), Anthoxanthum (−24)	−1.17a	0.93a	0.93a	9.11b	10.23***	1.27+
4	9.2	Andropogon (−80), Cynodon (+39), Anthoxanthum (+28)	0.10	0.50	−1.31	0.54	0.24	1.41*
5	7.9	Anthoxanthum (−83), Space (+43), Salvia (−21)	−0.35a	0.32a	−1.46a	4.19b	0.80	1.15
6	6.5	Danthonia (+73), Andropogon (−44), Cynodon (−27)	0.12ab	2.02a	−0.93ab	−3.11b	3.13*	0.87
7	6.1	Trifolium dubium (−83), Cynodon (+33), Danthonia (+32)	−0.13	0.03	0.06	1.19	0.23	1.18
8	5.5	Salvia (+84), Danthonia (−46), Anthoxanthum (−18)	−0.26	1.03	1.59	−0.30	0.73	1.13
9	2.5	Rumex (+79), Trifolium dubium (−32), Cynodon (−27)	−0.33a	−0.38a	1.81b	2.18	3.93***	1.32+
10	2.0	Hypochoeris (+87), Rumex (−41), Space (−11)	−0.17	0.87	0.04	0.40	0.63	1.58**

*** = $P < 0.001$; ** = $P < 0.01$; * = $P < 0.05$; + = $P < 0.1$.

positions (i.e. 20 cm–20 m). Scores of all the first ten principal components were positively correlated within a distance of 4 m (Fig. 11.7); in most cases negative autocorrelations did not appear consistently until transplants were 10 m apart.

Overall performance of *Anthoxanthum* ('the ecological environment')

The survival of transplanted tillers was comparable to that of the naturally occurring clonal parents of small size (Table 11.4). The fecundity of surviving plants generally increased with time, except for the last census, and was somewhat higher for the experimental than for the clonal parents (Table 11.4). Lifetime reproductive output showed a highly skewed distribution, with most individuals, even if they survived through the first winter, having no or very few inflorescences (Table 11.5). Naturally occurring clonal parent plants showed a similar highly skewed distribution of lifetime reproductive output (Table 11.5).

Because of this highly skewed distribution, statistical analysis could not be carried by conventional analysis of variance (using genotype and vegetation type as class variables and reproductive output as the dependent variable). Instead, individuals were assigned to reproductive output classes (0, 1, 2–9, >9 inflorescences), and this was used as a class variable in an analysis of variance to examine whether the 'dependent' variable, vegetation type, was significantly different among genotype, reproductive output class, and whether there was interaction of genotype and reproductive output class. This latter effect indicates if different genotypes with contrasting reproductive outputs, 'produced' different vegetation types: since genotypes were assigned to vegetation types at random, the reverse inference is that different genotypes perform differently in different vegetation types. Analyses were carried out using the GLM procedure of SAS (Ray, 1982b).

One-way analysis of variance on arcsin square root transformed percentage cover of each species individually showed (Table 11.1) that the reproductive output of *Anthoxanthum* was negatively correlated with increasing cover of *Panicum sphaerocarpon*, *Danthonia* and *Solidago* and positively with increasing cover of *Cynodon*, *Andropogon*, *Rumex*, *Aira*, *Carex* and *Oxalis*. Duncan's multiple range test revealed no case where significantly higher reproductive values were associated with intermediate levels of a particular species. The results of analyses on the principal component scores were consistent with the analyses on the individual species (Table 11.2). Components 1 and 6 showed large effects; these components were strongly correlated with the abundance of *Panicum sphaerocarpon* and *Danthonia*. Components 3 and 9 also showed large effects and were correlated with the abundance of 'space', *Cynodon* and *Rumex*.

Table 11.3 (a) Percentage cover scores for each species in five clusters from a cluster analysis (see text) on the first ten principal component scores. The value for the species with the greatest score out of all the clusters is underlined. (b) F-values and significance levels for cluster A and D for 'effect of' lifetime reproductive output class (R), for interaction of reproductive output class and genotype (G × R) on frequency of each species. (c) The F-value and significance levels for the three-way interaction of cluster, reproductive output-class, and genotype (C × R × G).

Species	(a) Percentage cover in cluster					(b) F-value for effect				(c) F-value for effect C × R × G
						Cluster A		Cluster D		
	A	B	C	D	E	R	R × G	R	R × G	
Panicum sphaerocarpon	2.87	1.61	2.05	<u>54.05</u>	13.44	8.43**	0.62	0.10	3.04***	2.82***
Space	<u>20.95</u>	8.37	11.26	10.89	8.34	1.88	0.47	2.34	0.63	0.66
Plantago	<u>15.33</u>	9.72	5.09	6.00	<u>46.48</u>	0.36	0.93	0.46	2.10*	1.22
Cynodon	8.28	<u>31.97</u>	3.42	3.33	4.49	2.47	1.19	2.70	1.97*	1.04
Anthoxanthum	<u>16.98</u>	3.04	2.38	5.57	4.50	2.52	1.48	0.95	1.07	1.20
Trifolium dubium	4.56	<u>22.42</u>	3.51	1.64	5.77	0.27	0.82	0.65	0.61	0.23
Andropogon	3.87	1.17	<u>58.70</u>	2.13	4.42	1.12	1.14	2.69	0.39	1.10
Salvia	7.04	1.45	4.71	2.69	<u>8.05</u>	1.32	0.25	0.11	0.92	0.56
Rumex	5.34	<u>7.97</u>	2.05	0.72	<u>3.12</u>	2.71	2.16*	0.03	0.69	1.56*

Danthonia	1.60	0.13	0.61	7.79	2.08	0.36	0.49	0.00	0.80	1.10
Poa	2.93	5.55	2.21	0.96	4.17	0.64	1.80*	0.81	2.36**	1.52
Hypochoeris	2.80	1.04	0.20	2.39	0.16	0.02	0.95	0.95	0.70	0.95
Aira	2.18	0.62	0.31	0.05	0.30	1.27	0.81	0.20	1.05	0.61
Carex	1.08	1.76	0.73	0.15	0.21	3.77+	1.05	0.42	0.32	0.76
Paspalum	0.91	0.78	0.10	0.07	0.59	0.13	1.10	0.50	0.38	0.75
Oxalis	0.74	1.11	0.52	0.03	0.08	0.42	0.63	0.08	0.13	0.49
Trifolium repens	0.48	0.22	0.72	0.14	0.47	2.30	2.29**	7.77**	3.48***	1.13
Trioidia	0.75	0.15	0.00	0.02	0.13	0.65	0.98	–	–	–
Trifolium arvense	0.38	0.36	0.95	0.03	0.16	1.53	0.90	0.08	0.13	0.61
Solidago	0.16	0.00	0.00	0.57	0.00	0.17	0.90	6.77**	2.59**	2.79**
Moss	0.21	0.02	0.05	0.20	0.36	0.32	1.42	0.06	0.94	0.60
Panicum anceps	0.11	0.05	0.12	0.31	0.23	0.02	0.97	1.23	2.20*	2.22*
Number in each cluster	427	160	54	355	104					
Mean reproductive output of Anthoxanthum in each cluster	4.05	9.86	7.28	0.47	7.96					

*** = $P < 0.001$; ** = $P < 0.01$; * = $P < 0.05$; + = $P < 0.1$.

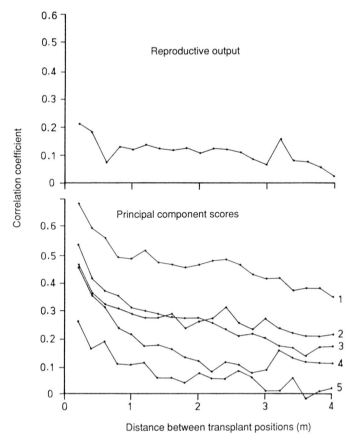

Fig. 11.7 Autocorrelograms showing the correlations between principal component scores (for the first five principal components), and correlations between log lifetime reproductive output at successively increasing distances between transect positions: data are means of calculations performed separately on each transect.

Anthoxanthum transplants showed the lowest performance in cluster D, representing the vegetation type dominated by *Panicum sphaerocarpon* (Table 11.3). They showed high and approximately equal performance in vegetation clusters B, C and E even though these clusters represented quite different vegetation types dominated by *Cynodon*, *Andropogon* and *Plantago*, respectively. *Anthoxanthum* showed intermediate performance in cluster A, which was a vegetation type with abundant space and a high frequency of resident *Anthoxanthum*.

Reproductive output per individual transplant was positively autocorrelated over 4 m, but the overall level of autocorrelation was less than that for the principal component scores (Fig. 11.7).

Table 11.4 Survival (1_x) and fecundity (m_x = inflorescence number) of experimentally planted and naturally occurring individuals of two size classes (large > 2 inflos; small ≤ 2 inflos). Numbers are based on 1100 experimental, and 16 large and 23 small naturally occurring individuals. The naturally occurring individuals were of unknown age and the 1_x, m_x values therefore represent depletion data and not age-specific data. Blanks indicate no census.

| | | | Naturally occurring | | | |
| | Experimental | | Small | | Large | |
Date	1_x	m_x	1_x	m_x	1_x	m_x
May 78	–	–	1	1.17	1	11.31
Nov 78	1	0	–	–	–	–
May 79	0.568	0.37	0.435	1.20	0.875	1.36
May 80	0.239	8.84	0.348	2.63	0.750	3.25
May 81	0.091	18.82	0.217	9.20	0.250	4.50
May 82	0.048	5.15	–	–	–	–

Relative performance of *Anthoxanthum* genotypes ('the selective environment')

Two-way analysis of variance on percentage cover, using genotype and reproductive output category as class variables, showed that the only abundant species that gave a significant genotype–environment interaction

Table 11.5 Distribution of life-time fecundity (inflorescence number) in experimental and naturally occurring individuals. Non-flowering individuals were not included in the census of naturally occurring plants.

| | Number of individuals | |
Fecundity class	Experimental	Naturally occurring
0	824	–
1	118	13
2	22	3
3	11	6
4	10	1
5	8	0
6–9	21	5
10–50	61	10
51–100	18	0
>100	7	1

effect was *Cynodon*, although several of the rarer species also did so (Table 11.1*c*). In only one case did the main effect of genotype approach significance (*Trifolium repens*, $P = 0.043$); in all other cases genotype effects were non-significant, as is to be expected since genotypes were planted randomly without regard to abundance of associate species. A two-way G-test on genotype and reproductive output class was also not significant, indicating that the overall performance of the different genotypes was not different.

Genotype–vegetation type interaction effects were also examined with regard to the principal component scores (Table 11.2*d*). Slight effects were present for components 3 and 4, both of which were correlated with *Cynodon* abundance. The strongest effect was for component 10, which was correlated with *Hypochoeris*; however, this component only explained 2% of the variation among quadrats.

We also tested for the presence of genotype–vegetation interactions within the two most abundant vegetation clusters, A and D. Only two reproductive classes were distinguished (≤ 1 and ≥ 1) and only those genotypes (13 in all) which were replicated at least twice in each cluster and reproductive class were included in the analysis. The results (Table 11.3*b*) showed that the interactions differed among the vegetation clusters. Thus within Cluster D where the vegetation was dominated by *Panicum sphaerocarpon*, there was a highly significant interaction with frequency of *Panicum sphaerocarpon*. In Cluster A, where *Panicum sphaerocarpon* was rarer, the frequency of *Panicum sphaerocarpon* had a strong negative relationship with reproductive output of *Anthoxanthum*, but there was no evidence for a genotype–environment interaction. There was a significant three-way interaction of reproductive output category, genotype, and cluster, so confirming that the interaction of genotype and reproductive output was different in the two clusters (Table 11.3*c*). Significant three-way interactions were also found for two rare species, *Solidago* and *Panicum anceps*. Other interactions not evident in the overall analysis were also seen when the communities were partitioned. Thus significant interaction effects were seen in *Poa* where before they were absent. In *Rumex*, interactions seen in the overall data were now restricted to Cluster A, characterized by a high frequency of *Rumex*.

Interpretation

The external measures of the environment (the percentage cover data on associated species) presents a picture of an environment that is highly heterogeneous. Both individual grids and transect positions were very different as far as the frequency of the individual species and principal component scores were concerned. This heterogeneous external environment was also ecologically heterogeneous for *Anthoxanthum*. Over the field as a whole, *Anthoxanthum* tillers showed marked differences in performance, with particular regions resulting in high reproductive output, others in extremely low or even zero output. However, the external heterogeneity (assessed by the vegetation composition) is only partly reflected as ecological heterogeneity (assessed by the response of *Anthoxanthum*). Thus *Anthoxanthum* was totally insensitive to the frequency of particular species in the community, and performed quite equivalently in divergent vegetation types as revealed by a cluster analysis. There is a further non-correspondence of the selective and ecological environment. In the overall analysis, it is notable that while there is abundant evidence that the different vegetation types affect the reproductive output of *Anthoxanthum*, there is much less evidence that they do so differentially for different genotypes. The selective environment appears rather homogeneous with significant effects being present for only one common species, *Cynodon dactylon* (or for principal components which are strongly correlated with the abundance of this species). Interactions are present more frequently for rare species (and for one minor principal component) and this could reflect the fact that these rare species indicate particular community types. This is confirmed by the result that when the vegetation in the field as a whole was differentiated into different vegetation clusters, not only were interactions evident among clusters, but also there were new interactions within each cluster. This shows that *Anthoxanthum* genotypes show differential response to particular community components in only some regions of the field or in some community types and not in others. Unfortunately, even though the experiment was relatively large, there was inadequate replication of particular genotypes to examine the small-scale variation in depth.

While these results demonstrate that different genotypes perform differently in different vegetation types and that therefore a necessary precondition favoring an open breeding system is met, it is clearly difficult to translate these results into a quantitative prediction about the short-term advantages of genetically variable progeny. From a number of other lines of evidence, it appears unlikely that changes in the large-scale spatial environment of the field (as examined in this study) are responsible for individual advantages to sexual reproduction. First, seed dispersal distances are short: almost no seed is dispersed beyond 4 m except perhaps during extremely rare events (Antonovics and Ellstrand, 1984; Kelley, 1985). The scale of vegetation heterogeneity in this field is such that seeds are more likely to fall within vegetation that is similar to the parental vegetation. Second, although genotype–environment interaction effects were present, they were generally small. Clones did not suddenly die or fail to reproduce at particular distances from the parents, and moreover all clones performed more or less similarly in particular regions of the field. The contrasting scenario would have been results similar to a large-scale reciprocal transplant experiment (e.g. Clausen and Hiesey, 1958) only compressed within the scale of a hectare; that is, where some genotypes had zero fitness in some regions, high fitness in others. Third, in a related experiment, Antonovics and Ellstrand (1986) planted parental clones and seed-derived progeny clones from those same parents at different distances from the parent. Advantages of the seed-derived progeny clones was only evident at 20–40 m from the parent, well beyond the normal dispersal distance.

The experiment carried out here cannot exclude the possibility that small-scale heterogeneity is important. Indeed, several lines of evidence suggest this. In particular, when the plant community was further sub-divided into vegetation clusters, additional genotype–environment interactions appeared. The sensitivity of such analyses becomes less and less as the community is partitioned into more and more subdivisions: nevertheless, there were more significant genotype–environment interactions following subdivision than in the overall analysis. However, as has been pointed out by a number of authors (Maynard Smith, 1978) such heterogeneity within the dispersal range of an organism can only be a force in the maintenance of sexual reproduction, if there is in addition, either habitat selection such that appropriate genotypes 'choose' (grow into) habitats to which they are best suited, or if there is 'sib-competition' leading to the best genotype surviving in its own microsite and pre-empting the resources that may have been occupied by less fit genotypes. Thus local heterogeneity may be important for the maintenance of sexual reproduction, but this would be by mechanisms other than those tested here.

The experiment described here was carried out with tillers and therefore did not test what might be the most critical stages of the life cycle, the seeds and seedlings. Genotype–environment interactions were also found to occur in seedling germination and survival of *Anthoxanthum* in this same field (Schmitt and Antonovics, 1986a). However, because this was a separate experiment from the present one, it is not clear whether genotype–environment interactions at the seedling stage would act in a similar direction to those at the adult stage and so reinforce them.

11.7 INTEGRATION

While it has been possible to demonstrate that some of the preconditions favoring open breeding systems exist in this one population of *Anthoxanthum*, it is difficult to translate these results into a prediction about whether a more or less open genetic system would indeed be favored. To assess this we would need to know for an average, naturally produced progeny array, the probability that different genotypes encounter different 'environments'. We then need to know how their relative fitness changes in the different 'environments'; these fitness functions should be non-parallel and over-lapping (i.e. there should be genotype–environment interaction). And for comparative purposes we need to have similar information for progeny produced from the alternative breeding system, with which the present naturally occurring system is being compared.

It is important to note that the 'environment' as used above is the selective environment; explicit measurement of the external environment or ecological environment (while perhaps providing a clue as to pertinent causal forces) is in fact unnecessary. Thus if we ask whether environments are heterogeneous in space it is sufficient to demonstrate merely that there is genotype × spatial position interaction in fitness; no explicit measures of environmental parameters as normally understood by an ecologist are necessary. A similar reasoning would apply to temporal variation. Where ecological or external environments can be identified as causal in the observed spatial position or temporal effects, then this may permit prediction to other ecological situations. But by themselves such measures may actually confuse rather than clarify the measurement of the evolutionary forces. Thus it would have been simpler in our experiments to avoid the 37×1100 point cover measures, and have instead genotypes replicated at different distances from their parental origin, to look at the physical distance over which G × E effects were manifest.

While measurement of within family variance in fitness functions over space and time would provide the raw material for a predictive theory, we still lack that predictive theory at a quantitative level. Word-models show the credibility of various processes affecting genetic systems, but by avoiding the quantitative aspects, as well as by externalizing our conceptualization of the environment, they have failed to provide a framework on which empirical studies can be based.

There are in addition considerable technical difficulties in studying the relative fitness of different types of progeny arrays, or of genotypes within those arrays. Comparison of sexually and asexually produced seed requires use of either facultative apomicts (Bayer and Stebbins, 1983) or of seed artificially created by crossing doubled-haploids generated from pollen or anther culture. In comparing self- and cross-pollinated progeny arrays it is extremely difficult to see how levels of progeny array variance can be deconfounded from individual levels of heterozygosity. Variable and uniform progeny arrays can be used in 'phenomenological' experiments which simulate dispersal around a parent individual (Kelly, Antonovics and Schmitt, 1988). But to delve into the factors which may be important in causing differential fitness of progeny arrays presents even more problems. Thus to estimate fitness functions of particular genotypes within progeny arrays would require replication of those genotypes over time or space: at present this is only easily done using vegetative clones, thus by-passing life-history stages that could be critical. And while a quantitative genetics approach could circumvent some of this problem by calculating the fitness performance of lines derived from within sibship crosses, it is not entirely clear the extent to which additive vs. non-additive components of variance are important in adaptation of progeny arrays to unpredictable environments. Such an indirect quantitative genetics approach would also be extremely labor intensive.

The connection between genetic variance and environmental variance has been largely an intuitive rather than explicit concept, particularly as far as developing an understanding of how and when open genetic systems provide a short-term advantage in spatially or temporally varying environments. While a number of mathematical and word models have been presented, these models have been difficult to translate into experimental tests of particular hypotheses. Conceptually, there has been confusion between the external measurable environment, the environment as it influences the organism, and the selectively heterogeneous environment. Environments are only selectively heterogeneous if different genotypes have different relative fitnesses in those environments. The ingredients that would be required for genetic variation among a progeny array to be favored have been proposed theoretically, but even for the most obvious cases (e.g. simple spatial or temporal variation) we have no comprehensive quantitative theory couched in terms measurable by the ecological geneticist. Practically, we still need to develop tissue culture technology to obtain replicates of individual genotypes, and to translate such technology into a natural population context so the critical experiments can be performed. Stebbins' (1950) exhortation that the evolution of genetic systems is a subject 'separate enough to be studied by itself' is as true today as it was then. We have barely begun to scratch the surface of many of the major and critical questions in the evolutionary biology of plant breeding systems, and the issue is likely to remain a challenge that will require the integrated efforts of theoreticians, field biologists, and biotechnologists.

11.8 ACKNOWLEDGEMENTS

The work reported here was supported by NSF grants DEC-772561, DEB-8022369, and an Intramural Research Grant from the Academic Senate of the University of California at Riverside. Most of the data analysis was carried out while one of us (J.A.) was a Visiting Research Scientist at the Centre Louis Emberger, Centre d'Etudes Phytosociologiques et Ecologiques, Centre National de la Recherche Scientifique, Montpellier, France. J.A. wishes to thank Pierre Jacquard and Georges Valdeyron for their hospitality and the CNRS for financial support. Many people helped with the planting and censuses: we thank them all for their effort and patience. The faculty and students at the Kellogg Biological Station provided valuable discussion of many of the ideas presented here.

11.9 REFERENCES

Allen, A. T. (1982) *SAS Econometrics and Time Series User's Guide, 1982 edn.* SAS Institute Inc., Cary, North Carolina.

Antonovics, J., Clay, K. and Schmitt, J. (1987) The measurement of small-scale environmental heterogeneity using clonal transplants of *Anthoxanthum odoratum* and *Danthonia spicata. Oecologia*, **71**, 601–7.

Antonovics, J. and Ellstrand, N. C. (1984) Experimental studies of the evolutionary significance of sexual reproduction. I. A test of the frequency-dependent selection hypothesis. *Evolution*, **38**, 103–15.

Antonovics, J. and Ellstrand, N. C. (1986) The fate of dispersed progeny: experimental studies with *Anthoxanthum.* in *Genetic Differentiation and Dispersal in Plants* (eds P. Jacquard, G. Heim and J. Antonovics), Springer-Verlag, Berlin, pp. 369–81.

Antonovics, J. and Primack, R. B. (1982) Experimental ecological genetics in *Plantago* VI. The demography of seedling transplants of *P. lanceolata. J. Ecol.*, **70**, 55–75.

Barton, N. H. and Post, R. J. (1986) Sibling competition and the advantage of mixed families. *J. Theor. Biol.*, **120**, 381–7.

Bayer, R. J. and Stebbins, G. L. (1983) Distribution of sexual and apomictic populations of *Antennaria parlinii. Evolution*, **37**, 555–61.

Bell, G. (1982) *The Masterpiece of Nature: the Evolution and Genetics of Sexuality.* University of California Press, Berkeley.

Brandon, R. N. (1986) On the concept of the environment in the theory of natural selection, (in preparation.)

Bulmer, M. G. (1980) The sib-competition model for the maintenance of sex and recombination. *J. Theor. Biol.*, **82**, 335–45.

Clausen, J. and Hiesey, W. M. (1958) *Experimental Studies on the Nature of Species.* IV. *Genetic Structure of Ecological Races.* Carnegie Institution of Washington Publication No. 615.

Clements, F. E. and Goldsmith, G. W. (1924) *The Phytometer Method in Ecology.* Carnegie Institution of Washington. Publication No. 356.

Darlington, C. D. (1939) *The Evolution of Genetic Systems.* Cambridge University Press, Cambridge.

Ellstrand, N. C. and Antonovics, J. (1985) Experimental studies of the evolutionary significance of sexual reproduction II. A test of the density-dependent selection hypothesis. *Evolution*, **39**, 657–66.

Fisher, R. A. (1930) *The Genetical Theory of Natural Selection.* Oxford University Press, Oxford.

Fowler, N. L. (1981) Competition and coexistence in a North Carolina grassland II. The effects of the experimental removal of species. *J. Ecol.*, **69**, 843–54.

Fowler, N. L. and Antonovics, J. (1981) Competition and coexistence in a North Carolina grassland I. Patterns in undisturbed vegetation. *J. Ecol.*, **69**, 825–41.

Ghiselin, M. T. (1974) *The Economy of Nature and the Evolution of Sex*. University of California Press, Berkeley.

Gillespie, J. (1977) Natural selection for variances in offspring numbers: a new evolutionary principle. *Am. Natur.*, **111**, 1010–14.

Glesener, R. R. (1979) Recombination in a simulated predator–prey interaction. *Am. Zool.*, **19**, 763–71.

Grime, J. P. (1979) *Plant Strategies and Vegetation Processes*. Wiley, Chichester.

Grime, J. P. and Hunt, R. (1975) Relative growth rate: its range and adaptive significance in a local flora. *J. Ecol.*, **63**, 393–422.

Hamilton, W. D. (1980) Sex versus non-sex versus parasite. *Oikos*, **35**, 282–90.

Jaenike, J. (1978) An hypothesis to account for the maintenance of sex within populations. *Evol. Theory*, **3**, 191–4.

Kelley, S. E. (1985) The mechanism of sib competition for the maintenance of sexual reproduction in *Anthoxanthum* odoratum. Ph.D. Dissertation, Duke University, Durham, USA.

Kelley, S. E., Antonovics, J. and Schmidt, J. (1988) The evolution of sexual reproduction: a test of the short-term advantage hypothesis. *Nature* (in press).

Lacey, E. P., Real, L., Antonovics, J. and Heckel, D. G. (1983) Variance models in the study of life-histories. *Am. Natur.*, **122**, 114–31.

Levin, D. A. (1975) Pest pressure and recombination systems in plants. *Am. Natur.*, **109**, 432–51.

Levins, R. (1968) *Evolution in Changing Environments*. Princeton University Press.

Lloyd, D. G. (1980a) Benefits and handicaps of sexual reproduction. *Evol. Biol.*, **13**, 69–111.

Lloyd, D. G. (1980b) Alternative formulations of the intrinsic cost of sex. *NZ Gen. Soc. Newsletter*, no. 6.

Maynard Smith, J. (1978) *The Evolution of Sex*. Cambridge University Press, Cambridge.

Price, M. V. and Waser, N. M. (1982) Population structure, frequency-dependent selection, and the maintenance of sexual reproduction. *Evolution*, **36**, 35–43.

Ray, A. A. (1982a) *SAS User's Guide: Basics*, 1982 edn. SAS Institute Inc., Cary, North Carolina.

Ray, A. A. (1982b) *SAS User's Guide: Statistics*, 1982 edn. SAS Institute Inc., Cary, North Carolina.

Schmitt, J. and Antonovics, J. (1986a) Experimental studies of the evolutionary significance of sexual reproduction. III. Maternal and paternal effects during seedling establishment. *Evolution*, **40**, 817–29.

Schmitt, J. and Antonovics, J. (1986b) Experimental studies of the evolutionary significance of sexual reproduction. IV. Effect of neighbor relatedness and aphid infestation on seedling performance. *Evolution*, **40**, 830–6.

Shaw, R. G. (1986) Response to density in a wild population of the perennial herb *Salvia lyrata*: variation among families. *Evolution*, **40**, 492–505.

Stebbins, G. L. (1950) *Variation and Evolution in Plants*. Columbia University Press, New York.

Tachida, H. and Mukai, T. (1985) The genetic structure of natural populations of *Drosophila melanogaster*. XIX. Genotype–environment interaction in viability. *Genetics*, **111**, 43–55.

Taylor, P. D. (1979) An analytical model of a short-term advantage for sex. *J. Theor. Biol.*, **81**, 407–21.

Tooby, J. (1982) Pathogens, polymorphism, and the evolution of sex. *J. Theor. Biol.*, **97**, 557–76.

Treisman, M. (1976) The evolution of sexual reproduction: a model which assumes individual selection. *J. Theor. Biol.*, **60**, 247–69.

Uyenoyama, M. K. (1984) On the evolution of parthenogenesis: a genetic representation of the 'Cost of Meiosis'. *Evolution*, **38**, 87–102.

Via, S. and Lande, R. (1985) Genotype–environment interaction and the evolution of phenotypic plasticity. *Evolution*, **39**, 505–22.

Williams, G. C. (1975) *Sex and Evolution*. Princeton University Press, Princeton.

Williams, G. C. and Mitton, J. B. (1973) Why reproduce sexually? *J. Theor. Biol.*, **39**, 545–54.

Young, J. P. W. (1981) Sib competition can favor sex in two ways. *J. Theor. Biol.*, **88**, 755–6.

CHAPTER 12

Local differentiation and the breeding structure of plant populations

DONALD A. LEVIN

12.1 INTRODUCTION

Plant populations are replete with genetic variation which often forms patterns in space. The pattern may coincide with that of the environment as a result of disruptive selection favoring different genotypes in different environments. A genotypic mosaic may develop independent of the environmental mosaic as a result of genetic drift within an established population or founder effect early in the history of the population. Genetic differentiation occurs within the context of demographic processes, and thus is best understood within this context (Jain, 1975, 1976; Solbrig, 1980; Ennos, 1984; Bradshaw, 1984; Rice and Jain, 1985). It occurs while a population is growing in area and number, and changing in population density and patch structure. Genetic differentiation emerges in the face of gene flow, which itself is a function of plant density and patch structure.

The genetic structure of a population, i.e. the amount and organization of genetic variation in space, plays a dominant role in their evolution by influencing the consequences of interactions among conspecifics, levels of selection, the amount of genetic variation which is maintained, and the ability to exploit ecological opportunity (Wright, 1969, 1980; Wade, 1978, 1980; Wilson, 1979). If a population is composed of many discontinuous subpopulations which are partially isolated from each other and which are genetically differentiated, the store of variability in the population will be much greater than in a random panmictic population of the same size. In a shifting random differentiation of subpopulations, the number of substantially different multilocus gene complexes which arise will far exceed that in a continuous panmictic population. Some of the novel genotypes may be pre-adapted for the exploitation of new habitats or for changes in the present habitat. Given the ecological opportunity, the novel genotypes will be favored, and the subpopulation will prosper in the new environment. Dispersion from this to other subpopulations would afford a higher level of adaptation to the population as a whole. Under the shifting balance process in a subdivided population, the response to selection is greater per generation than with mass selection.

Although genetic structure has been described in a few populations, the balances among selection, genetic drift and migrations are poorly understood. Indeed data on single factors are few. Except where there is manifest genotype–environment covariation, we cannot differentiate between gene frequency mosaics fostered by selection or genetic drift. We do not know the extent to which gene flow dilutes or obliterates the genetic structure which otherwise would be present. The problem is that we try to understand the process from the product. Only if one factor (selection, genetic drift, migration) far prevails over the others, will a correct interpretation be made, because the interactions between the factors may be very complex, and a given pattern may arise in more than one way.

The focal point of this chapter is the role of the breeding structure in promoting or constraining the development of genetic structure within populations. I will first discuss pollen and seed dispersion in terms of Wright's neighborhood model, and then assess the maximum levels of random differentiation which could occur within continuous populations with known neighborhood properties. This is followed by a consideration of the effect of the spatial dynamics of population growth on the potential for random differentiation, a discussion of selection–migration balances in patchy environments, and the characteristics of clines based on observed gene dispersion. Finally, the rate of spread of advantageous genes based on observed gene dispersion is discussed.

12.2 BREEDING STRUCTURE OF POPULATIONS

It is useful to consider the breeding structure of populations in terms of Wright's (1943a, 1946) neighborhood model, because it provides insight into the number of individuals which form panmictic breeding units, and the area which these units occupy. In turn, we can obtain insight to the extent which different parts of a population are isolated by distance, and the potential for random and selective differentiation.

A neighborhood is defined as an area from which about 86% of the parents of some central individual may be treated as if drawn at random. One parent will lie outside this area ca. 12.5% of the time, and both parents will lie outside ca. 1.5% of the time. The exact proportion depends on the balance between pollen and seed dispersal variances (Crawford, 1984a). If mating and seed dispersal distances were at random, the neighborhood would encompass the entire population. With restricted dispersal distances, the area is very much less. The neighborhood area (A) is a circle of radius 2σ so that $A = 4\pi\sigma^2$, where σ^2 is the parent–offspring dispersal variance measured around a zero mean and relative to a single reference axis passing through the population. The neighborhood size is $N_e = Ad$, where d, the effective density, is approximately the density of simultaneously flowering plants, if the population is stable in age structure and stationary in numbers, and if the distribution per parent of offspring reaching reproductive maturity is Poisson. Deviation from these conditions undoubtedly exist, and the effective density usually will be less than the number of flowering plants within a neighborhood area.

Neighborhood size also may be viewed in terms of parentage. The neighborhood size is the reciprocal of the probability that two uniting gametes come from the same parent. N_e also is the reciprocal of the average probabilities that two plants right next to each other have the same female parent, the same male parent, or that the female parent of one plant is the male parent of the other or vice versa (van Dijk, 1985).

Let us consider how to estimate neighborhood parameters in plants following the procedure described by Crawford (1984a). The total parent–offspring dispersal variance is σ^2 (axial) $= 1/2t\sigma^2_p + \sigma^2_s$, where t is the outcrossing rate, σ^2_p is the pollen dispersal variance and σ^2_s is the seed dispersal variance. Pollen and seed dispersion are assumed to follow a normal distribution. The pollen dispersal variance may be roughly estimated from pollinator flight distances between plants or more precisely with pollen or genetic markers. Seed dispersal variance may be estimated *in situ* or under experimental conditions.

Representative neighborhood size and area estimates for several herbs are presented in Table 12.1. All but *Plantago lanceolata* and *Avena barbata* are insect-pollinated, and all but *Avena barbata* are predominantly outcrossing. The neighborhood area varies from ca. 4 m^2 in *Avena barbata* to 66 m^2 in *Liatris cylindracea*. The number of individuals in that area (neighborhood size) varies from less than 20 in two cases to over 600 in *Liatris cylindracea*.

The neighborhood concept is so formulated that gene migration over one neighborhood diameter occurs at a given rate per generation. Therefore, the greater the number of neighborhood diameters separating two subpopulations, the greater their isolation and the greater the potential for their genetic divergence. Neighborhood diameters are presented for a series of herbs in Table 12.1. For the most part, neighborhood diameters are less than 5.0 m. The predominantly inbreeding *Avena barbata* has a neighborhood diameter of ca 2 m. Neighborhood diameters are very small relative to population diameters. Accordingly, groups of plants in different parts of a population may be separated by tens to perhaps hundreds of neighbor-

Table 12.1 Observed neighborhood and expected inbreeding characteristics of various herbs.

Species	N_a (m²)	N_a diameter (m)	N_e	ΔF	F_{ST}^*	$\hat{F}_{ST}^\dagger$	Reference
Phlox pilosa	41	3.61	533	0.0009	0.089	0.045	Levin and Kerster, 1968
Liatris cylindracea	66	4.58	633	0.0007	0.068	0.037	Schaal and Levin, 1978
Liatris aspera	35	3.34	175	0.0028	0.249	0.125	Levin and Kerster, 1969
Viola pedata	48	3.91	432	0.0011	0.104	0.055	Beattie and Culver, 1979
Viola rostrata‡	25	2.82	167	0.0030	0.269	0.154	Beattie and Culver, 1979
Primula veris	30	3.09	7.4	0.0680	0.999	0.772	Richards and Ibrahim, 1978
Plantago lanceolata	10	1.78	17	0.0294	0.947	0.595	Bos, Harmens and Vrieling, 1986
Avena barbata	4	1.13	140	0.0036	0.301	0.152	Rai and Jain, 1982

* After 100 generations.
† At equilibrium, $m = 0.01$.
‡ Estimates do not take into account selfing by cleistogamic flowers.

hoods. The isolation afforded by even a few diameters is essentially complete, because groups two diameters apart would exchange genes at very low rates.

The neighborhood characteristics of a species are dependent upon numerous variables foremost of which is plant density (Levin and Kerster, 1969; Levin, 1981). In *Liatris aspera* neighborhood size is positively correlated with density, varying from 45 in the sparse population (1 plant/m^2) to 363 in the dense population (11 plants/m^2). On the other hand, neighborhood area is negatively correlated with density, varying from 45 m^2 in the sparse population to 33 m^2 in the dense population. Neighborhood structure varies with density, because pollen dispersion (estimated from interplant pollinator flight distances) is inversely related to density. In the sparse population, the axial variance is 2.30 vs. 0.36 in the dense population.

The neighborhood size and area of species may be smaller than estimated, because the variance of progeny numbers relative to the mean typically is greater than 1, as will be discussed below. On the other hand, there are several factors which may render them larger than estimated from pollen and seed dispersion (Levin, 1981):

1. Pollen dispersal distances based on pollinator flight distance do not allow for pollen carryover and flight directionality.
2. Pollen–pistil compatibility may be lower between near neighbors than moderately spaced plants.
3. Crosses between neighbors may be less fruitful than crosses between moderately spaced plants.
4. Seed and seedling mortality may be higher near the seed parent because of predation, disease and competition.
5. The tails of pollen and seed distributions may be underrepresented due to inadequate tracking techniques.

Given these considerations, the neighborhood values that I and others have calculated are only rough approximations, albeit valuable in comparative studies of gene flow.

12.3 RANDOM DIFFERENTIATION IN CONTINUOUS POPULATIONS

Wright (1943a, 1946, 1978) showed that populations distributed continuously over an area will undergo genetic subdivision, if gene dispersal within them is sufficiently restricted, as a result of the isolation by distance. The smaller the neighborhood, the greater the genetic subdivision within the population, and correlatively the greater the level of inbreeding and homozygosity. Rohlf and Schnell (1971), Turner, Stephens and Anderson

(1982), and Sokal and Wartenberg (1983) conducted computer simulations of Wright's neighborhood (or isolation by distance) model, and observed rapid establishment of spatial patterns in gene frequencies which persisted for many generations. Gene dispersion in these models was assumed to be much more restricted than observed in natural populations. Thus it is of interest to determine what level of subdivision might develop within populations of the species for which gene dispersion is known.

For the sake of simplicity and to obtain estimates of the maximum amount of inbreeding and random differentiation, I will assume that a population is composed of an infinite number of totally isolated non-overlapping neighborhoods of the same genetic composition, and that there is no selection or mutation. We may estimate the rate of increase of the inbreeding coefficient in one generation as $\Delta F = 1/2N_e$ (Falconer, 1981) relative to the distance that remains to reach complete inbreeding. Estimates of ΔF are presented for several species in Table 12.1. For the most part, these numbers are very small, indicating that the genetic structure of populations changes very little from one generation to another. However, change is cumulative, and after several generations neighborhoods not only become more homozygous on average, but also diverge in gene frequencies. This divergence may be described in terms of the standardized variance of gene frequencies $F_{ST} = \sigma_p^2/\bar{p}\bar{q}$ (Wright, 1943a, 1965). F_{ST} is the inbreeding coefficient of a subpopulation (neighborhood in this case) relative to the total population. F_{ST} may be estimated from ΔF from the following equation $F_{ST} = 1 - (1 - \Delta F)^t$, where t is the number of generations. Possible values range from zero (no differentiation) to 1.0 (maximum differentiation).

Estimates of F_{ST} after 100 generations of isolation are presented in Table 12.1. In some species (e.g. *Liatris cylindracea*. $F_{ST} = 0.07$), very little random differentiation would occur over 100 generations, whereas in others (e.g. *Primula veris*, $F_{ST} = 0.99$) differentiation approaches the extreme. Suppose that neighborhoods are a collection of totally isolated subpopulations, but received the same level of immigration from a common gene pool. Wright (1931) showed that at equilibrium

$$\hat{F}_{ST} = 1/[4\,N_e m] + 1$$

where N_e is the size of the subpopulation and m is the immigration rate. The estimates of $\hat{F}_{ST}$ for $m = 0.01$ are presented in Table 12.1. Immigration from an extraneous source reduces the level of random differentiation between neighborhoods. Indeed, the level achieved with 1% immigration is quite modest; the $\hat{F}_{ST}$ values for most species are less than 0.15. Were the immigration rates higher, the $\hat{F}_{ST}$ values would be lower.

These estimates of F_{ST} with and without migration are maxima. Actual F_{ST} values will be much smaller, because neighborhoods in natural populations are not isolated from one another. Therefore, in most of the species considered, little differentiation is likely to accrue from isolation by distance in continuous populations.

12.4 RANDOM DIFFERENTIATION IN DISCONTINUOUS POPULATIONS

The neighborhood or isolation by distance model assumed spatial continuity and constant density. These assumptions are not very realistic, because spatial structure usually develops in plant populations. During the colonization of a site, scattered individuals often act as nuclei around which patches of plants subsequently develop. The patches then increase in size, and may eventually coalesce (Grieg-Smith, 1964; Brereton, 1971; Yarranton and Morrison, 1974; Bannister, 1965). Patches may persist for only a few generations, contracting or dissipating and being replaced by others elsewhere in the population (Leith, 1960; Grubb, 1977; Turkington and Harper, 1979, Thompson, 1978; Platt, 1975).

If the patches remain discontinuous for several generations, they may undergo random genetic differentiation, the magnitude of which is an inverse function of the number of plants per patch (Wright, 1931; Chakraborty and Nei, 1977; Nei, Maruyama and Chakraborty, 1975). Random differentiation will be accompanied by a loss of heterozygosity and average genetic variation within patches. The level of differentiation and loss of variation will be amplified if the number of plants within patches periodically contracts (Wright, 1938; Motro and Thomson, 1982).

Effective patch size

The patch size appropriate for population genetic consideration is the genetically effective size. It is not the number of plants or number of breeding plants. Wright (1931, 1938) defined the effective size as the number of breeding individuals in an idealized population that would show the same amount of dispersion of allele frequencies under genetic drift or the same amount of inbreeding as the population in question. The effective size that deals with genetic drift is the variance effective size. The key to the variance effective size of a group of individuals is the distribution of progeny numbers. The greater this distribution deviates from a Poisson distribution in which the mean and variance of progeny numbers are the same, the greater the level of genetic drift between parents and progeny, and the greater the level of random differentiation between patches (Wright, 1931). The relationships between fertility distributions, effective size and genetic drift in plants have been considered by a number of workers during the past decade (Levin, 1978; Levin and Wilson, 1978; Crawford, 1984b; Heywood, 1986). I will review here some theory and observations relating to these issues.

Heywood (1986) analyzed the mean and variance of seed and fruit (= progeny) numbers for populations of 34 monoecious annual species as presented in the literature. The ratio of the variance of seed number to the mean number ($V_k/\bar{k}$) exceeded 10 in most species and 100 in some. The distribution of progeny numbers thus was in stark contrast to the Poisson distribution ($V_k/\bar{k} = 1$) (Karlin and McGregor, 1968).

The effective size of patches may be estimated from the following expression formulated by Kimura and Crow (1963):

$$N_e = \frac{N\bar{k} - 1}{1 - F + (1 + F)V_k/\bar{k}}$$

where N is the number of breeding individuals, $\bar{k}$ is the mean number of seeds to which plants contribute gametes, V_k is the variance of seed number, and F is the fixation index. F = O with random mating, and F = 1 with self-fertilization.

Although we lack information on the number of breeding individuals per patch in the aforementioned 34 species, we can still estimate N_e relative to N. Heywood (1986) found that if patch size were constant ($\bar{k} = 2$), F = O and differences in progeny numbers were not heritable, the average N_e/N ratio for all species would be 0.41. The N_e/N ratios would be reduced by inbreeding and by the heritability of progeny numbers. If F = 1 and $\bar{k} = 2$, the average N_e/N ratio would decline to 0.29. If the heritability were 0.5, $\bar{k} = 2$ and F = O, the ratio would be 0.23.

The effective size of patches determines their penchant for genetic drift. A moderate amount of drift would be expected if N_e was less than 200, and a considerable amount would be expected if N_e was less than 20 (Wright, 1969). Populations need not be very small for drift to be important, because the N_e/N ratio may be very much less than 1 (e.g. 0.20 in *Linum catharticum* and *Chaenorhinum minus*, Heywood, 1986). Accordingly, patches of 100 plants will be subject to the same random gene frequency changes and loss of gene diversity as patches of 20 plants which have a ratio of the progeny variance to mean of 1.0.

The entire population is subject to genetic drift as are single patches therein. If no patch extinctions occur, the effective population size is the product of the mean effective sizes of patches and the number of patches. However, if the colonization–extinction rate is high, the effective size of the total population is reduced substantially (Maruyama and Kimura, 1980). Accordingly, the shorter the duration of patches within a long-lived population, the less will be the amount of genetic diversity within the total population, and the greater the random divergence of populations by genetic drift.

Patch dynamics

The genetic structure which develops within a population through stochastic processes depends not only on the effective patch size, but also upon the subsequent pattern of population growth in space. A population may grow from one or more patches. If the population expands gradually from a single patch, little structure would ensue. On the other hand, if the population expands from one original patch and other patches derived from it through successive colonization, considerable structure may emerge. The fewer the number of founders and the less closely they are related, the greater the level of genetic differentiation among patches (colonies). Genetic diversity in the initial patch is a prerequisite for interpatch divergence. The impact of stochastic processes on patch divergence is a positive function of diversity. A population may expand from multiple patches which were established independently of one another. The amount of subsequent structure depends on the amount and apportionment of the diversity within and among patches.

The distance between patches has a bearing on their genetic divergence, because the amount of migration between patches is an inverse function of the distance between them. Even a light pollen and seed rain from extraneous sources may erode genetic differences between small populations (Antonovics, 1968; Levin, 1984). Patches that were less than 50 m apart probably would not be free to diverge by stochastic processes, unless their effective sizes were very small (Levin and Kerster, 1974). A number of studies have inferred that migration may rapidly regenerate gene diversity in populations which have passed through bottlenecks (Jain, 1979; Brown and Marshall, 1981; Harding and Barnes, 1977; Clegg and Allard, 1972).

Random differentiation early in the history of a population is not a prerequisite for its subsequent development. Populations growing from one or a few patches, or from multiple patches with similar genetic compositions, may later develop structure as a result of low gene dispersion and inbreeding, as discussed above. However, the magnitude of the structure depends on the time of its initiation. The earlier it is initiated, the greater the effect of random drift, and the more pronounced the resultant pattern.

In the absence of selection, the mosaic of gene frequencies which develops through stochastic processes is likely to persist for tens or even hundreds of generations even in a continuous, stable population so long as gene dispersion is narrow (Endler, 1977; Turner, Stephens and Anderson, 1982; Sokal and Wartenberg, 1983). Structure also may remain intact as a population undergoes numerical fluctuations if there exists a long-lived seed pool. The seed pool maintains the genetic continuity of populations in time, serving as a source of 'immigrants' from past generations (Gottlieb, 1974; Templeton and Levin, 1979). If a seed pool is lacking, periodic contraction and expansion of patches or local extinction—recolonization episodes may cause the gene frequency surface to change, because additional opportunities occur for sampling error.

One highly probable case for local random differentiation is described by the spatial distribution of flower polymorphism in the outcrossing annual *Linanthus parryae* (Epling and Dobzhansky, 1942; Epling, Lewis and Ball, 1960). The percentage of blue (vs. white) morphs in 260 regularly spaced (every 10 ft.) quadrats along a half-mile transect in the Mojave Desert varied from 1% to more than 80%. Adjacent quadrats had an average gene frequency correlation of 0.66. Half the correlation was lost at 300 ft. The statistical and spatial distribution of morph frequencies were consistent with expectations for gene frequency dispersion in a system of small isolated populations (Wright, 1943b, 1978). The pattern of variation was independent of that of the environment. Presumably the gene frequency heterogeneity is maintained through sampling error associated with the contraction and expansion of local patches, and/or local extinction and recolonization. Most gene flow through pollen and seeds appears to be over a few meters or less, so that drift-generated variation patterns are likely to persist for many generations. The estimated neighborhood area is about 10 m^2. A large seed bank ostensibly buffers the pattern against gene flow.

Microdifferentiation independent of the pattern of environmental variation also occurs in the annuals *Avena barbata* (Rai and Jain, 1982), *Impatiens pallida* (Schemske, 1984) and *Armeria maritima* (Lefebvre, 1985). Gene frequency heterogeneity probably developed during the establishment of populations or periods of contraction thereafter, and persisted because pollen and seeds were narrowly dispersed. *Avena barbata* and *Impatiens pallida* are predominantly self-fertilizing; *Armeria maritima* is an outcrosser. Although two of the four examples involved outcrossers, Wright (1946, 1949b) demonstrated that random genetic differentiation is much more likely to develop in selfers than in outcrossers, because selfing (or any form of inbreeding) reduces the effective size of populations and patches therein, and reduces gene flow.

Populations of late successional or climax species are less prone to random microdifferentiation than those of weedy species, because populations of the former are more stable in time, and gene flow therein tends to be over greater distances (Levin and Kerster, 1974; Jain, 1975; Venable and Levin, 1983). Weakly differentiated patches independent of the pattern of the environment have been described in several climax species (*Liatris cylindracea*, Schaal, 1975; *Pinus sylvestris*, Tigerstedt *et al.*, 1982; *Pinus ponderosa*, Linhart *et al.*, 1981; *Larix laricina*, Park and Fowler, 1982; *Picea abies*, Brunel and Rodolphe, 1985).

Opposition by selection

The pattern of genetic substructure generated by stochastic processes during the early stages of population growth subsequently may be altered by natural selection. If patch survivorship, growth and emigration were functions of genetic diversity, gene frequency heterogeneity would erode in time. In essence, the population would be subject to stabilizing interdeme selection, because the genetically disparate (and thus extreme) demes would be at a disadvantage in producing emigrants. Variation-dependent fitness is evident in *Trifolium hirtum*. The most successful patches in time and size might have relatively high gene diversity and heterozygosity (Jain and Martins, 1979; Rice and Jain, 1985).

Gene frequency heterogeneity also may erode in time if patch fitness is dependent upon multilocus genotypes. As formulated by Wright (1931) in his shifting balance theory, wide stochastic variability may lead to the assemblage of novel multilocus genotypes, far superior to other genotypes in the patch and elsewhere in the population. By selection these genotypes would increase in the assemblage, and confer unto it unusually high fitness relative to other assemblages. This assemblage then grows in size, and by excess dispersion shifts allele frequencies in other assemblages until superior genotypes prevail therein as well. Wright refers to the third phase of this process as interdeme selection. It is directional selection, because one kind of population is replacing the others. In the process, gene frequency heterogeneity is reduced. In a heterogeneous environment, selection may amplify some of the drift-generated differences between assemblages. Selection may also yield a gene frequency mosaic in accord with the environmental variable in a population initially lacking genetic subdivision (Bradshaw, 1972; Snaydon, 1980; Allard *et al.*, 1972).

12.5 INFERENCES ABOUT MIGRATION RATES FROM GENETIC STRUCTURE

The level of gene frequency heterogeneity between subpopulations provides some insight about migration between them. This can be done within the context of Wright's (1943a) island model, in which a subpopulation receives immigrants at a rate m chosen at random from other subpopulations. Assuming allele neutrality and an equilibrium between migration and genetic drift, the number of migrants per generation can be estimated from

$$(Nm)_{est} = (1/F_{ST} - 1)/4$$

where N is the effective population size and m is the migration rate per generation. Schaal (1975) reported $F_{ST} = 0.07$ for allozyme alleles in adjacent subpopulations of *Liatris cylindracea*. If the population were in equilibrium, there was no selection, and immigration was from a common gene pool, $Nm = 3.32$, i.e. each subpopulation would receive 3.32 immigrants per generation from the population as a whole. The mean number of reproductive plants per subpopulation was approximately 90, which gives $m = 0.037$. Bos, Harmens and Vrieling (1986) performed a similar set of calculations for allozyme alleles in scattered subpopulations of *Plantago lanceolata*. They reported $F_{ST} = 0.04$, and from it estimated $Nm = 6$, with $N = 73.3$, $m = 0.08$.

The *Nm* estimates for *Liatris* and *Plantago* assume an island model of population structure. However, population structure is probably much closer to the stepping-stone model, wherein immigrants are from adjacent subdivisions. Crow and Aoki (1984) showed that if a population has very many subdivisions, one immigrant chosen randomly from the rest of the population has about twice the genetic effect as one from a neighboring group. Accordingly, a better approximation of *Nm* is obtained if the island model estimate is multiplied by 2. Then in *Liatris* $Nm = 6.6$ and $m = 0.07$. In *Plantago*, $Nm = 12.0$ and $m = 0.16$.

12.6 GENE FLOW AND SELECTIVE DIFFERENTIATION

The breeding structure of a population may influence its response to disruptive selection. The relationship between gene flow and selection in shaping local variation patterns in a heterogeneous environment, and the spatial scale of such variation are best considered in terms of a one-dimensional gene flow scale l, the square root of the mean squared dispersal distance (Fisher, 1950; Slatkin, 1973; Endler, 1977). The quantity l includes no assumptions about the normality of gene dispersal. The gene flow distance, l, is related to the neighborhood radius by $r = l\sqrt{2}$. The smaller that l is the greater the isolation by distance between two subpopulations, and the more rapidly and fully these aggregates may respond to disruptive selection.

Knowing l, we may calculate the minimum distance over which a population may respond to selection, assuming an abrupt selective change between adjacent environments as in Endler's (1977) ecotone model. This is referred to as the characteristic scale length of variation of gene frequencies (l_c) and is defined as $l_c = l/\sqrt{s}$, where s is the maximum difference in fitness between homozygotes in two environments (Slatkin, 1973). Regardless of the spatial heterogeneity in selection pressures, gene frequencies would not vary significantly over a distance less than this length. If the environment changes on a scale less than this length, gene frequencies will respond to selection intensities averaged over the characteristic length. When the scale of variation in the environment is greater than this length, populations can

Table 12.2   Predicted cline widths in several herbaceous species*.

Species	s = 0.10		s = 0.50	
	Δ = 0.10m	Δ = 100m	Δ = 0.10m	Δ = 100m
Phlox pilosa	9.43	48.24	4.21	28.32
Liatris cylindracea	10.08	50.40	4.51	29.64
Liatris aspera	7.46	41.32	3.33	24.27
Viola pedata	8.72	44.76	3.90	26.93
Primula veris	6.89	45.81	3.08	23.05
Primula vulgaris	2.81	21.64	1.26	12.78
Plantago lanceolata	3.98	27.30	1.78	13.27
Avena barbata	2.53	20.24	1.13	12.58

Based on observed gene flow estimates (Table 12.1).

respond to a heterogeneous environment and differentiate into distinctive subpopulations (May, Endler and McMurtie, 1975; Endler, 1977). Endlers' ecotone model may be applied to plant populations, because considerable changes in soil chemistry, ecological associates, and levels of disturbance may occur over a distance of a few meters or less.

The characteristic scale length of gene frequencies for selection coefficients of 0.1 and 0.5 are presented for several species in Table 12.2. The characteristic scale length of gene frequencies ($s = 0.1$) usually is less than 10 meters. When the coefficient of selection is increased to 0.5, characteristic scale lengths become less than 5 m for most species. Thus gene flow does not preclude adaptation to moderate size environmental patches.

Strong selective differentials may occur in closely adjacent habitats. Selection coefficients exceeding 0.5 have been described in *Agrostis tenuis* (Jain and Bradshaw, 1966; McNeilly, 1968), *Anthoxanthum odoratum* (Hickey and McNeilly, 1975; Davies and Snaydon, 1976); *Plantago major* (Warwick and Briggs, 1980), and *Dryas octopetala* (McGraw and Antonovics, 1983). Selection coefficients between 0.1 and 0.5 have been reported in several species, in relation to many environmental variables (Bradshaw, 1972; Levin, 1984). If the gene flow distance (l) in these species were similar to those discussed above, populations could readily respond to strong disruptive selection, because their characteristic scale lengths would be only a few meters.

Clinal variation

The movement of genes between differentiated subpopulations will produce a cline. Suppose that genotypes AA, Aa and aa have fitness of $1 + s$, 1 and

$1 - s$ respectively, in one habitat, and $1 - s$, 1, and $1 + s$, respectively, in the other. If the transition between habitat patches occurs over a distance less than that of the characteristic scale length, the width of a cline (p = 1 to p = 0) at equilibrium is l_c (May, Endler and McMurtie, 1975). Using 0.10 as the value of s, the width of clines in most species would be between 15 and 25 meters (Table 12.2). In species with very restricted dispersal, clines may be as little as 7 m wide. The width of clines diminishes as the selection intensity increases. With $s = 0.5$, species would have cline widths of less than 10 m.

Jain and Bradshaw (1966) simulated the joint effects of selection and gene flow via pollen in a linear habitat. The width of clines increased when they increased the level of pollen flow between breeding units. The pollen flow distribution also affected the width. When the level of pollen flow was held constant, clines generated by leptokurtic pollen dispersion were wider than those by normal dispersion. Near-neighbor pollination produced the narrowest clines.

The width of clines also depends on the vehicle of gene flow. Under high selection intensities, pollen flow will generate wider clines than by seed flow, because nearly all of the immigrants would be killed prior to reproduction (Antonovics, 1968). At low selection intensities, the opposite will be true.

The width of clines is best understood in populations experiencing disruptive selection for heavy metal tolerance. The boundary between the contaminated mine soil and the normal pasture soil is often very sharp, the intermediate zone being a meter wide. Metal tolerant and intolerant subpopulations are bridged by clines of 10 meters or less in *Anthoxanthum oderatum* and *Agrostis tenuis* (Jain and Bradshaw, 1966). Clines in the order of 10 m also link morphologically differentiated subpopulations of *Agrostis stolonifera* growing on cliffs and the adjacent pasture (Aston and Bradshaw, 1966) and subpopulations of *Potentilla erecta* growing in contrasting vegetation (Watson, 1969). Unfortunately, nothing is known about gene dispersion in these species.

McNeilly's (1968) study of *Agrostis tenuis* in and adjacent to a copper mine in North Wales demonstrates the effect of gene dispersion on cline width. Adult populations at the upwind end of the mine changed from predominantly tolerant to non-tolerant plants over a distance of 1 m at the mine:pasture interface. Populations at the downwind end of the mine changed from predominantly tolerant to non-tolerant plants over a distance of 100 m. Pasture plants in the upwind populations received much less pollen from mine plants than did downwind pasture plants.

Consider next clines along an environmental gradient. Suppose that one homozygote is best fit for one environment and the other homozygote the other, and that the fitness of homozygotes varied over a 100 m gradient. Also suppose that genotypes AA, Aa and aa have fitness of $1 + s$, 1 and

$1 - s$ respectively, at one end of the 100 m gradient, and $1 - s$, 1 and $1 + s$, respectively at the other end. The width of the cline (p = 0 to p = 1.0) is $2.40 \, (1_c^2 \, \Delta)^{1/3}$, where Δ is the transition distance over which both homozygote fitnesses are changing relative to one another (May, Endler and McMurtie, 1975). For the case where $s = 0.10$, the width for the species in question would be between 20 and 50 m (Table 12.2). This illustration shows that the cline width will be narrower than that of the ecotone, and that substantial differentiation may occur over moderate distance along an ecotone in the face of gene flow.

A prime example of clinal variation along an ecotone involves *Eucalyptus urnigera* on Mount Wellington, Tasmania. Below 800 m, the population is composed exclusively of green adults, and above 1000 m only of glaucous adults. The intervening cline occurs over 800 m in ground distance (Barber and Jackson, 1957; Barber, 1965). The cline in seedling frequencies is shallower than is the cline for adults. If the adult cline is in equilibrium, it would follow that glaucous juveniles were being eliminated at lower elevations and green juveniles at higher elevations. The selective advantage of glaucous leaves at higher altitudes is due to their water repellancy, which renders them more resistant to freezing than wettable green leaves (Thomas and Barber, 1974). The selective advantage of the alternate leaf morphs in their favored environments probably exceeds 0.50, because gene flow is quite extensive. Seeds and pollen may travel 100 m or more (Barber, 1965). The difference between leaf types is heritable; crossing studies suggested expression is controlled by a few major genes.

The presence of a cline may signify a selective differential which varies in space. However, it is not necessarily so. Clines may form in unselected characters, because of pleiotropy, linkage or developmental and functional correlations with a character which is being selected (Hedrick and Holden, 1979; Barton, 1979, 1983). Clinal variation in heavy metal tolerance in *Anthoxanthum odoratum* is accompanied by clinal variation in morphology, phenology and self-fertility (Antonovics and Bradshaw, 1970). From observation alone, it is impossible to establish which traits other than metal tolerance are being selected.

A cline may also arise from immigration via pollen, because it has the greatest impact on the population border facing the source (Jones and Brooks, 1950; Knowles and Ghosh, 1968; Green and Jones, 1953; Levin and Kerster, 1975). A cline will form regardless of the selective value of the character, as long as the gene frequencies in the donor population differ from those in the recipient. Immigration most likely explains the variation pattern in male-sterility genotypes in *Plantago lanceolata*. Van Damme and Graveland (1985) found a population in which one genotype was present only near the border, in some areas accounting for 50% of the plants. This genotype was common in nearby populations.

Complex variation patterns

Populations grow in and adapt to a fine-scale environmental mosaic. Two of the prime examples of microadaptation involve selfers. Nevo *et al.* (1986) recently described allozyme variation in *Hordeum spontaneum* in relation to six microniches in an oak forest. Each microniche, characterized by several edaphic variables, occurred several times in the overall mosaic pattern in which they were embedded. There was significant multilocus allozyme frequency – microniche covariation. Similar results were obtained in *Avena barbata* when allozyme frequencies were analyzed in relation to a hillside microniche mosaic (Hamrick and Allard, 1972; Allard *et al.*, 1972; Hamrick and Holden, 1979). Genotype–environment covariances in *Hordeum* and *Avena* ostensibly are the result of diversifying selection. The possibility of a spurious correlation is very low, because each type of microniche was examined at several positions in the complex environmental fabric. The target of selection remains to be determined. Hedrick and Holden (1979) have shown that the strong genotype–environment associations found in the hillside population in *Avena barbata* could be transient patterns generated by genetic hitchhiking of selectively neutral allozymes with alleles at other loci on which selection acts. The same may be said of the *Hordeum spontaneum* population, because in highly autogamous species allozymes may hitchhike with any locus in the genome (Hedrick, 1982). The outcrossing rates in the two species is about 2%.

The differentiation of populations occurs over time. Unfortunately, the *Hordeum* and *Avena* populations are unlikely to provide much insight into this process, because they have been in place for at least tens of generations, and probably are near selection–migration equilibria. Thus little change is expected from one generation to another within or among patches. Indeed, most natural populations growing in a stable patchy environment probably are near selection–migration equilibria.

Local differentiation is best studied in an ecological mosaic created within the past few decades. The Park Grass Experiment at Rothamsted is ideal for this purpose. Various fertilizer treatments have been imposed annually since 1856, and liming treatments have been imposed at four-year intervals since 1903. *Anthoxanthum odoratum* subpopulations of this species have diverged genetically in response to the soil treatments over a distance of 30 m in the face of gene flow by pollen and seed. The subpopulations differed in a number of morphological characteristics including plant height and posture, yield, seasonal pattern of growth, reproductive strategy, and disease susceptibility (Snaydon and Davies, 1972). The subpopulations also differed physiologically, in response to several soil factors (Davies and Snaydon, 1974). Moreover, when reciprocally transplanted, plants survived longer and grew faster on their native plot (Davies and Snaydon, 1976).

There is thus considerable evidence that genetic change was adaptive. Snaydon and Davies (1982) analyzed subpopulations of A. *odoratum* from plots which had received no lime for 100 years and from adjacent plots, which were limed six years previously but, had otherwise been treated similarly. Considerable genetic change had occurred for height, yield and disease resistance six years (for 2–3 generations) after the new treatment was imposed. The half-life of the species on the study plots was 1–3 years.

12.7   THE SPREAD OF ADVANTAGEOUS GENES

If an advantageous gene appears through mutation or migration, it will increase locally at the expense of the other genes, and then spread throughout the population. The velocity of the wave of increase of advantageous gene frequency is a positive function of the level of gene dispersion, as is the length of the wave, or the distance between designated gene frequencies (Fisher, 1937). The advance of advantageous genes or the velocity of the wave of gene increase per generation along a continuous linear population can be described as $v = \sigma(2s)^{1/2}$, where σ is the standard deviation of gene flow distance and s is the selective advantage estimated.

The velocity of this wave was estimated for several species from the data on gene dispersion presented earlier (Table 12.3). No dominance is assumed. If the mutant had a 10% advantage ($s = 10$), the wave would move between 0.25 m (*Avena barbata*) and 1.03 m (*Liatris cylindracea*) per generation. The mean rate across species was 0.68 m per generation. The rate of advance would be about 2.3 times greater if the advantageous gene

Table 12.3   Expected rate of advance of advantageous genes in a linear population.

Species	Rate (m/generation)	
	$s = 0.10$	$s = 0.50$
Phlox pilosa	0.81	1.81
Liatris cylindracea	1.03	2.29
Liatris aspera	0.75	1.66
Viola pedata	0.88	1.95
*Viola rostrata**	0.64	1.41
Primula veris	0.69	1.54
Plantago lanceolata	0.40	0.89
Avena barbata	0.25	0.56

*Estimates do not take into account selfing by cleistogamic flowers.

had a 50% advantage. Most species occur in two-dimensional populations rather than in the linear one considered thus far. The rate of advance is roughly 1.5 times greater in a two-dimensional habitat (Cavalli-Sforza and Bodmer, 1971).

Fisher's formulation deals with a stable continuum, and does not take into account the spatial dynamics of populations during their growth. The models of Slatkin (1976) and Ammerman and Cavalli-Sforza (1984) suggest that the expansion of patches, the founding of new patches, and migration between patches during the growth process would make the rate of spread greater than would be the case in a stable continuum.

All things considered, it appears that an advantageous gene is unlikely to spread more than a few meters per generation under the most favorable conditions. Thus the impact of a new advantageous gene probably will be limited in space unless a population is very long-lived or small.

Thus far I have referred to the radiation of a gene from a single point. If a given gene was introduced several times by immigration, it would radiate from several focal points. The more such points there were, the higher the rate of increase in gene frequency, and the greater the areal preeminence of the gene at any point in time. The rate of response to selection is dependent on the number of focal points, because when gene dispersion is over short distances, neighboring plants may share the same allele at the locus in question. Encounters between like genotypes are not selective events. The more restricted is gene dispersion, the greater will be the proportion of neighbor encounters that are between like genotypes, and the more dependent will be the response to selection on the number of focal points.

12.8 CONCLUSIONS

The nature and degree of genetic subdivision within populations depend on the balances among genetic drift, selection and migration. These balances may shift because of changes in the size, number and proximity of patches, as a population grows or as it undergoes periodic contraction and expansion. Therefore, the pattern fashioned over one time frame may be eroded and replaced by another. However, patterns are slow to change, so that which we see today may not reflect the present interactions among selection, genetic drift and migration.

The breeding structure of a population places constraints upon the magnitude of differentiation achieved by selection or genetic drift. The level of differentiation will be an inverse function of gene dispersion, all else being equal. Since all else is not equal, we should not expect a close association between neighborhood area and gene frequency heterogeneity in natural populations. Jain (1979) compared the range of pollen and seed dispersal, and the amount of genetic variation between patches in six grassland annuals of California. The species with the most intrapopulation differentiation had the broadest gene dispersion. The species with the narrowest gene dispersion had one of the lower levels of intrapopulation differentiation.

Although neighborhood area cannot be used for predicting the specific character of the gene frequency mosaic, it can provide estimates of the minimum distances over which differentiation can occur, and the degree of isolation by distance. The more restricted the dispersal of genes in space, the finer the scale of the genetic patchwork may be, and the less will be the exchange of genes between patches a given distance apart.

The neighborhood area (or amount of gene flow over a given distance) is in part an adaptation to the scale of environmental heterogeneity. It is determined by pollen and seed dispersal, both of which are adaptations for successful parenting and progeny placement, respectively, in a heterogeneous environment. Based on an analytical model of Levins (1964), I propose that the optimal neighborhood area (or amount of gene flow) depends on the slope of the environmental gradient; the steeper the slope the smaller the optimal area. Where the environment and gene frequency surface change rapidly, the seeds produced by a plant will be best fathered by another in the same environment, and thus one located nearby; and the progeny of a plant are best distributed near the seed parent. Thus we would expect species with small neighborhood areas to occur in habitats which would change substantially over short distances, whereas those with large neighborhood areas are expected in habitats which on average change little in space.

Populations with broad gene dispersion will be less able to adapt to small-scale environmental heterogeneity by genetic differentiation than species with narrow gene dispersion. Correlatively, the progeny of plants with narrow gene dispersion will be better adapted to their immediate environment than progeny of plants with broad dispersion. The highest level of localized adaptation would be achieved through self-fertilization and very local seed dispersal. The more widely pollen and seed are dispersed, the weaker the correlation between the microhabitats of parents and progeny, and the greater is the advantage of phenotypic plasticity as an alternate means of adaptation.

12.9 REFERENCES

Allard, R. W., Babbel, G. R., Clegg, M. T. and Kahler, A. L. (1972) Evidence for coadaptation in *Avena barbata*. *Proc. Natl. Acad. Sci. USA*, **69**, 3043–8.

Ammerman, A. J. and Cavalli-Sforza, L. L. (1984) *The Neolithic Transition and the Genetics of Populations in Europe*. Princeton University Press, Princton, N.J.

•Antonovics, J. (1968) Evolution in closely adjacent populations. VI. Manifold effects of gene flow. *Heredity*, **23**, 507–27.

Antonovics, J. and Bradshaw, A. D. (1970) Evolution in closely adjacent populations. VII. Clinal patterns at a mine boundary. *Heredity*, **23**, 507–24.

324   *Local differentiation of populations*

Aston, J. L. and Bradshaw, A. D. (1966) Evolution in closely adjacent plant populations. II. *Agrostis stolonifera* in maritime habitats. *Heredity*, 21, 649–64.

Bannister, M. H. (1965) Variation in the breeding system of *Pinus radiata*. in *The Genetics of Colonizing Species* (eds H. G. Baker and G. L. Stebbins), Academic Press, New York, pp. 353–72.

Barber, H. N. (1965) Selection in natural populations. *Heredity*, 20, 551–72.

Barber, H. N. and Jackson, W. D. (1957) Natural selection in action in *Eucalyptus*. *Nature*, 179, 1267–9.

Barton, N. H. (1979) Gene flow past a cline. *Heredity*, 43, 333–9.

Barton, N. H. (1983) Multilocus clines. *Evolution*, 37, 454–71.

Beattie, A. J. and Culver, D. C. (1979) Neighborhood size in *Viola*. *Evolution*, 33, 1226–9.

Bos, M., Harmens, H. and Vrieling, K. (1986) Gene flow in *Plantago* 1. Gene flow and neighborhood size in *P. lanceolata*. *Heredity*, 56, 43–54.

Bradshaw, A. D. (1972) Some of the evolutionary consequences of being a plant. *Evol. Biol.*, 5, 25–47.

Bradshaw, A. D. (1984) The importance of evolutionary ideas in ecology – and vice versa. in *Evolutionary Ecology* (ed. B. Shorrocks), Blackwell, London, pp. 1–25.

Brereton, A. J. (1971) The structure of species populations in the initial stages of salt-marsh succession. *J. Ecol.*, 59, 321–38.

Brown, A. D. H. and Marshall, D. R. (1981) Evolutionary changes accompanying colonization in plants. *Evolution Today, Proc. 2nd Int. Congr. Syst. Evol. Biol.* Vancouver, BC, pp. 351–63.

Brunel, D. and Rodolphe, F. (1985) Genetic neighborhood structure in a population of *Picea abies*. *Theor. Appl. Genet.*, 71, 101–10.

Cavalli-Sforza, L. L. and Bodmer, W. F. (1971) *The Genetics of Human Populations*. Freeman, San Francisco.

Chakraborty, R. and Nei, M. (1977) Bottleneck effects on average heterozygosity and genetic distance with stepwise mutation model. *Evolution*, 31, 347–56.

Clegg, M. T. and Allard, R. W. (1972) Patterns of genetic differentiation in the slender wild oat species, *Avena barbata*. *Proc. Natl. Acad. Sci. USA*, 69, 1820–4.

Crawford, T. J. (1984a) The estimation of neighborhood parameters in plant populations. *Heredity*, 52, 272–83.

Crawford, T. J. (1984b) What is a population? in *Evolutionary Ecology* (ed. B. Shorrocks), Blackwell, Oxford, pp. 135–73.

Crow, J. F. and Aoki, K. (1984) Group selection for a polygenic behavior trait: estimating the degree of population subdivision. *Proc. Natl. Acad. Sci. USA*, 81, 6073–7.

Davies, M. S. and Snaydon, R. W. (1973) Physiological differences among populations of *Anthoxanthum odoratum* L. collected from the Park Grass Experiment, Rothamsted. I. Response to calcium, and II. Response to aluminum. *J. Appl. Ecol.*, 10, 33–55.

Davies, M. S. and Snaydon, R. W. (1974) Physiological differences among populations *Anthoxanthum odoratum* L. collected from the Park Grass Experiment, Rothamsted. III. Response to phosphate. *J. Appl. Ecol.*, 11, 699–708.

Davies, M. S. and Snaydon, R. W. (1976) Rapid population differentiation in a mosaic environment. III. Measures of selection pressures. *Heredity*, 36, 59–66.

Endler, J. A. (1977) *Geographic Variation, Speciation and Clines*. Princeton University Press, Princeton, NJ.

Ennos, R. A. (1984) Maintenance of genetic variation in plant populations. *Evol. Biol.*, **15**, 129–55.

Epling, C. and Dobzhansky, T. (1942) Genetics of natural populations. VI. Microgeographical races in *Linanthus parryae*. *Genetics*, **27**, 317–32.

Epling, C., Lewis, H. and Ball, F. M. (1960) The breeding group and seed storage: A study in population dynamics. *Evolution*, **14**, 238–55.

Falconer, D. S. (1981) *Introduction to Quantitative Genetics*. 2nd edn, Longman, New York.

Fisher, R. A. (1937) The wave of advance of advantageous genes. *Ann. Eugenics*, **7**, 355–69.

Fisher, R. A. (1950) Gene frequencies in a cline determined by selection and diffusion. *Biometrics*, **6**, 353–61.

Gottlieb, L. D. (1974) Genetic stability in a peripheral isolate of *Stephanomeria exigua* ssp. *coronaria* that fluctuates in population size. *Genetics*, **76**, 551–6.

Green, J. M. and Jones, M. D. (1953) Isolation of cotton for seed increase. *Agron. J.*, **45**, 366–8.

Grieg-Smith, P. (1964) *Quantitative Plant Ecology* 2nd edn, Butterworth, London.

Grime, J. P. (1979) *Plant Strategies and Vegetation Processes*. Wiley, New York.

Grime, J. P. and Hunt, R. (1975) Relative growth rate: its range and adaptive significance in a local flora. *J. Ecol.*, **63**, 393–422.

Grubb, P. J. (1977) The maintenance of species richness in plant communities: the importance of the regeneration niche. *Biol. Rev.*, **52**, 107–45.

Hamrick, J. L. and Allard, R. W. (1972) Microgeographical variation in allozyme frequencies in *Avena barbata*. *Proc. Natl. Acad. Sci. USA*, **69**, 2100–4.

Hamrick, J. L. and Holden, L. R. (1979) Influence of microhabitat heterogeneity on gene frequency distribution and gametic phase disequilibrium in *Avena barbata*. *Evolution*, **33**, 521–33.

Harding, J. A. and Barnes, K. (1977) Genetics of *Lupinus*. X. Genetic variability, heterozygosity and outcrossing in colonial populations of *Lupinus succulentus*. *Evolution*, **31**, 247–55.

Hedrick, P. W. (1982) Genetic hitchhiking: A new factor in evolution. *BioScience*, **32**, 845–53.

Hedrick, P. W. and Holden, L. (1979) Hitch-hiking: an alternative to coadaptation for barley and slender wild oat examples. *Heredity*, **43**, 79–86.

Heywood, J. S. (1986) The effect of plant size variation on genetic drift in populations of annuals. *Am. Natur.*, **127**, 851–61.

Hickey, D. A. and McNeilly, T. (1975) Competition between metal tolerant and normal plant populations: a field experiment on normal soil. *Evolution*, **29**, 458–64.

Jain, S. K. (1975) Population structure and the effects of breeding system. in *Plant Genetic Resources: Today and Tomorrow* (eds O. Frankel and J. G. Hawkes), Cambridge University Press, Cambridge, pp. 15–36.

Jain, S. K. (1976) Patterns of survival and microevolution in plant populations. in *Population Genetics and Ecology* (eds S. Karlin and E. Nevo), Academic Press, New York, pp. 49–75.

Jain, S. K. (1979) Adaptive strategies: polymorphism, plasticity and homeostasis. in *Topics in Plant Population Biology* (eds O. T. Solbrig, S. K. Jain, G. B. Johnson and P. H. Raven), Cornell University Press, New York, pp. 160–87.

Jain, S. K. and Bradshaw, A. D. (1966) Evolutionary divergence among adjacent plant populations. I. Evidence and its theoretical analysis. *Heredity*, 21, 407–41.

Jain, S. K. and Martins, P. S. (1979) Ecological genetics of the colonizing ability of rose clover (*Trifolium hirtum* All.) *Am. J. Bot.*, 66, 361–6.

Jones, M. D. and Brooks, J. S. (1950) Effectiveness of distance and border rows in preventing outcrossing in corn. *Oklahoma Agric. Expt. Sta. Tech. Bull.* T-38.

Karlin, S. and McGregor, J. (1968) The role of the Poisson progeny distribution in population genetic models. *Math. BioSci.*, 2, 11–17.

Kimura, M. and Crow, J. F. (1963) The measurement of effective population number. *Evolution*, 17, 279–88.

Kimura, M. and Ohta, T. (1971) *Theoretical Aspects of Population Genetics.* Princeton University Press, Princeton, New Jersey.

Knowles, R. P. and Ghosh, A. W. (1968) Isolation requirements for smooth bromegrass, *Bromus inermis* Leyss. *Crop Sci.*, 3, 371–4.

Lefebvre, C. (1985) Morphological variation, breeding system and demography at populational and subpopulational levels in *Armeria maritima* (Mill.) Willd. in *Genetic Differentiation and Dispersal in Plants* (eds P. Jacquard, G. Heim, and J. Antonovics), Springer-Verlag, New York, pp. 129–39.

Leith, H. (1960) Patterns of change within grassland communities. in *The Biology of Weeds* (ed. J. L. Harper), Blackwell, Oxford.

Levin, D. A. (1978) Some genetic consequences of being a plant, in *The Interface of Genetics and Ecology* (ed. P. Brussard) Academic Press, New York, pp. 189–219.

Levin, D. A. (1981) Dispersal versus gene flow in plants. *Ann. Missouri Bot. Gard.*, 68, 233–53.

Levin, D. A. (1984) Immigration in plants: an exercise in the subjunctive. in *Perspectives on Plant Population Ecology* (eds. R. Dirzo and J. Sarukhan), Sinauer, Sunderland, MA, pp. 242–60.

Levin, D. A. and Kerster, H. W. (1968) Local gene dispersal in *Phlox. Evolution*, 22, 130–9.

Levin, D. A. and Kerster, H. W. (1969) Density-dependent gene dispersal in *Liatris. Am. Natur.*, 103, 61–74.

Levin, D. A. and Kerster, H. W. (1974) Gene flow in seed plants. *Evol. Biol.*, 7, 139–220.

Levin, D. A. and Kerster, H. W. (1975) The effect of gene dispersal on the dynamics and statistics of gene substitution in plants. *Heredity*, 35, 317–36.

Levin, D. A. and Wilson, J. B. (1978) The genetic implications of ecological adaptations in plants. in *Structure and Functioning of Plant Populations* (eds A. H. J. Freysen and J. W. Woldendorp), North Holland, New York, pp. 75–98.

Levins, R. (1964) The theory of fitness in a heterogeneous environment. IV. The adaptive significance of gene flow. *Evolution*, 18, 635–8.

Linhart, Y. B., Mitton, J. B., Sturgeon, K. B. and Davis, M. L. (1981) Genetic variation in time and space in a population of Ponderosa pine. *Heredity*, 46, 407–26.

McGraw, J. B. and Antonovics, J. (1983) Experimental ecology of *Dryas octopetala* ecotypes. I. Ecotypic differentiation and life cycle stages of selection. *J. Ecol.*, **71**, 879–97.

McNeilly, T. (1968) Evolution in closely adjacent plant populations. III. *Agrostis tenuis* on a small copper mine. *Heredity*, **23**, 99–108.

Maruyama, T. and Kimura, M. (1980) Genetic variability and effective population size when local extinction and recolonization are frequent. *Proc. Natl. Acad. Sci. USA*, **77**, 6710–14.

May, R. M., Endler, J. A. and McMurtie, R. E. (1975) Gene frequency clines in the presence of selection opposed by genetic drift. *Am. Natur.*, **109**, 659–76.

Motro, U. and Thompson, G. (1982) On heterozygosity and the effective size of populations subject to size changes. *Evolution*, **36**, 1059–66.

Nei, M., Maruyama, T. and Chakraborty, R. (1975) The bottleneck effect and genetic variability in populations. *Evolution*, **29**, 1–10.

Nevo, E., Beiles, A., Kaplan, D., Golenberg, E. M., Olsvig-Whittaker, L. and Naveh, Z. (1986) Natural selection of allozyme polymorphisms: A microsite test revealing ecological differentiation in wild barley. *Evolution*, **40**, 13–20.

Park, Y. S. and Fowler, D. P. (1982) Effects of inbreeding and genetic variances in a natural population of Tamarack (*Larix laricina* (Du Roi) K. Koch) in eastern Canada. *Silvae Genet.*, **31**, 21–6.

Platt, W. J. (1975) The colonization and formation of equilibrium plant associations on badger disturbances in a tall-grass prairie. *Ecol. Monogr.*, **45**, 285–305.

Rai, K. N. and Jain, S. K. (1982) Population biology of *Avena*. IX. Gene flow and neighborhood size in relation to microgeographic variation in *Avena barbata*. *Oecologia*, **53**, 399–405.

Richards, A. J. and Ibrahim, H. (1978) The estimation of neighborhood size in two populations of *Primula veris*. in *The Pollination of Flower by Insects* (ed. A. J. Richards), Academic Press, New York, pp. 165–74.

Rice, K. and Jain, S. (1985) Plant populations genetics and evolution in disturbed environments. in *Natural Disturbance and Patch Dynamics* (eds S. T. A. Pickett and P. S. White), Academic Press, New York, pp. 287–303.

Rohlf, F. J. and Schnell, G. D. (1971) An investigation of the isolation by distance model. *Am. Natur.*, **105**, 295–324.

Schaal, B. A. (1975) Population structure and local differentiation in *Liatris cylindracea. Am. Natur.*, **109**, 511–28.

Schaal, B. A. and Levin, D. A. (1978) Morphological differentiation and neighborhood size in *Liatris cylindracea*. *Am. J. Bot.*, **65**, 923–8.

Schemske, D. W. (1984) Population structure and local selection in *Impatiens pallida* (Balsaminaceae), a selfing annual. *Evolution*, **38**, 817–32.

Slatkin, M. (1973) Gene flow and selection in a cline. *Genetics*, **75**, 733–56.

Slatkin, M. (1976) The rate of spread of an advantageous allele in a subdivided population. in *Population Genetics and Ecology* (eds S. Karlin and E. Nevo), Academic Press, New York, pp. 767–79.

Snaydon, R. W. (1980) Plant demography in agricultural systems. in *Demography and Evolution in Plant Populations* (ed. O. T. Solbrig), Blackwell, London, pp. 131–60.

Snaydon, R. W. and Davies, M. S. (1972) Rapid population differentiation in a mosaic environment. II. Morphological variation in *Anthoxanthum odoratum*. *Evolution*, **26**, 390–405.

Snaydon, R. W. and Davies, M. S. (1976) Rapid population differentiation in a mosaic environment. IV. Populations of *Anthoxanthum odoratum* at sharp boundaries. *Heredity*, **37**, 9–25.

Snaydon, R. W. and Davies, M. S. (1982) Rapid divergence of plant populations in response to recent changes in soil conditions. *Evolution*, **36**, 289–97.

Sokal, R. and Wartenberg, D. E. (1983) A test of spatial autocorrelation analysis using an isolation-by-distance model. *Genetics*, **105**, 219–37.

Solbrig, O. T. (ed.) (1980) *Demography and Evolution in Plant Populations*. Blackwell Scientific, London.

Templeton, A. R. and Levin, D. A. (1979) Evolutionary consequences of seed pools. *Am. Natur.*, **114**, 1–22.

Thomas, D. A. and Barber, H. N. (1974) Studies on leaf characteristics of a cline of *Eucalyptus urnigera* from Mount Wellington, Tasmania. I. Water repellency and the freezing of leaves. *Aust. J. Bot.*, **22**, 501–12.

Thompson, J. N. (1978) Within-patch structure and dynamics in *Pastinaca sativa* and resource availability to a specialized herbivore. *Ecology*, **59**, 443–8.

Tigerstedt, P. M. A., Rudin, D., Niemela, T. and Tammisola, J. (1982) Competition and neighboring effect in a naturally regenerating population of Scots pine. *Silvae Fenn.*, **16**, 122–9.

Turkington, R. and Harper, J. L. (1979) The growth distribution and neighbour relationships of *Trifolium repens* in a permanent pasture. I. Ordination, pattern and contact. *J. Ecol.*, **67**, 201–18.

Turner, M. E., Stephens, J. C. and Anderson, W. W. (1982) Homozygosity and patch structure in plant populations as a result of nearest-neighbor pollination. *Proc. Natl. Acad. Sci. USA*, **79**, 203–7.

Van Damme, J. M. M. and Graveland, J. (1985) Does cytoplasmic variation in *Plantago lanceolata* contribute to ecological differentiation? in *Genetic Differentiation and Dispersal in Plants* (eds P. Jacquard, G. Heim, and J. Antonovics), Springer-Verlag, New York, pp. 67–79.

van Dijk, H. (1985) The estimation of gene flow parameters from a continuous population structure. In *Genetic Differentiation and Dispersal in Plants* (eds P. Jacquard, G. Heim and J. Antonovics), Springer Verlag, New York, pp. 311–25.

Venable, D. L. and Levin, D. A. (1983) Morphological dispersal structures in relation to growth form in the Compositae. *Plant Syst. Evol.*, **143**, 1–16.

Wade, M. J. (1978) A critical review of the models of group selection. *Quart. Rev. Biol.*, **53**, 101–14.

Wade, M. J. (1980) An experimental study of kin selection. *Evolution*, **34**, 844–55.

Warwick, S. I. and Briggs, D. (1980) The genecology of lawn weeds. VI. The adaptive significance of variation in *Achillea millefolium* as investigated by transplant experiments. *New Phytol.*, **85**, 451–60.

Watson, P. (1969) Evolution in closely adjacent plant populations. VI. An entomophilous species, *Potentilla erecta*, in two contrasting environments. *Heredity*, **24**, 407–22.

Wilson, D. S. (1979) *The Natural Selection of Populations and Communities*. Benjamin Cummins, Menlo Park.

Wright, S. (1931) Evolution in Mendelian populations. *Genetics*, **16**, 97–159.

Wright, S. (1938) Size of population and breeding structure in relation to evolution. *Science*, **87**, 430–1.

Wright, S. (1943a) Isolation by distance. *Genetics*, **28**, 114–28.

Wright, S. (1943b) An analysis of local variability of flower color in *Linanthus parryae*. *Genetics*, **28**, 139–56.

Wright, S. (1946) Isolation by distance under diverse systems of mating. *Genetics*, **31**, 39–59.

Wright, S. (1949a) Adaptation and selection. in *Genetics, Paleontology and Evolution* (eds G. L. Jepsen, E. Mayr and G. G. Simpson), Princeton University Press, Princeton, NJ, pp. 365–89.

Wright, S. (1949b) Population structure in evolution. *Proc. Am. Phil. Soc.*, **93**, 471–8.

Wright, S. (1965) The interpretation of population structure by F-statistics with regard to systems of mating. *Evolution*, **19**, 395–420.

Wright, S. (1969) *Evolution and Genetics of Populations*. Vol. 2. University of Chicago Press, Chicago.

Wright, S. (1978) *Evolution and the Genetics of Populations*. Vol. 4. University of Chicago.

Wright, S. (1980) Genic and organismic selection. *Evolution*, **34**, 825–43.

Yarranton, G. A. and Morrison, R. G. (1974) Spatial dynamics of a primary succession: Nucleation. *J. Ecol.*, **62**, 417–28.

Editors' commentary on Part 5

Asexuality and breeding systems occupy the center stage in the theater of plant evolutionary research. Darwin (1876) discussed the great diversity of outbreeding and inbreeding systems in plants primarily in terms of the role of hybrid vigor in selection for devices such as self-sterility and dioecy and of reproductive assurance under stress environments favoring self-fertilization. Soon after the rediscovery of Mendel's work, the role of genetic recombination in producing new gene combinations (new phenotypes) under outbreeding added an important dimension to the Darwinian views. The development of breeding methods for crop improvement, combined with the rapid advances in cytogenetics and population/biometrical genetics, led C. D. Darlington, K. Mather and R. A. Fisher, among others, to propose the concept of genetic systems, in which longevity, population size, outcrossing rate, chromosome number and various other features of a species' genetic makeup control the rates of release of new variation due to recombination (Grant, 1981). Accordingly, asexuality and self-fertilization having much lowered rates of recombination were labelled as closed, and outbreeding as open systems respectively; their adaptive features are assumed to depend upon the levels and turnover rate of genetic variability in their populations (Stebbins, 1958). Numerous studies describing the genetic structure of plant populations (i.e. variation within versus between populations), are based on pooled data on polymorphic loci and quantitative traits. Moreover, following Darwinian views as well as numerous biosystematic studies dealing with speciation mechanisms, a consensus developed that inbreeders are descended from outbreeding relatives under conditions of colonization or ecological specialization. Stebbins provided numerous examples from his pioneering work in *Elymus*, *Dactylis*, *Bromus*, and other grass genera in addition to his classic synthesis of numerous studies in Asteraceae, and researches continuing over a 50 year period! Most such studies, however, did not identify the specificity of individual traits that might be critically responsible in evolving adaptation to particular environmental factors. Genetic analyses of specific phenotypic features as well as recombination mechanisms were largely ignored.

It is evident from the Hardy–Weinberg theorem and numerous other theoretical foundations of population genetics that mating systems determine the basic rules of gene transmission. Therefore, genes modifying mating systems and thereby their evolutionary dynamics are of particular interest. Also, many aspects of reproductive ecology focus on the resource allocation and seed output which covary with the mating patterns. For example, cleistogamy and gynodioecy contribute to the relative rates of selfing and outcrossing, and also allow for variable allocation to male and female functions to maximize seed number and quality per individual. Evidence for the evolution of mating systems involves a four-part argument: (a) variation in mating system results in certain patterns of variation within and between populations due to differing levels of heterozygosity and genetic recombination; (b) natural selection sorts out individuals with higher Darwinian fitness depending on these variation components within populations (or some spatial subunit); (c) a part of this selective process involves maximization of reproductive success, given environmental heterogeneity that either induces or requires changes in mating patterns (Lloyd and Bawa, 1984); and finally, (d) these selective changes alter genetic makeup of populations through heritable changes in the mating system *per se*. Karlin and Lessard (1986) elegantly show the critical role of specific genetic models of sex determination (autosomal, sex-linked, extranuclear; one or many loci; multiple allelism) in evolutionary outcomes. Numerous details of population structure, pollination biology and seed set process are included to specify various experimental approaches; however, we already have an excellent start toward research on these four issues in the cases of sex-ratio variation, heterostyly, gynodioecy and cleistogamy. Note that selection under (b) and (c) might rarely be treated separately so that ecological and genetic models for the evolution of sex-determination and various mating systems are totally interdependent.

In this section chapters by Antonovics, Ellstrand and Brandon, by Levin, and by Clegg and Epperson that bring out the diversity of approaches are juxtaposed. In Chapter 11 Antonovics *et al.* discuss basic tests of models describing evolutionary advantages of sex; this chapter along with earlier writings of Antonovics emphasizes the importance of measuring selective forces using ecologically relevant phenotypic and environmental descriptors. In Chapter 12 Levin deals with the genetic and demographic factors affecting population structure and gene flow, both of which are an integral part of the reproductive system as well as models of microevolution. Thus, a reader is introduced to the ecological genetic literature in which demography, local versus global population structure, environmental heterogeneity and adaptive responses by individuals or populations are linked with the tools of population genetics. As a result, new empirical facts and new models are constantly demanded by their mutual feedback. This is elegantly shown in Chapter 10 by Clegg and Epperson who extend the mixed selfing and random mating model to incorporate unequal contributions of pollen parents due to pollinator preference and/or gametic selection, as suggested by their own observations in morning glory populations. Their work clearly establishes the utility of genetic markers (major gene polymorphisms) in the precise estimation of mating system and population substructure parameters.

There are many variations on the themes of reproductive assurance and colonizing success favoring selfers under certain restrictive environments. All mathematical or word-models require more rigorous population genetic thinking as well as more empirical work on the genetic and ecological aspects of variation. The chapters by Antonovics *et al.* and by Clegg and Epperson provide several interesting comparisons among the experimental approaches in ecological genetics. Clegg and Epperson show how measurements of outcrossing rates using appropriate genetic marker stocks provide a critical test of not only the role of higher selfing rate for the white flower class, but the role of population substructure and pollinator preference. Levin makes theoretical predictions about the scale of clinal variation, assuming certain values of migration and selection coefficients. Population genetic theory helps one advance both the formal treatment of gene frequency changes and search for the biological nature of selection. This is important to recognize in order to dispel skepticism about the merits of such theoretical underpinning in evolutionary research.

This is also evident in the chapter by Antonovics *et al.* who discuss their experimental approaches in which sexual and asexual progenies of individuals sampled from a population of *Anthoxanthum odoratum* are compared for relative survivorship and growth under a series of ingeneous field experiment designs. Environmental variables ranged from the relative spacing distance or neighborhood relations of sibs within the species diversity of a community. The role of sex is tested in relation to three propositions on the nature of selective process: (1) heterozygote advantage, where increased heterozygosity levels under outbreeding system is favored for one or more loci controlling a particular component of morphology or phenology; (2) sib-competition favoring new gene combinations, as related individuals from among the progenies of individual mothers remain dispersed locally in sibships (demes or neighborhoods) but continuously changing environmental conditions favor different genotypes than the parental ones; and (3) frequency-dependent advantage of rare genotypes, selected in a local context for such ecological reasons as escape from a disease or herbivore (Dirzo and Sarukhan, 1984). The results emphasize various difficult questions on the measure of a plant's response and on the interpretation of its adaptive nature. For example, one immediately finds a colonist's dilemma in terms of high dispersal to new uncertain environments versus failure to experiment in new areas due to poor dispersal.

Further, we should note several problems characteristic of ecological research. First, both genetic and non-genetic sources of variation often yield a wide range of values for each parameter of mating system or gene flow, as noted especially by Levin; moreover, different kinds of species and their habitat types might require divergent scales of experimental design. How do we simplify such data sets into a few values for testing a theory or in comparing related taxa? Second, each reductionist step in the experimental approaches might provide a statistical test of a subhypothesis but holistic aggregate of such tests might not permit 'strong inference' approach, especially since the interactions are necessarily ignored in such a procedure (cf. Levins and Lewontin, 1985). Third, most ecological experiments frequently involve very long-term observations.

All three points are illustrated in our studies on the origin of highly self-pollinated *Limnanthes floccosa* from its outcrossing relative *L. alba* through a series of colonizing events in xeric/marginal sites lacking insect pollinators (Jain, 1984; and unpublished data). Estimates of outcrossing rates vary widely both within and between populations of *L. alba*; measurements of inbreeding depression and self-fertility also gave a wide range of estimates, and so did relative colonizing success in artificially founded colonies observed over a six-year period. Quantitative genetic variation in founded colonies suggested varying amounts of variance reduction due to random drift. We still do not have a clear determination of the role of inbreeding in colonizing ability, or any simple paradigm for incorporating the observed variability levels in each of these parameters. Further, parallel long-term studies are needed to verify the role of specific genetic consequence of modified breeding systems during the early colonizing phases of *L. floccosa*. It must be recognized that population biology of different species researched under differing circumstances often provide some comparative likelihood scenarios in favor of one particular model or a mix of several models. This is in contrast to the use of molecular methods, for example, in phylogenetic hypothesis-testing (as discussed in Part I) which might frequently provide clearcut answers.

The primary question is whether and how do outbreeding species differ from the asexuals or inbreeders in their adaptability to changing environments. Most population genetic models in this area assume: (a) localized dispersal of pollen and seed; (b) heterogeneity of abiotic environment in terms of various nutrients, moisture, etc. and/or biotic factors such as pathogen or herbivore densities, some of which fluctuate over different generations; (c) role of linked gene systems, such that sex and recombination produce a low frequency of novel genotypes regularly; (d) selection enhancing local adaptedness temporarily, but this is constantly dependent upon the higher fitness of newly arising, hitchhiked gene combinations and various stochastic processes; and (e) in turn these forces somehow favoring individuals with increased sexual reproduction. Maynard Smith (1978) provided an excellent review of theory emphasizing frequency-dependent nature of selection and the role of multilocus variation.

These models might be inadequate in defining the process of change toward sexual reproduction in specific genetic terms, or in describing the demographic sequence of adaptive events. Many ecologists might also find the definition of environmental parameters too superficial, but it must be recognized that models are caricatures of reality, and that different models enhance precision at the cost of generality. Population genetic models help us develop a paradigm step-by-step for answering the 'big' questions on the causes of evolution of sex, dioecy, selfing, or gender allocation. It follows naturally that the measures of fitness be prescribed clearly in relation to the genetic analyses of variation in specific traits, and that phenotypic evolution be interpreted more discretely in terms of certain selective forces generated by the environmental changes. One would, of course, quickly recognize the enormous challenge in carrying out the needed large variety of experiments with appropriate controls of genetic and ecological variables separately and jointly and the difficult nature of scientific inference based on such studies.

Briefly, the research design has the following parts: (a) Genetic variation for factors affecting the mating systems *per se* and for the traits of form and function that would be related to the variance in fitness. Major gene polymorphisms and the so-called polygenic traits are neither independent in their heredity nor should they have to be independent in any evolutionary change. In fact, recent developments in the theory of multilocus organiza- tion of variation, and availability of new methods for extensive genome mapping suggest a joint use of simply inherited and biometrical traits. (b) Spatial and temporal deme structure, in which dispersability, progeny size, consanguineous matings and microhabitat factors play an important role. (c) Environmental heterogeneity, which in particular relates to the nature and intensity of selection. Items (b) and (c) provide the demographic and physiological sides of the triad with (a) genetics as the starting point in the source of variation and (b)–(c) together describing the parameters of selection, migration and random drift. Measurement of selective forces in the wild clearly involves some of the most difficult conceptual and methodological problems about the definitions of relative fitnesses and whether or how various fitness components are to be aggregated (Endler, 1986). Genetic research, on the other hand, has already been initiated in several related pairs of outbreeder–selfer species. Wyatt, and Rick (Chap- ters 5 and 6) discuss the evolution of selfers under the unfavorable condi- tions of pollinator paucity, and xeric or otherwise stress conditions and under the circumstances requiring high reproductive efficiency (viz. domes- tication and colonization of islands or new habitats). As noted earlier, research on two *Limnanthes* species has involved initial description of their ecology and genetic structures, and crosses to study the inheritance of both mating system and life-history features (Jain, 1984; unpubl. data). A large series of marked genetic stocks (population dynamics of major gene polymorphisms used for ecologically significant traits as well as markers) are being used for founding new populations in nature. Monitoring of natural and founded populations over several years will record the relative success of various genotypes. Many other details are not yet incorporated such as local deme structure and the role of seed dispersers or predators.

Research on mating systems and asexual reproduction relies on theories which provide a rigorous statement of assumptions about the genetic and demographic properties of a species, and develops equations for the evolutionary processes. Predictions about adaptive outcomes are made about the role of parameters describing sexuality, selfing rate, seed dispersability, seed dormancy, flower color polymorphism, root-shoot allocation, generation length, and so on. Traits affecting mating systems involve both major genes and many modifiers which affect rates of recombination in their linked gene systems and, therefore, we need to have well-mapped genomes for population genetic research in this area.

Many studies have also begun to deal with population structure in relation to interdeme versus intrademe selection. For example, a series of models has shown how inbreeding within demes on one hand and certain patterns of limited migration among demes on the other would allow a mixed selfing and random mating system to evolve (Uyenoyama, 1986; Holsinger, 1986), whereas a single population model might predict either complete selfing or panmixia to be the alternative evolutionary outcomes. Both theory and intuition tell us that reproductive assurance of asexuals and inbreeders has to be extended to include the fate of progeny after dispersal under varying environments. We also know now the importance of more accurate description of the breeding system itself. Both heterozygote advantage and frequency-dependent selection are also the subject of numerous field experiments. Here, one can also modify genetic systems through the use of specific mutants or naturally occurring variants and follow the ecological consequences using certain marker loci and other measures of genetic variation.

REFERENCES

Charlesworth, B. and Charlesworth, D. (1979) The evolutionary genetics of sexual systems in flowering plants. *Proc. R. Soc. London B.*, **205**, 513–30.
Darwin, C. (1876) *The Effects of Cross- and Self-fertilization in the Vegetable Kingdom.* Murray, London.
Dirzo, R. and Sarukhan, J. (eds) (1984) *Perspectives on Plant Population Ecology.* Sinauer, Sunderland.
Endler, J. A. (1986) *Natural Selection in the Wild.* Princeton University Press, Princeton.

Ennos, R. (1983) Maintenance of genetic variation in plant populations. *Evol. Biol.*, **16**, 129–55.

Grant, V. (1981) *Plant Speciation*, 2nd edn. Columbia University Press, New York.

Holsinger, K. E. (1986) Dispersal and plant mating systems: The evolution of self-fertilization in subdivided populations. *Evolution*, **40**, 405–13.

Jain, S. K. (1984) Breeding systems and the dynamics of plant populations. in *Genetics: New Frontiers* IV, (eds V. L. Chopra, B. C. Joshi, R. P. Sharma and H. C. Bansal), Oxford & IBH, New Delhi, pp. 291–316.

Jain, S. K. Evolution of inbreeding in genus *Limnanthes*. (in preparation)

Karlin, S. and Lessard, S. (1986) *Theoretical Studies on Sex Ratio Evolution*. Princeton University Press, Princeton.

Levins, R. and Lewontin, R. C. (1985) *The Dialectical Biologist*. Harvard University Press, Harvard.

Lloyd, D. G. and Bawa, K. S. (1984) Modification of the gender of seed plants in varying conditions. *Evol. Biol.*, **17**, 255–338.

Maynard Smith, J. (1978) *The Evolution of Sex*. Cambridge University Press, Cambridge.

Stebbins, G. L. (1958) Longevity, habitat, and release of genetic variability in the higher plants. *Cold Spring Harbor Symp. Quant. Biol.*, **23**, 365–78.

Uyenoyama, M. K. (1986) Inbreeding and the cost of meiosis: The evolution of selfing in populations practicing biparental inbreeding. *Evolution*, **40**, 388–404.

PART 6

Life histories in a community context

CHAPTER 13

Vegetational mosaics, plant–animal interactions and resources for plant growth

R. L. JEFFERIES

13.1 INTRODUCTION

Herbivores shape both the structure and dynamics of terrestrial and aquatic communities, a topic of considerable interest to Stebbins (1981). They act as agents of natural selection in the evolution of life histories of plants. Species richness, their abundance and the physical environment are all affected by herbivore activity (Harper, 1977; Rosenthal and Janzen, 1979; Crawley, 1983). Frequently, spatial and temporal heterogeneity in vegetation is a direct consequence of foraging and disturbance by animals (Paine and Levin 1981; Sousa, 1984; Pickett and White, 1985). As will be indicated later, a herbivore may act as a predator (removal of plants or seeds), a parasite (partial reduction of plant biomass), or it may promote some form of mutualistic association with the same plant species in a community, depending on seasonal conditions (Crawley, 1983).

Plant–herbivore interactions cannot be considered in isolation from the ecosystem in which they occur. They affect primary and secondary production, energy flow and nutrient cycling (Mooney and Godron, 1983; Sprugel, 1984; Vitousek, 1984). An operational model used to describe the direct and indirect effects of the interactions is based on a series of feedback loops in which recurrent disturbances brought about by a herbivore produce shifts in the availability of resources and in the size and frequency of patches which may be recolonized. Together with the effects of removal of plant tissues on the growth of plants, the alterations in the availability of resources may result in short- and long-term demographic and genetic changes to populations and the coevolution of forage species and herbivores.

The processes that produce these different responses of individuals, populations and communities to herbivory are poorly understood. Neglected areas of study include the time that elapses before the effects of herbivory on plants are manifest, the need to study the processes at the ecosystem level, and the influence of events beyond the immediate habitat on the outcome of the plant–herbivore interactions. In this chapter I shall examine three postulated effects of herbivory on graminoid plants that bear on the operational model and the neglected topics outlined above. Graminoid plants are commonly morphologically reduced and consist of short-lived repeating modular morphological units with intercalary meristems. Clonal reproduction is well developed. Aside from physical defenses, such as silification and lignification, they are generally depauperate in secondary chemical defenses (McNaughton, 1983a and b). Graminoid plants produce few toxins; they rely largely on physical defenses that limit damage by herbivores. They appear to have undergone adaptive radiation in response to grazing mammals and insects (Stebbins, 1981; McNaughton, 1983a).

The first postulate is that the net primary production of graminoid plants is altered by grazing. A lively debate has ensued as to whether herbivory can lead to increased production of forage plants beyond that of ungrazed plants (the herbivore-optimization model; Dyer, 1975; Owen and Wiegert, 1976, 1981, 1982a,b, 1984; Stenseth, 1978, 1983; McNaughton, 1979a,b; Dyer *et al.*, 1982; Silvertown, 1982; Belsky, 1986). Evidence in support of the claim comes from studies of graminoid communities in temperate and tropical latitudes dominated by patch-forming rhizomatous and stoloniferous plants, and from studies of grazing of coral reef algae by fish and sea urchins (Ogden and Lobel, 1978) and grazing of sea grass communities by turtles (Bjorndal, 1980). Stenseth (1978, 1983) and Belsky (1986) claim that much of the evidence in support of the model is circumstantial and that some of the results are confounded by poor experimental designs, failure to take into account possible changes in below-ground biomass, and unsatisfactory methods to determine net primary production. Some of the confusion stems from whether an increase in net primary production operates at the level of the individual, population or community. The increase may occur at the level of both the individual and the community, if reduced competition from grazed plants allows undamaged plants to grow more rapidly (Crawley, 1983).

The second postulate is that in intensely grazed swards there is selection for genotypes with faster growth rates, which can produce new tissues rapidly following defoliation. It is well known that intensive grazing selects for graminoid plants with prostrate leaves and shoots, reduced leaf width and shortened internode lengths (Stapledon, 1928; McNaughton, 1979a; Detling and Painter, 1983; Crawley, 1983; McNaughton, 1984). Although there is some evidence of genetic differentiation of populations of grasses in pasture mosaics in response to grazing (Kemp, 1937; Detling and Painter, 1983), the extent of this population differentiation in relation to the growth rate of plants in different types of grazed graminoid communities is poorly documented.

The third postulate is that a consumer creates patches in a plant community, thereby altering the supply of resources for the surviving plants. Not all species are equally tolerant of grazing. By foraging at intervals in an area, the consumer may 'reset the successional clock', so that the hierarchies of competitive abilities of the component species in a natural stand are adjusted and readjusted along a temporal pattern of vegetational cycling in favour of the preferred forage species. However, if the periodicity and intensity of foraging is increased somehow by an excessive rate of immigration of consumers from elsewhere, it is predicted that even these preferred species will be destroyed.

Each of the three postulates involves a consideration of the responses of individuals, populations and communities to the direct effects of herbivory. Plant–animal interactions also influence the rates of different processes at the ecosystem level (i.e. decomposition, nutrient cycling), which, in turn, act as agents of natural selection in the evolution of plant populations. I shall draw heavily on results from our studies on the responses of plants in coastal marshes in the Arctic to foraging by Lesser Snow Geese (*Chen caerulescens caerulescens*) in order to address these questions.

13.2 LESSER SNOW GEESE AND THE EFFECTS OF FORAGING ON THE VEGETATION OF SALT MARSHES AND FRESHWATER MARSHES

A salt marsh at La Pérouse Bay, Manitoba on the shores of Hudson Bay is the site of a breeding colony of an estimated 7000 pairs of Lesser Snow Geese. The dominant vegetation of the salt marsh on which the birds feed in summer consists of two perennials, *Puccinellia phryganodes* (Trin.) Scribn. and Merr., a stoloniferous grass and *Carex subspathacea* Wormsk. a rhizomatous sedge. Dicotyledonous plants present at low frequency include *Plantago maritima* L., *Potentilla egedii* Wormsk. and *Ranunculus cymbalaria* Pursh. Landward of the salt marsh are extensive freshwater sedge meadows of the Hudson Bay lowlands in which the birds also feed. The meadows are dominated by *Carex aquatilis* Wahl. and *Carex X flavicans* Nyl. The demand of the birds for forage is high. More than 15 000 goslings increase in weight in seven weeks from about 75–80 g at hatch to 1500–1800 g when fully fledged in an environment in which the growth of vegetation is nitrogen limited (Cooke *et al.*, 1982; Cargill and Jefferies, 1984a). Grazing by geese occurs in nearly all salt marshes along the southern and western shores of Hudson Bay. However, there is no record of regular summer grazing in a salt marsh in Button Bay, just to the west of the town of Churchill. The vegetation of this marsh is dominated by *Puccinellia phryganodes*.

A positive feedback model (Fig. 13.1) accounts for the interaction between the grazer and the forage species in summer at La Pérouse Bay, such that grazing by a colony of geese results in a rapid gain in body weight of birds, an acceleration of the nitrogen cycle as soluble nitrogen is released in feces for plant growth, and the maintenance of high densities of the preferred forage species (Jefferies, Bazely and Cargill, 1984; Bazely and Jefferies 1986; Bazely and Jefferies, 1988). Low amounts of plant litter accumulate in grazed swards (c. 20 g m^{-2}) (Cargill and Jefferies, 1984b). Nitrogen-fixing cyanobacteria colonize the surface of sediments between grazed shoots of *Carex* and *Puccinellia*. The substantially higher input of nitrogen into grazed swards, compared with that into ungrazed swards, replaces nitrogen that is incorporated into the body tissues of the geese (Bazely and Jefferies, 1988).

In spring, before the onset of above-ground net primary production but after snow melt, the geese grub for roots and rhizomes of the graminoid plants in the salt marsh, creating patches of disturbed sediment (Table 13.1). In freshwater sedge meadows the geese eat the swollen basal portion of shoots of Carices and discard the remainder. The degree of destruction of vegetation is dependent on the density of geese at a site and the duration of the grubbing.

POSITIVE FEEDBACK RESPONSES

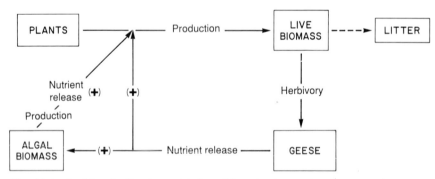

Fig. 13.1 Positive feedback model describing the interaction between the grazer (lesser snow geese) and the graminoid forage species of an arctic salt marsh. Grazing results in a rapid gain in body weight of birds and increased production of the preferred forage species.

Table 13.1 Summary of disturbances brought about by Lesser Snow Geese in spring and summer in salt- and freshwater marshes along the west and south Hudson Bay coasts.

	Salt-marsh grazing flats	Freshwater sedge communities
Dominant graminoids	*Puccinellia phryganodes* *Carex subspathacea*	*Carex aquatilis* *Carex* × *flavicans*

	SPRING	
Type of disturbance	(1) Small-scale grubbing of sward, patch size c. 1 m². Recolonization from inward growth of plants at edge of patch and by surviving tillers in grubbed areas.	(3) Shoot pulling: probably resulting in establishment of moss carpets where intense pulling activity occurs around the edges of small ponds.
	(2) Extensive grubbing by large numbers of birds during successive springs leads to the formation of a series of peat barrens, each c. 300 × 300 m. Erosion of peat occurs, followed by exposure of mineral substratum. Former plant communities appear unable to reestablish.	(4) In dry sedge communities both (1) and (2) occur. On the west Hudson Bay coast a coastal strip 30 km × 3 km has been partially or severely disturbed.

	SUMMER	
	(5) Intensive grazing maintains patches between individual plants of *Puccinellia* and *Carex* (c. 0.5 cm²). These are colonized by cyanobacteria. High rates of nitrogen fixation in grazed swards. Grazing lawn of graminoid plants kept in an actively growing state during summer, as a result of nutrients supplied in patches of feces.	(6) Periodic grazing of leaves of *Carex aquatilis*, upper part of leaf blade removed.

In recent years there has been a large increase in the size of the breeding colony of Lesser Snow Geese at La Pérouse Bay from 1200 ± 200 pairs in 1968 to 7000 ± 500 pairs in 1986 (Cooke *et al.*, 1982; Cooke and Rockwell, unpublished data). Elsewhere the west Hudson Bay population, which comprises an estimated 40% of the total eastern Arctic population has recovered from a drop in numbers, so that it now stands at 218 000 pairs, a size similar to that in 1973 (195 000 pairs) (Kerbes 1975, 1982, unpublished data for 1985 survey). Large breeding colonies of Lesser Snow Geese are also present on Southhampton Island (>50 000 nests) and Baffin Island in the region of the Foxe basin (>100 000 nests) (Boyd, Smith and Cooch, 1982). At least some of these birds use the Churchill area as a staging ground during the spring migration.

The McConnell River breeding colony, the largest of the west Hudson Bay colonies has declined from 163 000 pairs in 1973 to 132 000 in 1985 (Kerbes, 1975, 1982, unpublished data). Surveys of vegetation in the region of the McConnell estuary in 1985 and 1986 indicate that sedge meadow communities that were abundant in 1957 (based on evidence from photographs and field notes of Dr F. G. Cooch) have been destroyed, and the habitats degraded, exposing the underlying peat (Kerbes, Kotanen and Jefferies, unpublished data). The tidal flats in the estuary of the McConnell are also devoid of vegetation. The field evidence indicates that the grubbing of graminoid plants and the shoot pulling of sedges in aquatic habitats by the geese are responsible for the majority of the damage. Although the birds display some fidelity towards a nesting area, there is evidence, as given above, and from La Pérouse Bay, that the locations of nesting sites at high densities within a colony change over a number of years (Cooke *et al.*, 1982; Hik, 1986). Although the causes are not fully understood, it is invariably correlated with the destruction of a habitat associated with spring foraging prior to egg-laying and during the incubation period. Once the young hatch in these degraded systems there is insufficient grazing pasture to sustain large numbers of birds. At both La Pérouse Bay and along the west coast of Hudson Bay evidence from banding records and from aerial photography indicates that families move up to 50 km in search of grazing pasture after hatch (Kerbes, 1982; McLaren and McLaren 1982; Rockwell, unpublished data for 1985).

In Table 13.1 a summary of the disturbances by Lesser Snow Geese to the vegetation of salt- and freshwater marshes is given, and in Fig. 13.2 the effects of these disturbances on the differentiation of plant populations and on the species composition of plant communities of the salt marsh are presented as a flow diagram. The processes depicted in the diagram are discussed in subsequent sections.

13.3 THE HERBIVORE OPTIMIZATION MODEL AND THE NET PRIMARY PRODUCTION OF VEGETATION

There are three alternative hypotheses of the effects of herbivores on plant growth (Fig. 13.3; Dyer, 1975; McNaughton, 1979a, 1983b; Hilbert *et al.*, 1981). The first hypothesis is that plant growth declines as the intensity of

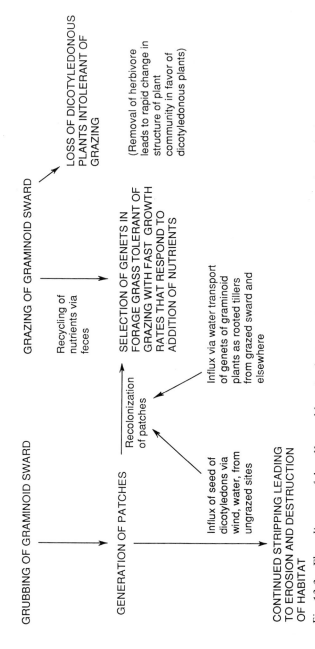

Fig. 13.2 Flow diagram of the effects of foraging by Lesser Snow Geese on an arctic salt-marsh plant community.

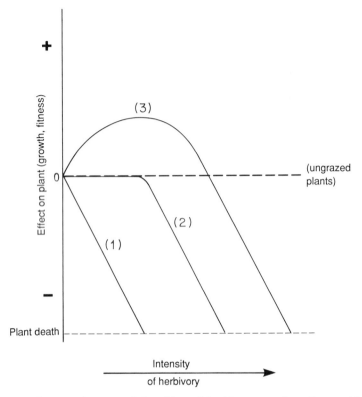

Fig. 13.3 Three predictions of the effect of herbivory on plant fitness. (1) Plant fitness declines as the intensity of herbivory increases; (2) plants are able to compensate up to a particular level of herbivory, and then fitness declines as the intensity of herbivory increases; (3) as the intensity of grazing increases, fitness of plants is increased at moderate levels of herbivory above that of ungrazed plants. At higher intensities fitness declines. (Modified after Dyer (1975), McNaughton (1979a) and Dyer *et al.* (1982).)

herbivory increases. Herbivory is detrimental to plant growth. In the second hypothesis, plants are able to maintain growth up to a certain level of herbivory, but at higher intensities growth declines. Of particular interest is the third hypothesis (the herbivore-optimization model), which states that net primary production of grazed plants may exceed that of ungrazed (control) plants at low to moderate levels of herbivory. Intrinsic mechanisms that operate at the level of the individual, which may account for an increase in net primary production of grazed plants, include increased photosynthetic rates of residual tissue, mobilization of stored reserves, activation of remaining meristems by growth-promoting substances, and a decrease in the mortality of leaves. Extrinsic mechanisms that involve processes at the ecosystem level that may lead to an increase in production include the recycling of nutrients from faeces and urine (McNaughton, 1979a).

The ecological time frame (i.e. within a growing season or between growing seasons) over which these changes in net primary production are supposed to occur is rarely discussed, either in relation to foraging patterns of consumers, the phenologies of plants, or the different mechanisms which may account for the increase in production.

In our study, measurements were made of changes in the dry weight of the standing crop in permanent exclosures (ungrazed sites) established at the beginning of each growing season, and of the regrowth (based on dry weight) of *Puccinellia* and *Carex* swards in short-term exclosures immediately after the geese had grazed an area. The cumulative production in ungrazed areas was estimated by summing increments of dry weight of standing crop between each sample date. In grazed areas, increments in growth for consecutive periods of regrowth in a series of short-term exclosures were summed to give an estimate of net above-ground primary production (NAPP). Grazing significantly increased NAPP of the two species by 30% in 1979 and 80% in 1980 (Cargill and Jefferies, 1984b). In 1982 and 1983, respectively, the NAPP of grazed *Puccinellia* swards was 46% and 106% higher than that of ungrazed *Puccinellia* swards (Bazely and Jefferies 1988) (Fig. 13.4). The percentages are minimal estimates, as we were unable to measure the regrowth of vegetation late in the season, except in 1983.

A major consequence of post-hatch summer grazing is that plant tissue is converted into goose biomass or feces; little accumulates as live plant biomass or litter. Since much of the nitrogen in feces is soluble (Cargill, 1981; Bazely and Jefferies 1985) and since droppings accumulate at high densities throughout grazed swards, considerable impact on the NAPP of vegetation is to be expected. Bazely and Jefferies (1985) demonstrated that addition of fresh feces to experimental plots at densities comparable to those recorded in the salt marsh resulted in significant increases in standing crop compared with untreated plots within a season.

The geese therefore act on this ecosystem by increasing the rate of turnover of nitrogen. This is sufficient to enhance production of the preferred forage species in a nitrogen-deficient environment at a time when it is most required. In other ecosystems the growth response of grazed plants of some species to fertilization may be delayed until the following season, forcing a herbivore to forage over a wide area. For example, in the High Arctic, *Eriophorum triste*, a species that is grazed heavily by musk oxen in summer, fails to show a growth response to fertilization until the following season (Henry, Freedman and Svoboda 1987). Enhanced production within a season is therefore sensitive to the intrinsic ability of plants to show a

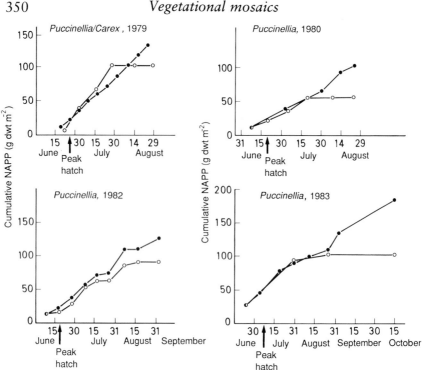

Fig. 13.4 Cumulative net above-ground biomass (d.wt. g m⁻²) of grazed (•) and ungrazed (○) plots of *Puccinellia phryganodes/Carex subspathacea* in 1979 and of *Puccinellia phryganodes* in 1980, 1982 and 1983 at La Pérouse Bay, Manitoba. Time of peak hatch of goslings of the Lesser Snow Goose is indicated by an arrow. (Number of plots = 6.)

rapid growth response to addition of nutrients, the prevailing background levels of available nutrients for plant growth and the timing of herbivory.

Although the action of the geese at La Pérouse Bay results in a higher cumulative production of above-ground biomass in grazed swards compared with that of ungrazed swards, the results provide no information on the relationship between grazing intensity and NAPP. (No differences were detected between treatments in below-ground biomass; Cargill and Jefferies, 1984b.)

David Hik carried out an experiment in which captive goslings grazed the first flush of spring growth of *Puccinellia* and *Carex* in experimental plots for different intervals of time. Above-ground biomass in all plots was measured at the start of the experiment and after the period of grazing. The feces remained in the plots.

The results show that at 24 and 36 days after grazing, the above-ground biomass (Fig. 13.5) was highest in plots where the period of grazing was intermediate in length, such that the standing crop (and the NAPP) far exceeded values for ungrazed plots. The rapid recycling of nutrients, particularly soluble nitrogen, from feces accounted for the increase in production. As the grazing period was increased beyond intermediate levels, there was a decline in biomass. The decline was probably a consequence of the

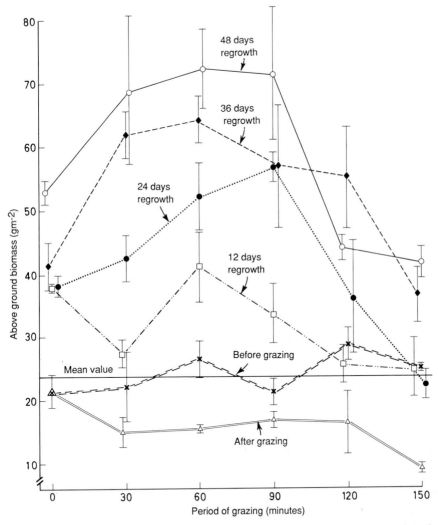

Fig. 13.5 Levels of standing crop (d.wt g m^{-2}) in experimental plots of *P. phryganodes* and *C. subspathacea* which have been grazed by goslings of the Lesser Snow Goose for different intervals of time. X, before grazing; $\triangle$, immediately after grazing; $\square$, 12 days regrowth; $\bullet$, 24 days regrowth; $\blacklozenge$, 36 days regrowth; $\circ$, 48 days regrowth. (Values are means ± SEM, $n = 4$.) (ANOVA: Treatments $F_{5,108} = 11.7$, $P < 0.0001$; Time $F_{5,108} = 67.1$, $P < 0.0001$).

toxic effect of ammonia present in the numerous droppings (Bazely, 1984). The general form of the experimental curves at 24 and 36 days follows that of hypothesis three as predicted by the herbivore-optimization model. As far as we are aware, this is the first time such a relationship has been demonstrated experimentally in the field, although McNaughton, Wallace and Coughenour (1983) have shown that clipping of a sedge under laboratory conditions can result in a higher productivity compared with that of unclipped leaves.

We wish to emphasize, however, that grazing *per se* has a detrimental effect on the growth of *Puccinellia*. Helen Sadul has grown individual tillers of the grass in pots in an experimental garden at La Pérouse Bay. The plants produced a considerable number of prostrate axillary shoots (i.e. morphological modular units). Leaves of main and axillary shoots were 'grazed' (i.e. partially clipped or removed), based on the results of detailed demographic studies of the leaves of grazed plants in the salt marsh. In the absence of the addition of nutrients, the cumulative births of new leaves and axillary shoots at the end of the summer were significantly lower than those of ungrazed plants.

The studies emphasize the plurality of the effects of 'grazing' and the necessity of including studies of nutrient cycling in investigations of plant–herbivore interactions.

13.4 POPULATION DIFFERENTIATION OF *PUCCINELLIA PHRYGANODES*

The results from the field experiments show that the herbivore actively accelerates the rates of breakdown and decomposition of plant material, so that pulses of soluble inorganic nitrogen are made available for plant growth throughout most of the summer. In essence, the geese convert capital into cash flow, in order to maintain growth!

The plant–herbivore interaction is 'successful' because of the ability of both *Puccinellia* and *Carex* to show a rapid growth response to the availability of nutrients in feces. Detailed demographic studies were carried out in which individual leaves were marked with dots of India ink and shoots were identified by toothpicks sunk into the sediment adjacent to the shoots. The data indicated that in 1983 grazed plots of *P. phryganodes* produced an average of 6.1 new leaves on main tillers and 4.3 new leaves on axillary tillers between 'melt' and just before 'freeze up'. Corresponding values for ungrazed plots were 5.9 leaves and 3.4 leaves respectively. Life expectancies of leaves produced during the first part of the season ranged from 24.9 to 34.1 days for ungrazed and partially grazed leaves in grazed plots and from 31.2 to 38.2 days for leaves in ungrazed plots (Bazely and Jefferies, 1988). A demographic study of leaf turnover in grazed and ungrazed plots of *Carex subspathacea* in 1984 produced similar results (Kotanen and Jefferies, 1987). Shoots in grazed plots produced an average of 8.2 new leaves compared with an average of 5.7 leaves per shoot in ungrazed plots. Values for life expectancies of ungrazed or partially grazed leaves in grazed plots were 29.3 to 35.0 days. Corresponding values for leaves in ungrazed plots were 44.6 to 45.8 days. A very rapid turnover of all leaves not removed by grazing occurred in grazed stands of both graminoid species in this arctic salt marsh, where the growing season was about 110 days. The mean numbers of axillary tillers produced per main tiller of *Puccinellia* between June and September 1982 and 1983 were 1.0 and 2.03 respectively for grazed plots and 0.54 and 1.47 for ungrazed plots. Unlike the stoloniferous grass, there was no burst of axillary shoot production associated with new leaves in the rhizomatous sedge. Responses within a season to the effects of grazing and input of nutrients from feces involved changes in the leaf population of *Carex* and the leaf and shoot populations of *Puccinellia*, reflecting the different growth responses of the two forage species to grazing.

The statement of Harper (1978) that 'a single genet of *Trifolium repens* lives long and [that] during its life the parts may wander through a sward as disconnected stolon fragments (like a terrestrial *Lemna*!)' is applicable to *Puccinellia*. However, there are important differences between *Trifolium repens* and this circumpolar grass of salt marshes. Aside from the propensity of genets of both species to spread through swards, ice rafting and the grubbing of *Puccinellia* roots by geese in early spring are two important processes that lead to the dispersal of ramets along shorelines. In addition, the grass is 'asexual'; it has never been known to set seed (cf. Sørensen, 1953; Dore and McNeill, 1980), although intensive collections have not been made throughout the entire geographical range of the species. Bowden (1961) has reported that plants from eastern North America are sterile triploids ($2n = 21$), a finding that we have confirmed. Jefferies and Gottlieb (1983) predicted that in the absence of sexual reproduction the level of genetic variation would be low, both within and between populations. Examination of electrophoretic mobilities of isozymes of 12 enzyme systems in plants from three widely separated populations in Arctic Canada, but including plants from La Pérouse Bay, indicated a high level of variability both within and between the populations. The source of this genetic variability was not determined. The occasional production of viable gametes and events of somatic mutation, similar to those described by Breese, Hayward and Thomas (1965), Whitham and Slobodchikoff (1981), and Hayward (1985) for *Lolium perenne*, may be the sources.

Helen Sadul has examined if there is evidence for different biotypes between populations of this 'asexual grass', and if there has been selection for plants that show a rapid growth response to grazing and addition of nutrients from feces. Such evidence would provide a genecological basis for interpreting the demographic and production data (Cargill and Jefferies, 1984b; Bazely and Jefferies, 1988; Katanen and Jefferies, 1987). In order to determine if selection of biotypes has taken place in response to the overall effects of foraging, 50 turves of *P. phryganodes* were collected at random at La Pérouse Bay from each of four sites at which grazing intensities differed based on field observations. Considerable summer grazing of *Puccinellia* swards occurred in the sites in the outer islands (I) and on the salt-marsh flats (F). No grazing had taken place for five years at the third site (E), an exclosure in the upper marsh, although prior to the erection of the fence, the sward was grazed intensively. The fourth site was just beyond the strand line in willow tundra (W), where a sward of *Puccinellia* persisted in a saline pocket. Grazing was very infrequent at this site and was only observed in August. With the exception of the exclosure site (5 × 5 m), the turves were collected in an area approximately 50 × 50 m. Individual tillers selected at random from each turf were grown in pots in the experimental garden in a uniform, nutrient-rich soil. Demographic methods were used to record the growth of tillers and to detect differences in growth between populations. In particular, births and deaths of leaves and shoots and lengths of main and axillary shoots were monitored, and leaf life-spans calculated. Two hundred tillers were transported to Toronto where the electrophoretic mobilities of isozymes of six enzymes were measured. Additional plants were collected from three sites (low, mid, high levels) in a tidal marsh at Button Bay to the west of Churchill and 30 km from La Pérouse Bay. There was no history of regular summer grazing by Lesser Snow Geese at these sites. Electrophoretic phenotypes were characterized in a total of 150 individuals from this Bay, based on sampling procedures outlined above. The growth of individuals was also compared with that of plants from La Pérouse Bay in a greenhouse at Toronto.

The results of mean cumulative values for leaf births and deaths and the number of axillary shoots produced by plants from the four populations in the experimental garden at La Pérouse Bay are given in Table 13.2. The population from the site in the willow tundra (W) produced significantly fewer axillary shoots per plant, and the numbers of births and deaths per shoot were also significantly lower than corresponding values for other populations. With the exception of leaf births and deaths on the main shoot, populations from the outer islands, the salt marsh flats and the exclosure were not significantly different. At sites at which the grazing of the sward was intense, there appears to have been selection for biotypes that have a high turnover of leaves and the ability to produce a large number of axillary shoots, both quantitative measures of plant growth. In a separate experiment carried out in a greenhouse at Toronto, the numbers of axillary shoots and the total length of all shoots were measured in plants from two Button Bay

Table 13.2 Cumulative mean values for leaf births and leaf deaths per shoot and number of axillary shoots per main shoot of plants of four populations of *Puccinellia phryganodes* from four localities at La Pérouse Bay subject to different grazing pressures. The plants were grown in an experimental garden at La Pérouse Bay during the summer of 1984. W = willow tundra population (rarely grazed); F = tidal flat population (intense grazing); E = population from an exclosure on tidal flats (no grazing for five seasons); I = outer island population in the Bay (intense grazing).

Demographic character	Mean values			
Leaf births per main shoot	W	E	F	I
	9.06	9.46	11.16	11.92
Leaf deaths per main shoot	W	E	F	I
	6.94	7.34	8.04	8.86
Leaf births per axillary shoot	W	E	F	I
	30.16	46.82	47.82	52.86
Leaf deaths per axillary shoot	W	E	F	I
	4.40	6.54	6.86	7.88
Number of axillary shoots per main shoot	W	E	F	I
	10.48	14.88	16.04	16.48

Bar indicates values not significantly different; ANOVA, Scheffe Method, $p > 0.05$.

populations and three populations (W,I,F,) from La Pérouse Bay. The plants from Button Bay produced on average significantly fewer axillary shoots, and the total length of all shoots was significantly lower than corresponding values for grazed populations.

Our interpretation of the results is that in the nitrogen-deficient ungrazed systems there is selection for genets with slow growth rates (Chapin, 1980), but where swards are grazed additional nutrients become available each summer, and selection favors genets with faster growth rates that produce numerous axillary shoots. We assume that in the exclosure insufficient time has elapsed for a significant change in the composition of the population to occur in the absence of grazing.

As in the previous study of electrophoretic variation in *P. phryganodes* (Jefferies and Gottlieb, 1983), a large amount of variability was detected both within and between the seven populations (three from Button Bay, four from La Pérouse Bay). For the six enzymes a total of 26 patterns was recognized. In plants from the La Pérouse Bay populations, ten sets of patterns of these six enzymes were observed. In contrast, only four sets were detected in plants from Button Bay, and of these, two were observed only in single individuals. Surprisingly, only two sets of patterns were common to plants from both marshes, again reflecting the large amount of electrophoretic variability present both within and between these local populations.

Aside from the origin of the genetic diversity, electrophoretic variation between individuals appeared to be more pronounced in populations from La Pérouse Bay than those from Button Bay. We suggest that this difference reflects the foraging activities of the geese. In early spring, geese feed on the roots and shoots of *Puccinellia*. It takes about one hour for an adult goose to strip an area about one metre in diameter. Besides the breeding population of an estimated 7000 pairs, migrants from other colonies located in the Hudson Bay region use La Pérouse Bay as a staging area. It is not difficult to appreciate that disturbance on this scale each year produces a large number of patches suitable for colonization. The marsh may be regarded as a mosaic of patches of *Puccinellia* of different ages. Some rooted shoots of the grass remain in the grubbed patches (Table 13.3), but many disturbed shoots are discarded by the geese and are dispersed to other areas by melt water. Tides that cover the marsh in late summer also disperse plants that may subsequently root in patches of bare sediment.

At Button Bay, summer grazing is absent or infrequent, and spring grubbing is very limited. In the absence of frequent disturbance, a clone may establish over a wide area of marsh. With the exception of two individuals, all other plants showed one of two sets of electrophoretic patterns.

In spite of the apparent absence of seed set in this grass, electrophoretic evidence revealed an unexpectedly high level of genetic variation. Although the genetic origins of this variation are unknown, frequent physical and biotic disturbances of the salt-marsh sward appear to maintain a supply of mature 'ready-made' rooted tillers of different genotypes from both grazed

Table 13.3 Total number of shoots of graminoid plants and dicotyledonous plants in grubbed, grazed and exclosed plots (10 × 10 cm, $n = 10$) on intertidal flats immediately after grubbing in mid-June and in early August 1986 (mean ± SE).

	Graminoid species		Dicotyledonous species	
	June	August	June	August
Grubbed plots	7.0 ± 1.0	15.0 ± 5.2	2.2 ± 0.7	1.0 ± 0.8
Ungrubbed plots				
Grazed plots	45.5 ± 5.0	45.0 ± 7.5	4.0 ± 2.0	4.8 ± 1.8
Exclosed plots	45.5 ± 5.0	45.8 ± 7.8	4.0 ± 2.0	4.1 ± 1.7

and highly grazed areas that are able to colonize patches created by the disturbances. Tillers remaining at the edges of and in the patches also spread rapidly in the absence of high densities of the species (Table 13.3). As the developing swards age, are grazed, and receive nutrients from feces, there appears to be selection for genotypes with fast growth rates. The grazed salt marsh represents a vegetational mosaic in which genotypes are being sorted and resorted as a result of disturbance and natural selection. The Lesser Snow Goose is an unwitting agent in the maintenance of intra- and interpopulation variation in *P. phryganodes*.

Populations of other clonal plants common in temperate grassland swards also show similar characteristics to *Puccinellia* (Burdon, 1980; Turkington and Burdon, 1983; Turkington, 1985). For example, populations of *Trifolium repens* differ in relative growth rate (Burdon and Harper 1980) and stolon length (Aarssen and Turkington 1985a,b). Tranthan (1983), quoted by Turkington (1985), found up to 50 electrophoretic phenotypes per m^2 of *T. repens*, and Aarssen and Turkington (1985a) reported a high level of variability within populations of *Holcus lanatus* and *Lolium perenne*. In populations of both *L. perenne* and *T. repens*, a decline in variance for different morphological and physiological characters appears to occur with increasing age of the pasture (Aarssen and Turkington, 1985b).

Turkington (1985) has suggested that as grassland swards age there is the potential for a single, successional clone to dominate large areas by eliminating less fit genotypes. The low number of electromorphs in the low- and mid-marsh at Button Bay, coupled with the near absence of disturbance by the geese is consistent with the suggestion.

13.5 RESOURCE AVAILABILITY, PATCH DYNAMICS AND COMMUNITY STRUCTURE

A positive feedback model accounts for the interaction between the grazer and the forage species in summer at La Pérouse Bay, such that grazing by a colony of geese results in a rapid gain in body weight of birds and the maintenance of high densities of the preferred forage species. We have estimated that the maximum amount of nitrogen removed annually by the geese is about 2 g m^{-2} (Cargill and Jefferies, 1984b). Most of the nitrogen either is incorporated into the body tissues of the geese between hatch and autumn migration, or is present as insoluble nitrogen in droppings. The latter are either carried out to sea on ebb tides or deposited on the strand line.

The export of nitrogen each year depletes the reservoir of this element within the marsh. The activities of the geese result in the accumulation of low amounts of plant litter (c. 20 g m^{-2}), so that small patches of sediment are visible between shoots of *Puccinellia* or *Carex*. The surface of these patches is colonized by species of *Oscillatoria* and *Lyngbya*. Certain species of these genera of cyanobacteria are known to fix atmospheric nitrogen under suitable conditions (Sprent, 1979; Postgate, 1982). Measurements of nitrogenase activity of surface sediments using the acetylene-reduction technique indicate that the activity is considerably higher in grazed swards than in ungrazed swards (Fig. 13.6; Bazely and Jefferies 1988). The organisms fix an estimated 1–2 g of nitrogen $m^{-2}y^{-1}$.

We believe that this input of nitrogen is essential to maintain the salt-marsh flats in their present state. The colonial foraging behavior of the geese involving intense cropping and the creation of patches approx 0.5 cm in diameter, which are colonized by cyanobacteria, ensures that the long-term input of nitrogen can be sustained.

As mentioned above, in early spring there is considerable grubbing by geese for roots and rhizomes of plants which are exposed at snow melt. Preliminary evidence indicates that it takes at least eight years before formerly grubbed areas are regrubbed by the geese. By creating the patches in spring and grazing the sward intensively, the consumer is continually resetting 'the successional clock', so that non-equilibrium conditions prevail. Species, such as *P. phryganodes* and *C. subspathacea*, display consid-

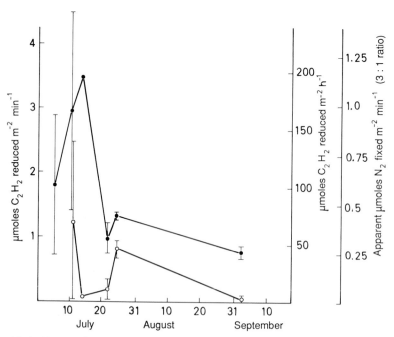

Fig. 13.6 Rates of acetylene reduction, as a measure of nitrogenase activity, in surface sediments of grazed (•) and ungrazed (○) plots of salt-marsh vegetation at La Pérouse Bay, Manitoba between early July and early September 1983. (After Bazely, 1984.)

erable resilience in their ability to recolonize these patches after a disturbance. However, the birds are generalists in their foraging behavior, as both monocots and dicots are eaten. The effect of grazing on numbers of individuals, production of ramets and reproductive success of the latter group of plants depends strongly on the growth habit of each species.

Helen Sadul has recorded that the removal by geese of stolons of *Potentilla egedii* Wormsk. and *Ranunculus cymbalaria* Pursh. resulted in no regrowth of existing stolons or their replacement within the season. As a consequence, the number of ramets established per individual for both species was significantly lower in grazed plots. Both *Potentilla egedii* and *Plantago maritima* L. produced erect inflorescences, unlike plants of *Ranunculus cymbalaria*, which were strongly prostrate. In the case of *Potentilla* and *Plantago*, few flowers reached maturity in grazed swards and the sizes of the populations decreased significantly relative to those in ungrazed plots over a two-year period. In contrast, no significant differences in numbers of mature flowers per individual, or in numbers of individuals of *Ranunculus* were detected between treatments, presumably a reflection of the inability of the grazer to forage efficiently at the immediate sediment surface. Prins, Ydenberg and Drent (1980) have also recorded heavy grazing of *Plantago maritima* by Brent Geese in Holland.

Unlike more lightly grazed systems, where species diversity may be high (Harper, 1969; Crawley, 1983), intense grazing favors *Puccinellia* and *Carex* at the expense of dicotyledonous plants. A similar situation in England has been reported by Gray and Scott (1977) where *Puccinellia maritima* replaces broad-leaved species in a salt marsh grazed by sheep.

The cessation of grazing in exclosures at La Pérouse Bay results in extensive stolon production by such species as *Potentilla egedii* and *Ranunculus cymbalaria*, which exploit bare patches between formerly grazed plants of *Carex* and *Puccinellia*. Rapid changes in species composition occur, associated with a decline in the frequency of *Puccinellia* and changes in the competitive hierarchy of species (Bazely and Jefferies, 1986). The composition and structure of the grazed community at La Pérouse Bay are therefore strongly dependent on the colonial feeding of the geese. When the fence of a two-year old exclosure was removed in the spring of 1983, very little grazing of the modified sward occurred in 1983 or 1984. In the ungrazed sward plant litter rapidly accumulated. Much of the live biomass consisted of stems and senescing tissue (Bazely and Jefferies, 1986). Once grazing pressure is relaxed and changes in the vegetation occur, it is unlikely that the grazers will show a high preference for such a plant community. The changes are effectively irreversible and represent a loss of preferred forage.

The results parallel those reported by McNaughton (1976, 1984) for the Serengeti, where large herds of wildebeest, buffalo and zebra graze the grasslands intensively, and as a result the concentration and quality of available food is enhanced. He suggests that a benefit of herd formation is increased foraging efficiency.

In freshwater sedge meadows in early spring, the geese eat the swollen basal portion of shoots of Carices and discard the remainder of the shoots. Peter Kotanen has shown that in the spring of 1986, the density of pulled shoots per 50×50 cm was a mean of 43.8 ± 3.4 SEM for stands of *Carex aquatilis* and a mean of 7.1 ± 1.5 SEM for stands of *Carex × flavicans*. We believe that sustained destruction on such a scale, together with summer grazing of leaves results in the establishment of a moss carpet, in which the only common angiosperms present are *Potentilla palustris* and *Hippuris tetraphylla*, both of which are avoided by the geese. Moss carpets may also develop on frost-heave mounds on the salt-marsh flats, where there has been frequent disturbance by geese and a large input of nutrients from fecal matter. Circumstantial evidence from elsewhere indicates that the presence of moss carpets at other lowland sites, especially around lakes and ponds in coastal areas, is a consequence of foraging by birds (Tikhomirov 1959; Bliss 1981; Giroux, Bedard and Bedard 1984; Prop, Eerden and Drent 1984). At present, we are 'grubbing and grazing' sedge communities, in an attempt to determine the length of time it takes to convert them into moss carpets under different levels of predation.

All these predatory activities of the geese occur at melt in early spring. The degree of destruction is dependent on the density of geese at a site and the time they remain there. If large numbers of migrants are held in an area in spring because of poor weather, the combined foraging activities of migrants and breeders may result in the elimination of individual plant species and abrupt changes in the composition of the communities over one or two seasons. Although it is difficult to be precise, the destruction occurs within eight years at a site where a nesting colony of Lesser Snow Geese is established (c. 15 nests per 100×100 m). Within an area, the nesting colony spreads outwards when numberse of birds are increasing, creating a 'doughnut' effect on the local graminoid communities.

At La Pérouse Bay, nesting sites and summer grazing areas are spatially separated. Geese rarely nest on the tidal flats! This partitioning of the use of plant communities in space and time would appear to ensure the maintenance of summer grazing pasture. Unfortunately the flats are also being degraded by spring grubbing. Limited observational evidence suggests that most of the birds responsible are migrants and not local birds. This requires further study, as it stresses the fact that the interactions between plants and consumers involve other goose populations aside from the breeding colony. The stability of the system may ultimately depend upon the number of itinerants.

It may be that the boom–bust cycles in numbers of geese, relocation of nesting sites and summer grazing areas, and the partial destruction and recovery of the wetlands is a long-standing phenomenon in the Arctic. Garoarsson and Sigurosson (1972) suggested that pink-footed geese in Iceland may have a similar effect on wetlands. The time scale of this 'pattern and process' may take more than a hundred years, and it has to be measured against the backdrop of isostatic uplift, as the Hudson Bay region is rising at a rate of about 1 meter/100 years (Hunter, 1970).

Numbers of Lesser Snow Geese in the eastern Canadian Arctic are unusually high at present. Although the reasons for this are not well understood, it is probably linked to changes in hunting practices and the use of agricultural crops as a food source (rice, corn and winter wheat) by the geese in the wintering and spring-staging areas of the United States and Canada. The likely effect of these changes is that many more geese are returning to breed in excellent condition. Events 5000 km away on the wintering grounds may profoundly affect the fate of these wetland plant communities in the Arctic.

13.6 COMPARISONS WITH OTHER ECOSYSTEMS

Although comparisons may be made with similar phenomena in terrestrial ecosystems, it is in marine ecosystems that the closest parallels appear to occur, as many marine communities are effectively structured by a few strong interactions (Paine, 1980; Wharton and Mann, 1981). Paine (1979) has pointed out that few, if any, naturally occurring populations are immune to disruptions. Selective removal of dominant species or general disturbance 'drives' most ecosystems. Harper (1977) distinguishes between a 'catastrophe', which is rare but results in the near destruction of an entire population, and a 'disaster', which occurs more frequently but leads to a reduction in a population only at the local scale (i.e. creation of a patch). Although Sousa (1984) has raised difficulties with the use of the terms 'catastrophe' and 'disaster', because of the lack of objective criteria and the uncertainty of the relative influence of these events on evolutionary processes, the terms can be used to describe the action of the geese on the plant communities. Regular and predictable summer grazing maintains the *Puccinellia/Carex* sward and removes those dicotyledonous species that can overgrow the graminoid plants. In the absence of grazing, the species composition of the community rapidly changes at the expense of species, such as *P. phryganodes*. Frequent grubbing (disasters) at the local scale creates patches (1 m²) that are colonized by *Puccinellia* and *Carex*. The effect on plant communities may be catastrophic if large numbers of geese are held in an area in spring. Large areas (> 300 m × 300 m) become 'barrens', and the original plant communities may fail to reestablish. In these Arctic wetlands, much depends on the ability of perennial plants capable of extensive modular growth to colonize the barrens. As soil conditions close to the surface immediately after grubbing bear little resemblance to those prior to foraging, this may preclude the establishment of some species. Unlike the small grubbed patches, the peat barrens serve as foci for opportunists, such as *Salicornia europaea*, *Senecio congestus* and *Eleochaeris acicularis*.

In aquatic macrophyte communities, changes in the population density of a key predator may result in shifts in the balance of primary producers (Duggins, 1980). Mann (1972a,b, 1977) has shown that sea urchins have a major effect on the structure of algal kelp communities. Sea urchins (*Strongylocentrotus droebachiensis*) formed destructive grazing aggregations that produced a macrophyte-free area of more than 500 km in length along the coast of Nova Scotia. The increase in the size of the sea urchin population appears to be related to decreases in both crustacean predators and predatory fish (Wharton and Mann, 1981). The interaction between graminoid communities and geese is a parallel case, in that these herbivores exert a major effect on the structure of the plant community. As in the case of the decline of the crustaceans (lobsters), the increase in the numbers of Lesser Snow Geese in the last decade is probably related to the activities of man.

Plant distributions in coral reefs are strongly influenced by herbivorous fish and invertebrates (Ogden and Lobel, 1978; Huston, 1985). Like geese, herbivorous fish have a low assimilation efficiency with a high ingestion rate and a retention time of a few hours (Odum, 1970). Preferred plants include species of *Polysiphonia* and *Enteromorpha* (Randall, 1961). Cyanobacteria are abundant on reefs but are generally not eaten by fishes (Randall, 1961). Heavy grazing, characteristic of open reef surfaces, leads to the development of a very productive algal mat and 'holds the plant community in a high turnover, early successional state' (Ogden and Lobel, 1978). Although the nutrient budgets of coral reefs are incomplete, it is likely that enrichment from feces of fishes and invertebrates, together with the release of nitrogen compounds from cyanobacteria, enables high levels of primary production to be sustained (Ogden and Lobel, 1978; McDonald, 1985). These processes, which probably involve positive-feedback mechanisms, appear similar to the interactions between the salt-marsh plant community and the geese at La Pérouse Bay.

13.7 CONCLUSIONS

Grazing of salt-marsh vegetation by Lesser Snow Geese maintains the graminoid communities, thereby delaying the onset of vegetational change. The geese adjust the competitive hierarchy in favor of preferred forage species in the early part of each summer and provide a nutrient repayment (direct or indirect) to maintain the growth of the preferred graminoids. A positive-feedback model accounts for the interaction between the geese and the forage plants, whereby growth of the plants is sustained and adult geese and goslings achieve a rapid gain in body weight. On a larger spatial scale, because the turnover times for the movements of nutrients between and within abiotic and biotic compartments of arctic soils are long, disturbances are necessary to release nutrients for sustained biological production. The grazing animal regulates the release of soluble nutrients via feces at the soil surface, thereby bypassing processes that occur in sediments. This compartmental shift of nutrient release is one of a limited number of ways whereby nutrients are made available in arctic ecosystems for plant growth.

Under the influence of intense grazing and the addition of nutrients from feces, there appears to be selection for plants of *P. phryganodes* that produce large numbers of axillary tillers and that the leaves have a high turnover rate. Leaves of *C. subspathacea* also turnover rapidly, but there is no increased production of shoots in response to grazing. These changes in the growth of leaves and shoots of the two forage species account for the increase in above-ground net primary production, which occurs when plants are grazed and the sward received nutrients from feces.

Frequent grubbing by geese during the spring for roots and rhizomes creates patches in the salt-marsh sward. The grazed sward represents a vegetational mosaic in which genets of *Puccinellia* are sorted and resorted as a result of disturbance and natural selection. Because of the presence of high numbers of geese at some sites in recent years, the scale of the disturbance has resulted in the destruction of the plant communities. As this also leads to changes in edaphic conditions, it is unlikely that the former plant communities will reestablish.

From an evolutionary standpoint, these local extinctions of populations and communities may be part of a long-term process in the arctic coastal systems that result in the displacement of communities because of physical disturbances and isostatic uplift. Although isostatic uplift leads to the disappearance of these salt-marsh communities on a local scale (irrespective of grazing which only delays the onset of change) and their replacement with freshwater wetlands, emerging shorelines, in turn, provide new areas suitable for colonization. The traits that enable plants of a number of species to colonize developing shorelines and to withstand the rigors of the arctic coast (prostrate growth habit, extensive clonal growth, ability of shoot systems to root readily in soft substrate) are the very same traits that contribute to their survival when grazed. The activities of the herbivore reinforce at a local scale the other sorting processes of genets of *Puccinellia* already taking place. If summer grazing predominates over a number of years at the local site with the attendant input of nutrients, selection of genets with fast growth rates from the pool of 'ready-made' plants is predicted. Whether the genetic structure of the local grazed population is maintained probably depends on the frequency of disturbance relative to the intensity of grazing. If the selection pressure is relaxed (i.e. the geese no longer graze the site) we predict rapid changes in the genetic structure of the local population.

The ability of the herbivore to regulate the genetic structure of plant populations and to modify the dynamics of the plant communities at a range of spatial (from 0.5 cm^2 to 1 km^2), and temporal (within a growing season to decades) scales illustrates not only the strong interactive nature of the relationships between plants and consumer but also the dangers of generalizations. The emphasis has been placed on the need for a long-term empirical investigation (c.f. Rice and Jain, 1985), as I find myself uneasy about straying too far from well-documented observations and sound experimental field data.

Another important research requirement is to obtain quantitative information on the rates of change in numbers of genets/ramets in plant populations in response to the foraging activities of the herbivore, in order to provide evidence for the disturbance-selection model outlined above. The use of electrophoretic markers to identify genets, coupled with demographic studies of the fate of ramets, represents a possible approach to the problem. Quantitative changes in the genetic structure of these graminoid populations in response to grazing can be linked to increased population fitness and the positive-feedback processes between the plants and the herbivore. Measurement of rates of colonization of patches and rates of extinction, as a result of disturbances, are also essential in determining 'pattern and process' in these arctic coastal communities. These changes are superimposed upon changes in vegetation brought about by isostatic uplift. A challenge is to determine the sequence of vegetational changes and the time scales over which they occur under these non-equilibrium conditions. Existing theories of plant succession may be of limited application in describing and predicting such changes. Episodic events, whether caused by biotic or environmental agencies, set limits on the evolution of plant populations, as well as on both the pattern and the extent of development of arctic plant communities.

13.8 ACKNOWLEDGEMENTS

I thank Susan Cargill, Dawn Bazely, Helen Sadul, Peter Kotanen and David Hik, who generously allowed me to use their data. Together with Tony Davy, Rudi Drent and the editors, they made valuable suggestions for the improvement of the text. Much of the paper was written at Eskimo Point and at the Queen's University Biology Station at La Pérouse Bay on the Hudson Bay coast, where I had many useful discussions with Fred Cooke, Robert Rockwell, Dick Kerbes, Rick Bello, Rudolf Harmsen and students about the topics discussed in the chapter. Brenda Missen kindly typed the paper in record time.

13.9 REFERENCES

Aarssen, L. W. and Turkington, R. (1985a) Vegetation dynamics and neighbour associations in pasture-community evolution. *J. Ecol.*, **73**, 585–603.

Aarssen, L. W. and Turkington, R. (1985b) Biotic specialization between neighbouring genotypes in *Lolium perenne* and *Trifolium repens* from a permanent pasture. *J. Ecol.*, **73**, 605–14.

Bazely, D. R. (1984) *Responses of salt-marsh vegetation to grazing by lesser snow geese (Anser caerulescens caerulescens)*, M.Sc. Thesis, University of Toronto, Canada.

Bazely, D. R. and Jefferies, R. L. (1985) Goose faeces: a source of nitrogen for plant growth in a grazed salt marsh. *J. Appl. Ecol.*, **22**, 693–703.

Bazely, D. R. and Jefferies, R. L. (1986) Changes in the composition and standing crop of salt-marsh communities in response to the removal of a grazer. *J. Ecol.*, **74**, 693–706.

Bazely, D. R. and Jefferies, R. L. (1988) Lesser Snow Geese and the nitrogen economy of a grazed salt marsh. *J. Ecol.*, (in press).

Belsky, A. J. (1986) Does herbivory benefit plants? A review of the evidence. *Am. Natur.*, **127**, 870–92.

Bjorndal, K. Q. (1980) Nutrition and grazing behaviour of the green turtle *Chelonia myclas*. *Mar. Biol.*, **56**, 147–54.

Bliss, L. C. (1981) North American and Scandinavian tundras and polar deserts. in *Tundra Ecosystems: A Comparative Analysis* (eds L. C. Bliss, O. W. Heal and J. J. Moore), Cambridge University Press, London, pp. 8–24.

Bowden, K. M. (1961) Chromosome numbers and taxonomic notes on northern grasses. IV. Tribe Festuceae: *Poa* and *Puccinellia*. *Can. J. Bot.*, **39**, 123–8.

Boyd, H., Smith, G. E. J. and Cooch, F. G. (1982) *The Lesser Snow Geese of the Eastern Canadian Arctic*. Occasional Paper Number 46, Canadian Wildlife Service, Environment Canada, Ottawa.

Breese, E. L., Hayward, M. D. and Thomas, A. C. (1965) Somatic selection in perennial ryegrass. *Heredity*, **20**, 367–79.

Burdon, J. J. (1980) Intra-specific diversity in a natural population of *Trifolium repens*. *J. Ecol.*, **68**, 717–35.

Burdon, J. J. and Harper, J. L. (1980) Relative growth rates of individual members of a plant population. *J. Ecol.*, **68**, 953–7.

Cargill, S. M. (1981) *The effects of grazing by lesser snow geese on the vegetation of an Arctic salt marsh*. M.Sc. Thesis, University of Toronto, Toronto, Canada.

Cargill, S. M. and Jefferies, R. L. (1984a) Nutrient limitation of primary production in a sub-arctic salt marsh. *J. Appl. Ecol.*, **21**, 657–68.

Cargill, S. M. and Jefferies, R. L. (1984b) The effects of grazing by lesser snow geese on the vegetation of a sub-arctic salt marsh. *J. Appl. Ecol.*, **21**, 669–86.

Chapin, S. A. III (1980) The mineral nutrition of wild plants. *Ann. Rev. Ecol. Syst.*, **11**, 233–60.

Cooke, F., Abraham, K. F., Davies, J. C., Findlay, C. S., Healey, R. F., Sadura, A. and Segin, R. J. (1982) *The La Pérouse Bay Snow Goose Project – A 13-year Report*. Department of Biology, Queen's University, Kingston, Ontario, Canada.

Crawley, M. J. (1983) *Herbivory: The Dynamics of Animal–Plant Interactions*. Blackwell, Oxford.

Detling, J. K. and Painter, E. L. (1983) Defoliation responses of western wheat grass populations with diverse histories of prairie dog grazing. *Oecologia*, 57, 65–71.

Dore, W. G. and McNeill, J. (1980) *Grasses of Ontario*. Monograph 26, Research Branch, Agriculture Canada, Ottawa.

Duggins, D. O. (1980) Kelp beds and sea otters: an experimental approach. *Ecology*, 61, 447–53.

Dyer, M. I. (1975) The effects of Red-Winged Blackbirds (*Agelaius phoeniceus* L.) on biomass production of corn grains (*Zea mays* L.). *J. Appl. Ecol.*, 12, 719–26.

Dyer, M. I., Detling, D. C., Coleman, D. C. and Hilbert, D. W. (1982) The role of herbivores in grasslands. *Grasses and grasslands: systematics and ecology* (eds J. R. Estes, R. J. Tyre and J. N. Brunken), University of Oklahoma Press, Norman, pp. 255–95.

Garoarsson, A. and Sigurosson, J. B. (1972) *Skyrslaum rannsoknir a heidgoes i jorsaverum sumarid 1971*. Rannsoknir pessorvoru unnar fyir orkusrofnun af natturufrazdistofunislands, University of Iceland, Reykjavik.

Giroux, J.-F., Bedard, Y. and Bedard, J. (1984) Habitat use by Greater Snow Geese during the brood rearing period. *Arctic*, 37, 155–60.

Gray, A. J. and Scott, R. (1977) The ecology of Morcambe Bay: VII. The distribution of *Puccinellia maritima*. *Festuca rubra* and *Agrostis stolonifera* in the salt marshes. *J. Appl. Ecol.*, 14, 229–41.

Grime, J. P. (1979) *Plant Strategies and Vegetation Processes*. Wiley, Chichester.

Harper, J. L. (1969) The role of predation in vegetational diversity. *Diversity and Stability in Ecological Systems* (eds G. M. Woodwell and H. H. Smith), Brookhaven National Laboratory, Upton, NY, pp. 48–62.

Harper, J. L. (1977) *Population Biology of Plants*. Academic Press, London.

Harper, J. L. (1978) The demography of plants with clonal growth. *Structure and Functioning of Plant Populations* (eds A. H. J. Freyson and J. W Woldendorp), North Holland, Amsterdam, pp. 27–48.

Hayward, M. D. (1985) Adaption, differentiation and reproductive systems in *Lolium perenne*. *Genetic Differentiation and Dispersal in Plants*. NATO ASI Series, Vol. 65 (eds P. Jacquard *et al.*), Springer-Verlag, Berlin.

Henry, G., Freedman, B. and Svoboda, J. (1987) Effects of fertilization on three tundra plant communities of a polar desert oasis. *Can. J. Bot.*, (in press).

Hik, D. (1986) *Philopatry and dispersal in the Lesser Snow Goose: the role of reproductive success*. B.Sc. Thesis, Queen's University, Kingston, Ontario.

Hilbert, D. W., Swift, D. M., Detling, J. K. and Dyer, M. I. (1981) Relative growth rates and the grazing optimization hypothesis. *Oecologia*, 51, 14–18.

Hunter, G. T. (1970) Postglacial uplift at Fort Albany, James Bay. *Can. J. Earth Sci.*, 7, 547–8.

Huston, M. A. (1985) Patterns of species diversity on coral reefs. *Ann. Rev. Ecol. Syst.*, 16, 149–77.

Jefferies, R. L., Bazely, D. R. and Cargill, S. M. (1984) *Effects of grazing on tundra vegetation – a positive feedback model*. Second International Rangeland Congress, Adelaide, Australia. Working Papers, Section 1c, Dynamics of Range Ecosystems.

Jefferies, R. L. and Gottlieb, L. D. (1983) Genetic variation within and between populations of asexual plant *Puccinellia × phryganodes*. *Can. J. Bot.*, 61, 774–9.

Kemp, W. B. (1937) Natural selection within plant species as exemplified in a permanent pasture. *J. Hered.*, **28**, 329–33.

Kerbes, R. H. (1975) *The Nesting Population of Lesser Snow Geese in the Eastern Canadian Arctic.* A photographic inventory of June 1973. Report series Number 35, Canadian Wildlife Services, Environment Canada, Ottawa.

Kerbes, R. H. (1982) Lesser snow geese and their habitat on west Hudson Bay. *Natur. Can. (Rev. Ecol. Syst.)*, **109**, 905–11.

Kotanen, P. and Jefferies, R. L. (1987) The leaf and shoot demography of grazed and ungrazed plants of *Carex subspathacea. J. Ecol.*, **75**, 961–75.

Mann, K. H. (1972a) Ecological energetics of the seaweed zone in a marine bay on the Atlantic coast of Canada. I. Zonation of biomass of seaweeds. *Mar. Biol.*, **12**, 1–10.

Mann, K. H. (1972b) Ecological energetics of the seaweed zone in a marine bay on the Atlantic coast of Canada. II. Productivity of the seaweeds. *Mar. Biol.*, **14**, 199–209.

Mann, K. H. (1977) Destruction of kelp beds by sea urchins: a cyclical phenomenon or irreversible degradation? *Helgolander wissenschaftliche Meeresuntersuchungen*, **30**, 455–67.

McDonald, M. E. (1985) Growth enhancement of a grazing phytoplanktivorus fish and growth enhancement of other grazed alga. *Oecologia (Berlin)*, **67**, 132–6.

McLaren, P. L. and McLaren, M. A. (1982) Migration and summer distribution of lesser snow geese in interior Keewatin. *Wilson Bulletin*, **94**, 494–504.

McNaughton, S. J. (1976) Serengeti migratory wildebeest: facilitation of energy flow by grazing. *Science, NY*, **191**, 92–4.

McNaughton, S. J. (1979a) Grazing as an optimization process: grass-ungulate relationships in the Serengeti. *Am. Natur.*, **113**, 691–703.

McNaughton, S. J. (1979b) Grassland–herbivore dynamics. in *Serengeti: Dynamics of an Ecosystem* (eds A. R. E. Sinclair and M. Norton-Griffiths), University of Chicago Press, Chicago, pp. 82–103.

McNaughton, S. J. (1983a) Physiological and ecological implication of herbivory. in *Physiological Plant Ecology III. Responses to the Chemical and Biological Environment* (eds O. L. Lange, P. S. Nobel, C. B. Osmond and H. Ziegler), pp. 657–78.

McNaughton, S. J. (1983b) Compensatory plant growth as a response to herbivory. *Oikos*, **40**, 329–36.

McNaughton, S. J. (1984) Grazing lawns: Animals in herds, plant form, and coevolution. *Am. Natur.*, **124**, 863–86.

McNaughton, S. J. (1985) Ecology of a grazing ecosystem: the Serengeti. *Ecol. Monogr.*, **55**, 259–94.

McNaughton, S. J., Wallace, L. L. and Coughenour, M. B. (1983) Plant adaptation in an ecosystem context: Effects of defoliation, nitrogen, and water on growth of an African C_4 sedge. *Ecology*, **64**, 307–18.

Mooney, H. A. and Godron, M. (eds) (1983) *Disturbance and Ecosystems.* Springer-Verlag, Berlin.

Odum, W. E. (1970) Utilization of the direct grazing and plant detritus food chains by the striped millet, *Mugil cephalus. Marine Food Chains* (ed. J. H. Steele), Oliver and Boyd, Edinburgh, pp. 222–40.

Ogden, J. C. and Lobel, S. L. (1978) The role of herbivorous fishes and urchins in coral reef communities. *Env. Biol. Fish.*, **3**, 49–63.

Owen, D. F. and Wiegert, R. G. (1976) Do consumers maximize plant fitness? *Oikos*, **27**, 488–92.

Owen, D. F. and Wiegert, R. G. (1981) Mutualism between grasses and grazers: an evolutionary hypothesis. *Oikos*, **36**, 376–8.

Owen, D. F. and Wiegert, R. G. (1982a) Grasses and grazers: is there a mutualism? *Oikos*, **38**, 258–9.

Owen, D. F. and Wiegert, R. G. (1982b) Beating the walnut tree: more on grass/grazer mutualism. *Oikos*, **39**, 115–16.

Owen, D. F. and Wiegert, R. G. (1984) Aphids and plant fitness: 1984. *Oikos*, **43**, 403.

Paine, R. T. (1979) Disaster, catastrophe, and local persistence of the sea palm *Postelsia palmaeformis*. *Science*, **205**, 685–7.

Paine, R. T. (1980) Food webs: linkage, interaction strength and community infrastructure. *J. Anim. Ecol.*, **49**, 667–85.

Paine, R. T. and Levin, S. A. (1981) Intertidal landscapes: disturbance and the dynamics of pattern. *Ecol. Monogr.*, **51**, 145–78.

Pickett, S. T. A. and White, P. S. (eds) (1985) *The Ecology of Natural Disturbance and Patch Dynamics*. Academic Press, New York.

Postgate, J. R. (1982) *The Fundamentals of Nitrogen Fixation*. Cambridge University Press, London.

Prins, H. H., Ydenberg, R. C. and Drent, R. H. (1980) The interaction of Brent geese *Branta bernicla* and sea plantain *Plantago maritima* during spring staging. Field observations and experiments. *Acta Bot. Neerl.*, **29**, 585–96.

Prop, J., van Eerden, M. R. and Drent, R. H. (1984) Reproductive success of the Barnacle Goose *Branta leucopsis* in relation to food exploitation on the breeding grounds, western Spitsbergen. *Norsk polarinstitut Skritter*, **181**, 87–117.

Randall, J. E. (1961) A contribution to the biology of convict surgeon fish of the Hawaiian Islands, *Acanthurus triostegus sandvicensis*. *Pacific Sci.*, **15**, 215–72.

Rice, K. and Jain, S. (1985) Plant population genetics and evolution in disturbed environments. *The Ecology of Natural Disturbance and Patch Dynamics* (eds S. T. A. Pickett and P. S. White), Academic Press, New York, pp. 287–303.

Rosenthal, G. A. and Janzen, D. H. (1979) *Herbivores. Their Interaction with Secondary Plant Metabolites*. Academic Press, New York.

Silvertown, J. W. (1982) No evolved mutualism between grasses and grazers. *Oikos*, **38**, 253–9.

Sørensen, T. (1953) A revision of the Greenland species of *Puccinellia* Parl. *Meddelelser om Gronland*, **136**, 1–179.

Sousa, W. P. (1984) The role of disturbance in natural communities. *Ann. Rev. Ecol. Syst.*, **15**, 353–92.

Sprent, J. I. (1979) *The biology of nitrogen-fixing organisms*. McGraw-Hill, New York.

Sprugel, D. G. (1984) Density, biomass, productivity, and nutrient cycling changes during stand development in wave-regenerated balsam fir forests. *Ecol. Monogr.*, **54**, 165–86.

Stapledon, R. G. (1928) Cocksfoot grass (*Dactylis glomerata* L.) ecotypes in relation to the biotic factor. *J. Ecol.*, **16**, 71–104.

Stebbins, G. L. (1981) Coevolution of grasses and herbivores. *Ann. Missouri Bot. Gard.*, **68**, 75–86.

Stenseth, N. C. (1978) Do grazers maximize individual plant fitness? *Oikos*, **31**, 299–306.

Stenseth, N. C. (1983) Grasses, grazers, mutualism and coevolution: a comment about handwaving in ecology. *Oikos*, **41**, 152–3.

Tikhomirov, B. A. (1959) *Relationship of the Animal World and the Plant Cover of the Tundra*. Publication of the Botanical Institute, Academy of Science of the USSR, Moscow and Leningrad. (Trans. E. Issakoff and T. W. Barry; ed. W. A. Fuller), The Boreal Institute, Calgary, Canada.

Tranthan, P. (1983) Clonal interactions of *Trifolium repens* and *Lolium perenne*. PhD Thesis, University of Wales, Bangor.

Turkington, R. (1985) Variation and differentiation in populations of *Trifolium repens* in permanent pastures. *Studies on Plant Demography: A Festschrift for John L. Harper* (ed. J. White), Academic Press, London, pp. 69–82.

Turkington, R. and Burdon, J. J. (1983) Biology of Canadian weeds. 57. *Trifolium repens* L. *Can. J. Plant Sci.*, **63**, 243–66.

Vitousek, P. M. (1984) Litterfall, nutrient cycling and nutrient limitation in tropical forests. *Ecology*, **65**, 285–98.

Wharton, W. G. and Mann, K. H. (1981) Relationship between destructive grazing by the sea urchin, *Strongylocentrotus droebachiensis*, and the abundance of American lobster, *Homarus americanus*, on the Atlantic coast of Nova Scotia. *Can. J. Fish. Aquat. Sci.*, **38**, 1339–49.

Whitham, T. G. and Slobodchikoff, C. N. (1981) Evolution by individuals, plant–herbivore interactions, and mosaics of genetic variability: The adaptive significance of somatic mutations in plants. *Oecologia*, **49**, 287–92.

CHAPTER 14

The C-S-R model of primary plant strategies – origins, implications and tests

J. PHILIP GRIME

14.1 INTRODUCTION

In both evolutionary biology and in ecology much effort has been devoted to the search for generalizing principles. One approach has been to seek to develop a universal functional classification of organisms. On first inspection this may seem an impossible task because there are obvious differences between autotrophs, herbivores, carnivores and decomposers, and each of these groups is itself represented by an immense variety of taxa, life-forms and physiologies. However, there is now a considerable amount of evidence suggesting that beneath this diversity there is a common pattern of evolutionary and ecological specialization which is highly relevant to our understanding of the structure and dynamics of communities and ecosystems.

Evidence of widely recurring types of specialization in plants and animals appears comparatively early in the scientific literature (e.g. Macleod, 1894; Ramenskii, 1938) and has been the subject of numerous books and review articles (e.g. MacArthur and Wilson, 1967; Stearns, 1977; Southwood, 1977; Whittaker and Goodman, 1979; Parry, 1981; Greenslade, 1983). These sources will not be reviewed here, instead attention will be confined to botanical evidence, most of which refers to the vascular plant flora of the British Isles. Nevertheless, strong parallels can be observed between the patterns which will be described and those reported for a wide range of autotrophic and heterotrophic organisms (e.g. Downes, 1964; Southwood 1977; Raven, 1981; Shepherd, 1981; Dring, 1982; Pugh, 1980; Greenslade, 1983; Cooke and Rayner, 1984; Lee, 1985; Jarvinen, 1986).

14.2 DATA COLLECTION

Since 1960, a long-term research program has been conducted at the University of Sheffield with the objective of characterizing the field ecology and laboratory characteristics of the commoner herbaceous plants of the British Isles. This research relies essentially upon large-scale comparisons of the field and laboratory characteristics of groups of plants drawn from contrasted habitats. The rationale for the program has been described in a series of papers (Grime, 1965, 1966; Grime and Hodgson, 1969; Grime and Hunt, 1975) and may be summarized as follows:

1. Mechanisms excluding an organism, or reducing its abundance, in a particular type of habitat may be suggested on the basis of the differences in requirements or in tolerance which exist between it and the organisms which are more successful in the habitat.
2. Comparisons between species of contrasted ecology reveal many differences and it is difficult to determine which, if any, are of ecological significance. This may be resolved to some extent by confining attention to the more consistent differences between large numbers of species which are successful (or unsuccessful) in the habitat under study. The advantage of this research strategy, as explained by Clutton-Brock and Harvey (1979), is that 'as more species are considered, it becomes progressively more difficult to fit several adaptive hypotheses to the empirical facts'.
3. Where available, populations of the same species drawn from contrasted habitats may provide opportunities to examine variation with respect to a smaller number of potentially-critical characteristics. However, many of the features which differentiate between species of contrasted ecology are the product of extended periods of evolution and are not adequately reflected in micro-evolutionary change within species. It seems advisable to review evidence from intraspecific studies within the broader context provided by large-scale interspecific comparisons.
4. It is rarely profitable to examine variation in a single attribute without reference to other characteristics of the organisms under study. Ecological specialization is usually associated with correlated changes within a set of traits (Hutchinson, 1959; MacArthur and Wilson, 1967; Pianka, 1970).

The series of comparative studies conducted in Sheffield has produced a large body of data documenting various aspects of the biology of native plants of contrasted ecology. This information has been used in conjunction with other published sources to provide a monograph (Grime, Hodgson and Hunt, 1988) containing standardized accounts of the morphology, life-history, physiology and reproductive biology of 281 common species of

flowering plants and pteridophytes, together with more skeletal character-
izations of the essential biology of 234 less common species. It is this dataset
and the insights gained during its collection which provide the basis for the
plant strategy theories which are reviewed in this chapter.

14.3 THE INITIAL HYPOTHESIS

The ideas leading to the C-S-R model first appeared in 1973 in a debate in
Nature; the matter at issue was the role of competition in the control of
species diversity in herbaceous vegetation. Newman (1973) maintained that
on fertile soils low diversity in productive, relatively undisturbed vegetation
was the result of competition for light, whereas on infertile soils he
suggested that intense competition for limiting mineral nutrients was the
dominant process but did not lead to competitive exclusion because small-
scale heterogeneity in the rhizosphere permitted a degree of niche differentiation far in excess of that possible in the aerial environment. In contesting
this interpretation, Grime (1973a,b) used the experimental evidence of
Donald (1958) to substantiate the view (a) that competitive exclusion in
undisturbed productive vegetation is the result of intense competition above
and below ground and (b) that on infertile soils there is a general decline in
the intensity of competition as a consequence of the reduced stature and low
potential growth rates of the constituent species. It is interesting to note that
this argument did not extend to the role of vegetation disturbance; here
there was agreement that diversity could be increased where the disruptive
effects of climate or management debilitated potential dominants and
initiated micro-successional mosaics, a view elaborated by Whittaker and
Levin (1977), Huston (1979), Connell (1979) and Pickett (1980).

The views expressed in the debate with Newman evolved into the C-S-R
model (Grime, 1974), in which it was proposed that competition reached its
maximum intensity in circumstances where high productivity and lack of
disturbance permitted monopoly by large, fast-growing perennial plants.
Also of critical importance in the model was the idea that two different types
of phenomena (stress and disturbance) were capable of restricting competition. Stresses (most notably resource shortages) exerted this effect by
reducing the growth of potentially strong competitors whereas disturbance
prevented the development of robust phenotypes through repeated physical
damage to the developing plants. At moderate intensities the effects of stress
and disturbance (separately or more usually in some combination between
the two) was merely to restrict the vigor of potentially strong competitors
and to permit plants of lower competitive ability to co-exist with them. At
severe intensities, however, both stress and disturbance were said to be
capable of exerting distinctive forms of natural selection. The low productivity associated with severe and continuous stress was conducive to long-

lived, slow-growing evergreens, whereas in productive but frequently disturbed habitats, ephemerals were promoted.

This is an extremely simplified description of the C-S-R model at a rudimentary stage of its development. However, before proceeding further it will be helpful to clarify two aspects of the model; (1) the C-S-D equilibrium and (2) the selection processes associated with competition, stress and disturbance.

The C-S-D equilibrium

The C-S-R model proposes that the vegetation which develops in a particular place and at one point in time is the result of an equilibrium which is established between the intensities of stress (constraints on production), disturbance (physical damage to the vegetation) and competition (the attempt by neighbors to capture the same unit of resource). In this model stress and disturbance control the intensity of competition by restricting the density and vigor of the vegetation. In the short term this control is exerted through immediate impacts on the established plants. Over a longer time-span control occurs by modification of the species and genotypic composition of the vegetation through selective effects on extinctions and immigrations. Where stress and disturbance remain low, plants of exceedingly high competitive ability will eventually occupy the site and a drift towards monoculture will occur. Where severe stress and frequent disturbance coincide no vegetation is possible. The viable equilibria between stress, disturbance and competition occupy a triangular area, and the characteristics of the plants expected to occupy any particular position within the triangle are predictable from their location at a particular intersection between the three coordinates of the triangle (Grime, 1979).

In the real world, the C-S-D equilibrium varies from place to place even within a plant community and on a diurnal, seasonal and successional time-scale. For this reason communities often contain species of widely different biology. In modern, floristically depauperate landscapes many species may be under-dispersed and there may be long delays in adjustment of the species composition of the vegetation to changes in the intensities of stress or disturbance; here perhaps we may suspect that the role of intraspecific variation in the occupying species will be enlarged.

The selection processes associated with competition, stress and disturbance

Where productive, undisturbed habitats are colonized by robust, perennial plants of high potential growth rate the zones immediately above and below the ground surface are occupied by a dense, rapidly expanding biomass. In the aerial environment, a shaded stratum extends upward beneath a rapidly ascending layer of foliage; fatalities are conspicuous among those individuals which are outstripped by their neighbors and become trapped in the shaded zone. Here the high respiratory burden and etiolated tissues render the suppressed individuals particularly susceptible to pathogens (Vaartaja, 1952). However, fatalities in such dense vegetation are not simply the result of events above ground. Competition for light imposes a severe drain on the carbon and energy reserves of the plant, which may lead to starvation of the root system and its confinement to zones of nutrient depletion within the rhizosphere (Nye, 1966, 1969; Bhat and Nye, 1973; Fitter and Hay, 1981). Equally important however, is the demand for mineral nutrients and water imposed by the rapid growth and turnover of foliage characteristics of plants competing within an ascending canopy. The interdependence of root and shoot is therefore a crucial part of the analysis of the selection forces which operate upon competing plants. Elsewhere (Mahmoud and Grime, 1976) the implications of this scenario have been considered in relation to the phenomenon of co-variance in the resource 'foraging' abilities of leaves and roots. At this point, however, the essential argument is that competition generates severe spatial gradients in resources and selection is likely to favor strongly those genotypes in which high morphological plasticity facilitates escape from the depletion zones, sustains resource capture and maintains reproductive fitness.

A marked contrast is immediately evident when attention is turned to the selection processes operating in habitats where severe stress restricts plant production to a continuously low level. Here only slow rates of capture of the limiting resources are possible and both survival and reproduction depend crucially upon the capacity of the plant to remain viable through long periods in which little growth is possible; this may be expected to confer a selective advantage upon species in which there is an uncoupling of growth from resource intake (Grime, 1977). Protection of the tissues (and the investment of captured mineral nutrients which they represent) against herbivory is also conspicuous in plants of chronically unproductive habitats (Bryant and Kuropat, 1980; Coley, Bryant and Chapin, 1985).

Where frequent and severe disturbance becomes the dominant influence upon vegetation, natural selection is likely to favor those genotypes in which rapid growth and early reproduction increase the probability that sufficient offspring will be produced to allow survival and re-establishment of the population. There is little difficulty in recognizing the relevance of this pattern of natural selection to circumstances where fertile habitats are subject to frequent mechanical disturbance (e.g. arable fields, gardens and heavily trampled paths); here through man's activities brief opportunities are being created which can be exploited by plants with a condensed life-history. More careful analysis is required, however, where the agents of disturbance are climatic. Where the effect of summer desiccation is to lower the waterlevel of a pond, the resulting exposure of bare mud may create a temporary but productive habitat which can be colonized by ephemeral plants (Salisbury, 1967; Furness and Hall, 1981). An essentially similar opportunity for ephemerals occurs during the vernal phase in deciduous woodlands on fertile soils in North America (Baskin and Baskin, 1985) and a further example is provided in certain desert areas where soil fertility is sufficient to allow extremely rapid development of ephemerals following rain showers (Went, 1949). It is of vital significance in assessing the predictive value of the C-S-R theory to recognize that in all these instances exploitation by ephemerals coincides with short duration but relatively high productivity in the temporal niche. At those pond margins, woodlands and deserts where mineral nutrients are strongly limiting we may predict that the low quality of the 'growth window' will preclude exploitation by ephemerals; here conditions are more likely to favor the uncoupling of resource-capture from growth which is characteristic of stress-tolerators (e.g. pond margin *Isoetes* spp., woodland evergreens, desert succulents).

14.4 THE CONCEPT OF STRATEGIES

In preceding sections, it has been argued that the C-S-D equilibrium is of central importance in the determination of vegetation structure and it has been suggested that C, S and D impose distinctive forms of natural selection. It is now necessary to consider whether evolutionary responses to high intensities of C, S and D are sufficiently predictable to provide the corner-stones of a universal functional classification of plants. Following the theories of Stebbins (1974) it may be supposed that according to their evolutionary history taxa will respond in different ways to the same selection pressure and it could be argued that against the different climatic, edaphic and biotic backgrounds afforded by contrasted biomes, evolutionary responses to C, S and D will differ radically. With these complications in mind Grubb (1985) has suggested that only complex arrays can adequately describe the range of functional types actually occurring in nature. Ultimately, these discussions can be resolved only by further research of the kinds described later in this chapter. However, the following questions appear to be of critical importance in the current debate.

1. The existence of great variety in the characteristics of species and populations is not in dispute, nor is the relevance of this detailed variation to our understanding of the 'fine-tuning' of plant ecologies. The question at issue refers to a higher level of organization and concerns the existence (or not) of recurrent, predictable patterns of specialization, recognition of which allows a functional analysis of communities and ecosystems.

2. There can be little doubt that evolutionary histories can strongly influence paths of contemporary ecological specialization. The critical question to be addressed here is whether different evolutionary histories can lead to radically different solutions to the same ecological problem or do they merely determine the extent to which particular taxa approach the same basic solution.

From diverse schools of research there is already available a considerable amount of evidence in support of the view that with respect to basic characteristics of life-history and physiology paths of ecological and evolutionary specialization in response to C, S and D are remarkably stereotype. This has led to the suggestion (Grime, 1977, 1979) that three primary strategies exist, each recognizable from a characteristic set of traits. Use of the term 'strategy' to describe such patterns of specialization remains controversial. Some biologists (Hutchinson, 1959; MacArthur and Wilson, 1967; Pianka, 1970; Southwood, 1977, Maynard Smith, 1982; May and Seger, 1986) have used the term widely whilst others (Harper, 1982; Godwin, 1985) have taken strong exception to it. With its teleological and anthropomorphic connotations, the term is not ideal and it is understandable that some biologists have preferred to use neutral expressions such as 'set of traits' or 'syndrome' (Stebbins, 1974).

I do not feel a commitment to the term 'strategy' but retain its use as a mark of respect for those who first used the word. Their achievement was to recognize that organisms exhibit sets of co-adapted traits which are predictably related to their ecology. This simple concept has more than any other allowed ecology to begin its escape from a morass of parochial and undigested observations; against this the semantic objections to 'strategy' pale into insignificance.

Here a strategy is defined as 'a grouping of similar or analogous genetic characteristics which recurs widely among species or populations and causes them to exhibit similarities in ecology' (Grime, 1979) and a primary strategy is recognized as one engaging the fundamental activities of the organism (resource capture, growth, reproduction) and recurring widely in both animals and plants.

A distinctive feature of the strategy concepts reviewed in this paper is the separation of the strategy exhibited in the established (adult) phase from that of the regenerative (juvenile) phase. The need for this separation has become apparent through the work of several biologists including Stebbins (1951, 1971, 1974), Wilbur, Tinkle and Collins (1974), Grubb (1977), Gill (1978) and Grime (1979) who have all recognized the peculiar nature of the selection forces and design constraints which determine the characteristics of offspring. In plants this has led to the suggestion (Table 14.1) that there are five major types of regenerative strategies which differ in such features as parental investment, mobility and dormancy and confer different predictable sets of ecological capacities and limitations upon the organism. In many plants (and some animals) the same genotypes may be capable of regenerating by quite different mechanisms, leading to the hypothesis (Grime, 1979) that ecological amplitude is determined not only by genetic variability and phenotypic plasticity but also by regenerative flexibility, a function of the number of regenerative strategies.

A detailed discussion of the traits associated with the three primary strategies is available in Grime (1979) which also contains a description of the five regenerative strategies of Table 14.1 and their inter-relationships with C, S and R. Here only a brief account of the primary strategies will be attempted with particular emphasis upon flowering plants and upon those

Table 14.1 Five regenerative strategies of widespread occurrence in terrestrial vegetation.

Strategy	Functional characteristics	Conditions under which strategy appears to enjoy a selective advantage
Vegetative expansion (V)	New shoots vegetative in origin and remaining attached to parent plant until well established	Productive or unproductive habitats subject to low intensities of disturbance
Seasonal regeneration (S)	Independent offspring (seeds or vegetative propagules) produced in a single cohort	Habitats subjected to seasonally predictable disturbance by climate or biotic factors
Persistent seed or spore bank (B_s)	Viable but dormant seeds or spores present throughout the year; some persisting more than 12 months	Habitats subjected to temporally unpredictable disturbance
Numerous widely dispersed seeds or spores (W)	Offspring numerous and exceedingly buoyant in air; widely dispersed and often of limited persistence	Habitats subjected to spatially unpredictable disturbance or relatively inaccessible (cliffs, walls, tree trunks, etc.)
Persistent juveniles (B_j)	Offspring derived from an independent propagule but seedling or sporeling capable of long-term persistence in a juvenile state	Unproductive habitats subjected to low intensities of disturbance

attributes which are of particular importance or knowledge of which has been amplified by recent research.

Before proceeding further some comment is necessary concerning the comparatively large number of characters associated with the primary strategies. It might be considered unlikely that so many variable attributes of plants would conform to these patterns. However, it is the cardinal assertion of the C-S-R model of primary plant strategies that under the distinctive selection pressures associated with the extremes of C, S and D the range of adaptive possibilities is extremely constrained such that only particular evolutionary solutions (C, S and R) are viable. Further conformity arises from the fact that each of these solutions depends upon an integrated response involving most of the fundamental activities of the plant.

14.5 RUDERALS

Two plant characteristics in particular are relevant to analysis of the population dynamics and ecology of plants exploiting habitats subject to frequent and severe disturbance. The first is a potentially high relative growth rate (Grime and Hunt, 1975) during the seedling phase whilst the second is the early onset of flowering, a feature which often coincides with self-pollination and rapid maturation and release of seeds. In arable weeds and ephemeral plants of frequently and severely disturbed habitats such as paths, these characteristics undoubtedly confer resilience and explain the remarkably rapid population fluctuations observed in species such as *Matricaria matricarioides*, *Stellaria media*, *Polygonum aviculare* and *Senecio vulgare*. As documented by Salisbury (1942), it is also characteristic of the ruderal strategy for allocation to seed production to be sustained, as a proportion of the total biomass, in plants stunted by water or mineral nutrient stress or as a consequence of growth at high population density. The tendency for early and heavy allocation to reproduction evident in the life-history, physiology and breeding systems of ruderals also plays an integral part in their failure to exploit relatively undisturbed habitats. The early diversion of captured resources into flowers and seeds is not compatible with development of the extensive root and shoot systems necessary for dominance and extended occupation of productive, stable habitats nor is it conducive to survival in undisturbed but highly stressed environments where success is usually associated with very conservative patterns of resource utilization.

14.6 COMPETITORS

Reference to the list of plant attributes associated with the competitive strategy (Grime, 1979) reveals several which are obviously related to the capacity to monopolize resource capture in productive, relatively undisturbed environments. These include a high potential relative growth rate, tall stature and the tendency to form a consolidated growth form by vigorous lateral spread above and below ground. These characteristics are all evident in the large clonal herbs which we may describe as the 'classical competitors' of the British Isles (e.g. *Chamerion angustifolium, Urtica dioica, Reynoutria japonica, Phalaris arundinacea, Epilobium hirsutum, Petasites hybridus.* The high rates of resource capture achieved by these plants are also due to other less obvious but equally important characteristics. In temperate regions with a well-defined growing season these include the formation of substantial underground storage organs which fuel the initial surge in root and shoot growth in the spring and allow a very large peak in shoot biomass to be developed in summer (Al-Mufti *et al.*, 1977). Perhaps most important of all, however, is the high degree of morphological plasticity in the development of roots and shoots. This feature, coupled with the short life-span of individual leaves and fine roots, brings about a constant re-adjustment in the spatial distribution of the leaf canopy and the actively absorbing part of the root system during the growing season. The consequence of this constant projection of new leaves and roots into the resource-rich zones of the patchy environments created by competing plants is the phenomenon of 'active-foraging' (Grime, 1977, 1979) which has been analysed experimentally and has been shown to be superior in resource capture from productive environments to mechanisms involving less dynamic root and shoot systems (Crick, 1985; Grime, Crick and Rincon, 1986).

In comparison with ephemeral plants of frequently disturbed productive habitats (ruderals), species exhibiting the competitive strategy often show a temporary delay in the onset of seed production (Boot, Raynal and Grime, 1986). In terms of Darwinian fitness the advantage of this developmental pattern is not difficult to recognize in relation to the habitats normally exploited by these species; in undisturbed productive conditions rapid vegetative monopoly is an essential prelude to sustained reproductive output over many years. Equally, however, there can be little doubt that the delayed reproduction of the competitors is one of the major factors restricting their resilience and abundance in severely disturbed habitats.

The costs involved in the 'active foraging' for light, mineral nutrients and water characteristic of the competitive strategy are considerable in terms of the high rates of re-investment of captured resources in the construction of new leaves and roots and in their rapid senescence. To these costs we must add those associated with the high rates of herbivory experienced by the weakly defended tissues of many competitors (for evidence of this phenomenon and an evolutionary explanation for it see Grime (1979), Coley (1983), Coley, Bryant and Chapin (1985) and Edwards and Wratten

(1985)). We suspect therefore that there are severe penalties attached to 'active foraging' and that these severely restrict the success of the competitive strategy in chronically unproductive habitats.

14.7 STRESS TOLERATORS

Under the last heading it has been suggested that heavy expenditure of captured resources in new leaves and roots will be of selective advantage only in circumstances where active foraging gains access to large reserves of light energy, water and mineral nutrients. From this argument we may suspect that competitors will fail in habitats where productivity is low and resource availability is brief and unpredictable (e.g. light as sunflecks, mineral nutrients as short pulses from decomposition processes). In these circumstances conservation of captured resources is of primary significance and successful plants are likely to be those with the capacity to harvest and retain scarce resources in a continuously hostile physical environment. In keeping with this prediction, we find that the leaves of plants of unproductive environments tend to be comparatively long-lived, morphologically inflexible structures which are strongly defended against herbivory (Grime, 1979; Bryant and Kuropat, 1980; Coley, 1983; Cooper-Driver, 1985). Conservation of resources is also apparent in the low potential relative growth rates and delayed onset of reproduction in these plants (Grime and Hunt, 1975; Tamm, 1956; Inghe and Tamm, 1985).

Stress tolerators, despite their many common features of life history and physiology, are associated with a wide range of life forms and ecologies. In Britain, for example, there are stress tolerators characteristic of calcareous soils (*Koeleria macrantha, Primula veris*), acidic soils (*Juncus squarrosus, Nardus stricta*), droughted habitats (*Sedum acre, Thymus praecox*), wetland (*Carex panicea, Succisa pratensis*), shaded situations (*Sanicula europaea, Viola riviniana*) and metal-contaminated spoil (*Minuartia verna*). Quite clearly, there are important distinctions to be drawn within the general category 'stress tolerator' and the literature contains an abundance of references to evidence of mechanisms of tolerance specific to particular types of stress tolerators. This diversity among the ranks of the stress tolerators prompts the question, 'What is the nature of the selection force(s) responsible for the features common to stress tolerators?' Two possible answers to this question deserve consideration.

The first proposes that the stress tolerant traits (Grime, 1979) are an inevitable evolutionary response to chronically low productivity and are relatively independent of the nature of the stresses constraining production. The alternative explanation is to suggest that despite superficial differences all the habitats exploited by stress tolerators share a common underlying stress.

Further research is required to test these ideas but the balance of evidence (Grime, 1979; Chapin, 1980) is in favor of the latter hypothesis and it seems highly probable that the common underlying stress is low availability of mineral nutrients, especially phosphorus and nitrogen. The notion of limiting mineral nutrient elements as the ultimate determinants of the stress-tolerant strategy in plants is not incompatible with the concept of a recurring pattern of evolutionary specialization involving all major aspects of the plant. Most plant activities depend upon the level of supply of mineral nutrients and, as proposed by Mahmoud and Grime (1976), Grime (1977), Chapin (1980) and Atkinson and Farrar (1983), it seems likely that conservative mechanisms of mineral nutrient capture and utilization will invariably be associated with constraints on both carbon nutrition and reproductive activity.

14.8 IMPLICATIONS FOR COMMUNITY DYNAMICS

In a number of publications (Grime, 1979; Leps, Osbornova-Kosinova and Rejmanek, 1982; Grime, 1984, 1986, 1987) the C-S-R model has been used as the basis for predicting the mechanisms controlling the structure and dynamics of plant communities. Some of the main predictions are as follows:

1. Where vegetation in a productive terrestrial habitat suffers frequent and severe disturbance, the ruderal strategy will be strongly represented and the populations of bryophytes and herbaceous plants present will be too sparse and ephemeral to exert a significant degree of dominance. Unless disturbance is excessive, the potential for diversity will be relatively high. The ephemeral, weakly defended tissues of ruderals have low resistance to predation and to mechanical or climatic damage but resilience of the community is maintained by rapid growth rates, precocious reproduction and usually also by the development of persistent banks of seeds or spores. Reduction in the intensity of vegetation disturbance will allow dominance by larger perennials and will lead to rapid loss of diversity.
2. Plant communities developed in chronically unproductive but relatively undisturbed conditions will consist of stress-tolerant lichens, bryophytes, small herbs or low shrubs. Interactions between neighbors may determine which species are most abundant but dominance will be weakly developed and, except where the habitat is very skeletal, diversity may be relatively high. Resistance to predation and to physical disturbance will be considerable but resilience will be low.
3. Under productive, relatively undisturbed conditions, plant biomass will expand rapidly and dominance of the community will be achieved by herbs, shrubs or trees conforming to the competitive strategy. Domi-

nance may approach a state of local monoculture due to the rapid growth, dense clonal structure of some herbs, shrubs and trees and the resulting zones of resource depletion above and below ground. Competitive dominants may further reduce diversity as a result of the dynamic foraging behavior of their shoots and roots which will create a highly unpredictable and hazardous environment for subordinate plants of narrower niche width. In comparison with ruderal communities, there will be greater resistance to disturbance but the longer life-spans and delayed reproduction will lower resilience.

4. In relatively undisturbed habitats of modest biomass and low productivity (many shrublands, heathlands, grasslands and infertile mire) and in mature ecosystems of large biomass but low productivity (many primary forests) dominance will be attained by plants with predominantly stress-tolerant traits. Dominance will be achieved slowly through the capacity to harvest and retain resources and in accord with Odum's (1969) theory of ecosystem development mineral nutrients will be tightly recycled. The more constant environment and more predictable matrix of living and dead materials associated with stress-tolerant dominants will allow greater opportunity for subordinate plants including epiphytes. Resistance to disturbance by predation or climatic factors will be high but resilience of the community will be exceedingly low.

5. Except in the extreme conditions corresponding to the corners of the C-S-R model (see (1)–(3) above), it is unlikely that communities will consist of species of similar strategy. Spatial heterogeneity within a stand of vegetation may arise from differences in topography, soil and microclimate and will be imposed also by vegetation, fauna and microflora. Habitats also change seasonally and on a successional time scale. One effect of this variation will be to cause spatial and temporal changes in the C-S-D equilibrium. Analysis of the strategic diversity within a plant community will therefore provide clues to the mechanisms which permit co-existence and control the relative abundance of species.

6. Further insights into the control of species composition and relative abundance in communities are available from the hump-backed model (Grime, 1973a), a close relative of the C-S-R model (see Fig. 14.1). Two main predictions, discussed in Grime (1979), arise from the hump-backed model:

   a. Communities remain species-poor in vegetation subjected to high intensities of stress and/or disturbance (corresponding to the left-hand side of the model) and also in the quite different circumstances where dominance is permitted (on the right-hand side).

   b. Species-rich communities (in the 'hump') contain a majority of subordinate plants which are neither potential dominants nor capable of surviving in extreme habitats.

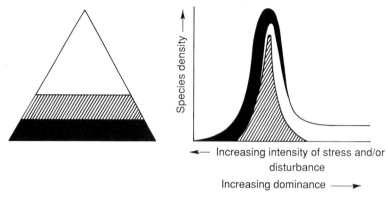

Fig. 14.1 Scheme describing the distribution of three floristic elements in the C-S-R model (left) and in the humped-back model (right). □ potential dominants; ■ plants highly adapted to extremely disturbed and/or unproductive conditions; ▨ subordinates.

The diagrams in Fig. 14.2 illustrate the potential of the C-S-R model to interpret various familiar successional phenomena. In each diagram the strategies of the dominant plants at particular points in time are indicated by the position of arrowed lines within the triangle. The passage of time in years during succession is represented by the numbers on each line and shoot biomass by the size of the circles.

Fig. 14.2(f) describes the course of secondary succession in a forest clearing on a moderately fertile soil in a temperate climate. Initially biomass development is rapid and there is a fairly swift replacement of species as rapidly growing herbs, shrubs and trees successively dominate the vegetation. Later the course of succession deflects towards the stress-tolerant corner of the triangle reflecting a gradual transition in dominance from species with high rates of resource capture and loss to those in which resources, particularly mineral nutrients, are efficiently retained.

In Fig. 14.2(g) secondary succession is portrayed for a site of lower productivity. The processes are essentially similar to those of Fig. 14.2(f) but the successional parabola is shallower and the plant biomass is reduced by the earlier onset of mineral nutrient limitation.

Primary succession on bare rock is examined in Fig. 14.2(h). Here the initial colonists are stress-tolerant lichens and bryophytes. Biomass development is exceedingly slow, the incursion of herbs and shrubs gradually occurring as soil formation takes place; this interpretation coincides with the facilitation model of Connell and Slatyer (1977).

Vegetation responses to major perturbations are also capable of description in strategic terms. The loop in Fig. 14.2(i) represents the cycle of

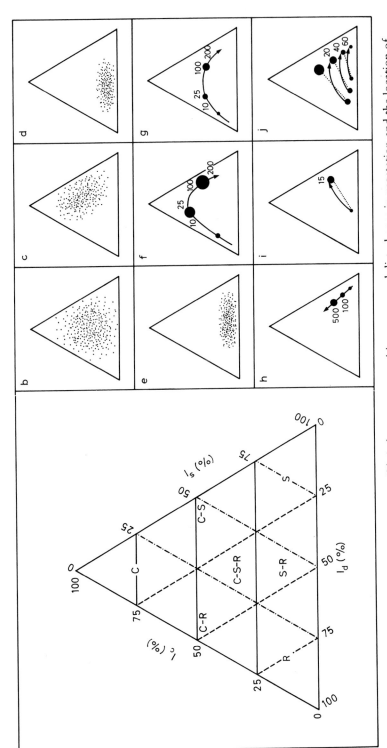

Fig. 14.2 (a) Model describing the various equilibria between competition, stress and disturbance in vegetation and the location of primary and secondary strategies. C, competitor; S, stress tolerator; R, ruderal; C-R, competitive-ruderal; S-R, stress tolerant ruderal; C-S, stress tolerant competitor; C-S-R, 'C-S-R- strategist'. I_c, intensity of competition (———); I_s, intensity of stress (– – –); I_d, intensity of disturbance (– – –). (b—e) The strategic range of four life forms: (b) herbs, (c) trees and shrubs, (d) bryophytes, (e) lichens. (f—j) Successional diagrams (see text).

vegetation change associated with rotational burning of *Calluna vulgaris* moorland in Northern Britain. A more complex sequence is depicted in Fig. 14.2(j) which attempts to explain the consequences of 'slash and burn' tropical agriculture where the declining mineral nutrient capital of the system may be expected to result in a series of arcs of progressively lower trajectory in successive cycles of vegetation destruction and recovery.

14.9 TESTS AND REFINEMENTS OF THE C-S-R MODEL

For the present, the most compelling evidence supporting the C-S-R model is the large number of very different ecological phenomena which appear to be explained by the theory. Efforts to falsify and, where necessary, refine the model are now required; several complementary approaches seem desirable.

Mathematical models

There is an urgent need for models which not only examine the influence of life history and reproductive schedules on fitness in defined circumstances, but also take account of resource capture and utilization. In particular there is a requirement for models analyzing the circumstances conducive to the extended life histories, slow growth rates, conservative patterns of resource use and heavy 'defence spending' said to be characteristic of adversity-selected organisms. A tentative contribution to this field is presented in Sibly and Grime (1986).

Intraspecific studies

Plants exploiting conditions corresponding to the extremities of the triangular model are expected to differ in genetic traits and are the products of natural selection over many generations in highly contrasted environments. According to their evolutionary history, families of flowering plants appear to differ considerably in their capacity to produce species with ruderal or competitive characteristics (Hodgson, 1986) and in the Pteridophyta a major impediment to adaptive radiation into the disturbed fertile habitats of agricultural land is discernible from the fact that only one ephemeral species (*Anogramma leptophylla*) has been recorded in the world flora (Grime, 1984).

This background is highly relevant to the choice of appropriate species and the development of realistic objectives in tests of the C-S-R model by selection experiments or comparison of populations of the same species collected from contrasted environments. It may be unreasonable to expect to find one species with genotypes corresponding to more than one corner of the C-S-R model. However, species which are predominantly C, S or R do not offer the best prospects for experimental tests. Among herbaceous plants, the most promising subjects appear to be species of relatively wide ecological amplitude and with characters intermediate between those associated with C, S and R (e.g. *Agrostis capillaris, Poa pratensis, Holcus lanatus, Deschampsia cespitosa, Trifolium repens, Plantago lanceolata*). All of these species are known to be genetically variable and would provide excellent experimental material. One insight into the potential of common wide-ranging herbaceous plants to allow tests of the C-S-R theory is available from studies of the common pasture grass *Poa annua* (Law, Bradshaw and Putwain, 1977; Law, 1979). These investigations have revealed variation between individuals with respect to the onset of flowering and have enabled an analysis of the trade-off between early allocation to reproduction and sustained vegetative vigor which is clearly an important component of the R-C dimension of the triangular model. There also exist a small number of pioneer studies which appear to explore intraspecific variation corresponding to the C-S axis of the model (Kruckeberg, 1954; Bocher, 1949, 1961).

Screening of individual traits

Standardized measurements of particular attributes on large numbers of species of contrasted ecology in the Sheffield region of Northern England (Grime and Hunt, 1975; Grime *et al.*, 1981) were involved in the development of the C-S-R theory. There is clearly scope for independent tests of aspects of the model by screening programs involving other strategy-related characteristics measured on plants drawn from other floras and following the rationale discussed earlier.

Multivariate screening and cluster analysis

Although measurement of individual traits provide limited tests of parts of the model, a more critical approach is to bring together information on diverse aspects of the biology of large numbers of plants drawn from contrasted habitats to seek evidence for or against the co-occurrence of attributes in the sets predicted by the C-S-R model. Although the scope for adequate tests using this approach is limited at present by the shortage of data, cluster analyses have been applied to a database for 273 common vascular plant species of the British flora and consisting of scores for 15 attributes of the established phase and nine relating to the regenerative phase. The results (Grime, Hunt and Krzanowski, 1987) have provided strong support for aspects of the C-S-R model.

Analyses based upon the characteristics of the established phase have revealed a large compact group of ephemeral plants with many of the predicted attributes of the ruderal strategy. Some stress tolerators have also emerged from the analysis as a coherent group recognizable as low-growing, evergreen plants with physically tough foliage but the remaining classes are heterogeneous ecologically and reflect the lack of physiological criteria and over-representation of morphological characters in the database.

Clustering of the regenerative attributes has produced four remarkably coherent groups. The first, including Pteridophytes and numerous Compositae and Onagraceae, corresponds quite clearly to the strategy of regeneration through the production of numerous buoyant seeds or spores (W in Table 14.1). The second contains a high proportion of species in which small seeds are known to be readily incorporated into a persistent buried seed bank (B$_s$ in Table 14.1, and including seed banks Types III and IV of Thompson and Grime, 1979). The two remaining classes to emerge from the analysis are dominated by comparatively large-seeded species and mainly correspond to two types of seasonal regeneration (S in Table 14.1). The first includes many grasses with elongated seeds, many with prominent awns, dormancy is lacking and there is a strong tendency for synchronous autumn germination (seed bank Type I, Thompson and Grime, 1979). Most of the species in the second group have a chilling requirement and germinate in the early spring (seed bank Type II, Thompson and Grime, 1979).

The outstanding feature of the results of the cluster analyses so far completed is the lack of consistent association between attributes of the established and regenerative phases. This appears to confirm the uncoupling predicted by C-S-R theory.

Use of controlled gradients of stress and disturbance

Plant responses to various intensities of stress and disturbance experienced either directly or through their determining effect on the intensity of competition for resources lie at the heart of the C-S-R model. Since both stress and disturbance are notoriously difficult to measure in natural environments, a simpler procedure is to examine the performance of plants of contrasted biology growing in isolation and in mixtures over crossed controlled gradients of mineral nutrient stress and vegetation disturbance.

An experiment evaluating this approach is currently in progress using seven grass species subjected to gradients of mineral nutrient supply and simulated grazing and trampling damage in square garden plots (Campbell, personal communication). Prior to the experiment each species has been classified with respect to strategy and predictions have been made of the zones of the plot which each species is expected to occupy in monoculture and in the presence of the other species. At this stage the most obvious feature of the experiment is the poor yield of all seven species in the zone experiencing both severe stress and frequent disturbance. Plant cover is confined in each plot to a triangular area, within which, as we might expect, maximum vegetative and reproductive vigor in all monocultures occurs in the productive, undisturbed sector. Detailed analysis of the modifying effect of interspecific competition on these patterns in the mixture treatments would be premature, but it is already evident that the strongest intensities of competitive exclusion coincide with low stress and infrequent disturbance.

14.10 ACKNOWLEDGEMENTS

I wish to record my thanks to all members of the Unit of Comparative Plant Ecology who participated in the studies described in this paper. I am particularly grateful to Mrs J. M. L. Mackey and Mrs N. Ruttle for help in preparing this chapter and to Mr B. D. Campbell for permission to refer to his unpublished work. Thanks are also due to the editors of this volume for their hospitality in Davis and their critical attention to this paper. This research was supported by the Natural Environment Research Council.

14.11 REFERENCES

Al-Mufti, M. M., Sydes, C. L., Furness, S. B., Grime, J. P. and Band, S. R. (1977) A quantitative analysis of shoot phenology and dominance in herbaceous vegetation. *J. Ecol.*, **65**, 759–91.

Atkinson, C. J. and Farrar, J. F. (1983) Allocation of photosynthetically-fixed carbon in *Festuca ovina* L. and *Nardus stricta* L. *New Phytol.*, **95**, 519–31.

Baskin, J. M. and Baskin, C. C. (1985) Life cycle ecology of annual plant species of cedar glades of Southeastern United States. in *The Population Structure of Vegetation* (ed. J. White), Junk, Dordrecht, pp. 371–98.

Bhat, K. K. and Nye, P. H. (1973) Diffusion of phosphate to plant roots in soil. I. Quantitative autoradiography of the depletion zone. *Plant Soil*, **38**, 161–75.

Bocher, T. W. (1949) Racial divergences in *Prunella vulgaris* in relation to habitat and climate. *New Phytol.*, **48**, 285–314.

Bocher, T. W. (1961) Experimental and cytological studies in plant species: VI. *Dactylis glomerata* and *Anthoxanthum odoratum*. *Bot. Tidskr.*, **56**, 314–55.

Boot, R., Raynal, D. J. and Grime, J. P. (1986) A comparative study of the influence of moisture stress on flowering in *Urtica dioica* and *Urtica urens*. *J. Ecol.*, **74**, 485–95.

Bryant, J. P. and Kuropat, P. J. (1980) Selection of winter forage by subarctic browsing vertebrates: the role of plant chemistry. *Ann. Rev. Ecol. Syst.*, **11**, 261–81.

Chapin, F. S. (1980) The mineral nutrition of wild plants. *Ann. Rev. Ecol. Syst.*, **11**, 233–60.

Clutton-Brock, T. H. and Harvey, P. H. (1979) Comparison and adaptation. *Proc. R. Soc. London B*, **205**, 547–65.

Coley, P. D. (1983) Herbivory and defensive characteristics of tree species in a lowland tropical forest. *Ecol. Monogr.*, **53**, 209–32.

Coley, P. D., Bryant, J. P. and Chapin, F. S. (1985) Resource availability and plant antiherbivore defence. *Science*, **230**, 895–9.

Connell, J. H. (1979) Tropical rain forests and coral reefs as open non-equilibrium systems, in *Population Dynamics* (eds R. M. Anderson, B. D. Turner and L. R. Taylor), Blackwell, Oxford, pp. 141–63.

Connell, J. H. and Slatyer, R. O. (1977) Mechanisms in natural communities and their role in community stability and organisation. *Am. Natur.*, **111**, 1119–45.

Cooke, R. C. and Rayner, A. D. M. (1984) *The Ecology of Saprotrophic Fungi: Towards a Predictive Approach*. Longman, London.

Cooper-Driver, G. (1985) Anti-predation strategies in pteridophytes – a biochemical approach. *Proc. R. Soc. Edin.*, **86B**, 397–402.

Crick, J. C. (1985) *The role of plasticity in resource acquisition by higher plants*. PhD thesis, University of Sheffield.

Donald, C. M. (1958) The interaction of competition for light and for nutrients. *Aust. J. Agric. Res.*, **9**, 421–32.

Downes, J. A. (1964) Arctic insects and their environment. *Can. Entomol.*, **96**, 279–307.

Dring, M. J. (1982) *The Biology of Marine Plants*. Arnold, London.

Edwards, P. J. and Wratten, S. D. (1985) Induced plant defences against insect grazing: fact or artefact? *Oikos*, **44**, 70–4.

Fitter, A. H. and Hay, R. K. M. (1981) *Environmental Physiology of Plants*. Academic Press, London.

Furness, S. B. and Hall, R. H. (1981) An explanation of the intermittent occurrence of *Physcomitrium sphaericum* (Hedw.) Brid. *J. Bryol.*, **11**, 733–42.

Gill, D. E. (1978) On selection at high population density. *Ecology*, **59**, 1289–91.

Godwin, H. (1985) *Cambridge and Clare*. Cambridge University Press, Cambridge.

Greenslade, P. J. M. (1983) Adversity selection and the habitat templet. *Am. Natur.*, **122**, 352–65.

Grime, J. P. (1965) Comparative experiments as a key to the ecology of flowering plants. *Ecology*, **45**, 513–15.

Grime, J. P. (1966) Shade avoidance and shade tolerance in flowering plants, in *Light as an Ecological Factor* (eds R. Bainbridge, G. C. Evans and O. Rackham), Blackwell, Oxford, pp. 281–301.

Grime, J. P. (1973a) Competitive exclusion in herbaceous vegetation. *Nature*, **242**, 344–7.

Grime, J. P. (1973b) Competition and diversity in herbaceous vegetation – a reply. *Nature*, **244**, 310–11.

Grime, J. P. (1974) Vegetation classification by reference to strategies. *Nature*, **250**, 26–31.

Grime, J. P. (1977) Evidence for the existence of three primary strategies in plants and its relevance to ecological and evolutionary theory. *Am. Natur.*, **111**, 1169–94.

Grime, J. P. (1979) *Plant Strategies and Vegetation Processes*. Wiley, Chichester.

Grime, J. P. (1981) An ecological approach to management, in *Amenity Grassland: an Ecological Perspective* (eds I. H. Rorison and R. Hunt). Wiley, London, pp. 13–55.

Grime, J. P. (1984) The ecology of species, families and communities of the contemporary British Flora. *New Phytol.*, **98**, 15–33.

Grime, J. P. (1985) Factors limiting the contribution of pteridophytes to a local flora. *Proc. R. Soc. Edin.*, **86B**, 403–21.

Grime, J. P. (1986) Manipulation of plant species and communities, in *Ecology and Design in Landscape* (eds A. D. Bradshaw, E. Thorpe and D. A. Goode), Blackwell, Oxford, pp. 175–94.

Grime, J. P. (1987) Dominant and subordinate components of plant communities – implications for succession, stability and diversity, in *Colonisation, Succession and Stability* (eds A. Gray, P. Edwards and M. Crawley), Blackwell, Oxford, pp. 413–28.

Grime, J. P., Crick, J. C. and Rincon, E. (1986) The ecological significance of plasticity, in *Plasticity in Plants* (eds D. H. Jennings and A. J. Trewavas), Company of Biologists, Cambridge, pp. 5–29.

Grime, J. P. and Hodgson, J. G. (1969) An investigation of the ecological significance of lime-chlorosis by means of large-scale comparative experiments. in *Ecological Aspects of the Mineral Nutrition of Plants* (ed. I. H. Rorison), Blackwell, Oxford, pp. 67–99.

Grime, J. P., Hodgson, J. G. and Hunt, R. (1988) *Comparative Plant Ecology: a Functional Approach to Common British Species*, London.

Grime, J. P. and Hunt, R. (1975) Relative growth-rate: its range and adaptive significance in a local flora. *J. Ecol.*, **63**, 393–422.

Grime, J. P., Hunt, R. and Krzanowski, W. J. (1987) Evolutionary physiological ecology of plants. in *Evolutionary Physiological Ecology* (ed. P. Calow), Cambridge University Press, Cambridge, pp. 105–25.

Grime, J. P., Mason, G., Curtis, A. V., Rodman, J., Band, S. R., Mowforth, M. A. G., Neal, A. M. and Shaw, S. (1981) A comparative study of germination characteristics in a local flora. *J. Ecol.*, **69**, 1017–59.

Grubb, P. J. (1977) The maintenance of species-richness in plant communities: the importance of the regeneration niche. *Biol. Rev.*, **52**, 107–45.

Grubb, P. J. (1985) Plant populations and vegetation in relation to habitat, disturbance and competition: problems of generalization. in *The Population Structure of Vegetation* (ed. J. White), Junk, Dordrecht, pp. 595–621.

Harper, J. L. (1982) After description. in *The Plant Community as a Working Mechanism* (ed. E. I. Newman), Special Publication No. 1 BES, Blackwell, Oxford, pp. 11–25.

Hodgson, J. G. (1986) Commonness and rarity in plants with special reference to the flora of the Sheffield region. III. Taxonomic and evolutionary aspects. *Biol. Conserv.*, **36**, 275–96.

Huston, M. (1979) A general hypothesis of species diversity. *Amer. Natur.*, **113**, 81–101.

Hutchinson, G. E. (1959) Homage to Santa Rosalia or why are there so many kinds of animals? *Am. Natur.*, **93**, 145–59.

Inghe, O. and Tamm, C. O. (1985) Survival and flowering of perennial herbs IV. *Oikos*, **45**, 400–20.

Jarvinen, A. (1986) Clutch size of passerines in harsh environments. *Oikos*, **46**, 365–71.

Kruckeberg, A. R. (1954) The ecology of serpentine soils: III. Plant species in relation to serpentine soils. *Ecology*, **35**, 267–74.

Law, R. (1979) The cost of reproduction in annual meadow-grass. *Am. Natur.*, **113**, 3–16.

Law, R., Bradshaw, A. D. and Putwain, P. D. (1977) Life history variation in *Poa annua*. *Evolution*, **31**, 233–46.

Lee, K. E. (1985) Ecological strategies. in *Earthworms, Their Ecology and Relationships with Soils and Land Use*. Academic Press, Sydney, pp. 102–31.

Leps, J. M., Osbornova-Kosinova, J. and Rejmanek, K. (1982) Community stability, complexity and species life-history strategies. *Vegetatio*, **50**, 53–63.

Nye, P. H. (1966) The effect of nutrient intensity and buffering power of a soil, and the absorbing power, size and root hairs of a root, on nutrient absorption by diffusion. *Plant Soil*, **25**, 81–105.

Nye, P. H. (1969) The soil model and its application to plant nutrition. in *Ecological Aspects of the Mineral Nutrition of Plants* (ed. I. H. Rorison), Blackwell, Oxford, pp. 105–14.

MacArthur, R. H. and Wilson, E. D. (1967) *The Theory of Island Biogeography*. Princeton University Press, Princeton NJ.

Macleod, J. (1894) Over de bevruchting der bloemen in het Kempisch gedeelte van Vlaanderen. *Deel II. Bot. Jaarboek*, **6**, 119–511.

Mahmoud, A. and Grime, J. P. (1976) An analysis of competitive ability in three perennial grasses. *New Phytol.*, **77**, 431–5.

May, R. M. and Seger, J. (1986) Ideas in ecology. *Am. Sci.*, **74**, 256–67.

Maynard Smith, J. (1982) *Evolution and the Theory of Games*. Cambridge University Press, Cambridge.

Newman, E. I. (1973) Competition and diversity in herbaceous vegetation. *Nature*, **244**, 310.

Odum, E. P. (1969) The strategy of ecosystem development. *Science*, **164**, 262–70.

Parry, G. D. (1981) The meanings of r- and K-selection. *Oecologia*, **48**, 260–4.

Pianka, E. R. (1970) On r- and K-selection. *Am. Natur.*, **104**, 592–7.

Pickett, S. T. A. (1980) Non-equilibrium co-existence of plants. *Bull. Torrey Bot. Club*, **107**, 238–48.

Pugh, G. J. F. (1980) Strategies in fungal ecology. *Trans. Br. Mycol. Soc.*, **75**, 1–14.

Ramenskii, L. G. (1938) *Introduction to the Geobotanical Study of Complex Vegetations*. Selkhozgiz, Moscow, 620 pp.

Raven, J. A. (1981) Nutritional strategies of submerged benthic plants: the acquisition of C, N and P by rhizophytes and haptophytes. *New Phytol.*, **88**, 1–30.

Salisbury, E. J. (1942) *The Reproductive Capacity of Plants*. Bell, London.

Salisbury, E. J. (1967) The reproduction and germination of *Limosella aquatica*. *Ann. Bot.*, **31**, 147–62.

Shepherd, S. A. (1981) Ecological strategies in a deep water red algal community. *Bot. Mar.*, **XXIV**, 457–63.

Sibly, R. M. and Grime, J. P. (1986) Strategies of resource capture by plants – evidence for adversity selection. *J. Theor. Biol.*, **118**, 247–50.

Southwood, T. R. E. (1977) Habitat, the templet for ecological strategies? *J. Anim. Ecol.*, **46**, 337–65.

Stearns, S. C. (1977) The evolution of life-history traits: a critique of the theory and a review of the data. *Ann. Rev. Ecol. Syst.*, **8**, 145–71.

Stebbins, G. L. (1951) Natural selection and the differentiation of angiosperm families. *Evolution*, **5**, 299–324.

Stebbins, G. L. (1971) Adaptive radiation of reproductive characters of angiosperms II. Seeds and seedlings. *Ann. Rev. Ecol. Syst.*, **2**, 237–60.

Stebbins, G. L. (1974) *Flowering Plants: Evolution Above the Species Level*. Arnold, London.

Tamm, C. O. (1956) Further observations on the survival and flowering of some perennial herbs. I. *Oikos*, **7**, 274–92.

Thompson, K. and Grime, J. P. (1979) Seasonal variation in the seed banks of herbaceous species in ten contrasting habitats. *J. Ecol.*, **67**, 893–922.

Vaartaja, O. (1952) Forest humus quality and light conditions as factors influencing damping off. *Phytopathology*, **42**, 501–6.

Went, F. W. (1949) Ecology of desert plants. II. The effect of rain and temperature on germination and growth. *Ecology*, **30**, 1–13.

Whittaker, R. H. and Levin, S. A. (1977) The role of mosaic phenomena in natural communities. *Theor. Pop. Biol.*, **12**, 117–39.

Whittaker, R. H. and Goodman, D. (1979) Classifying species according to their demographic strategy. I. Population fluctuations and environmental heterogeneity. *Am. Natur.*, **113**, 185–200.

Wilbur, H. M., Tinkle, D. W. and Collins, J. P. (1974) Environmental certainty, trophic level and resource availability in life history evolution. *Am. Natur.*, **108**, 805–17.

Editors' commentary on Part 6

In their emphasis on the distribution and abundance of species living together in a community, ecologists have naturally focused on comparative studies on population growth, demographic patterns of regulation, and species interactions. Natural selection is frequently invoked to explain various adaptive features of life histories as well as plant–animal coevolution. A life history is defined in terms of age-dependent birth and death rates with an emphasis on the first age of reproduction and the tradeoffs between growth (which is related to survivorship under competition, predation or nutrient scarcity) and reproduction. Accordingly, a population biologist measures selection intensities in terms of life cycle components of an aggregate fitness parameter, e.g. average reproductive value or Malthusian growth rate of a genotypic class. In an optimal life history, then, natural selection in a density-independent model maximizes the population growth rate (r), which also corresponds to an evolutionary stable strategy (Charlesworth, 1980), or maximizes carrying capacity (K) in a density-dependent environment. This is how 'r' and 'K' life history strategies are mathematically derived, and assuming genetic variation in birth and death rates, r- and K-selection could be simulated in a computer model (Anderson, 1971). (See Boyce (1984) for an updated and critical review.)

However, many attempts to test these models using limited assumptions of resource allocation and without adequate environmental description have not been successful. Temporal variation in life history traits further alters the predicted fitness relationships (Stearns, 1977). The concept of optimal life history is designed as a problem-solving algorithm under prescribed conditions of environmental heterogeneity and certain statistics of vital attributes. It can take many 'optimal' forms depending on the genetic and non-genetic constraints as well as random events. As Boyce (1984) noted, 'the precise nature of environmental variability is critical to understanding its potential impact on life history evolution; to collapse all environmental variability into one model of r- and K-selection is naive'. Empirical evidence from comparative ecology of related species supports the optimality arguments in general (e.g. Begon, Harper and Townsend, 1986). Since life-history traits bring together information from physiologial ecology and

395

community ecology, and we believe should include developmental biology and genetics to provide a complete fitness context, this pluralistic view of evolution seems attractive.

The chapters by Jefferies and Grime provide two different views of adaptation through life history evolution, and two levels of community organization. In Chapter 13 Jefferies discusses plant adaptation to herbivory in terms of an optimal model of grazing response in which the herbivore (Snow Geese) causes patchy disturbance in the vegetation, and nutrients are recycled rapidly through both the plant and the herbivore. Plant adaptation in such a community presumably involves life history evolution toward higher survivorship and recruitment success. Reproductive rates show high plastic response to grazing and competition. In Chapter 14 Grime lists characteristics that allowed him to define three strategies, namely competitive, stress tolerant and ruderal, which refer to the ability of plants to reproduce and persist in three kinds of habitat. The approach might be useful in making deductions and testing experimentally, as discussed by Grime, the patterns of survival, growth, reproduction, and resource allocation originating from one or more optimizing principles.

Fitness criteria have no single universal definition; different models emphasize different notions of ecological success and/or evolutionary stability (Lewontin, 1979). Accordingly, experiments on the role of life history variation in adaptive responses differ not only in the aggregate definitions of individual or population fitnesses but also in spatial–temporal scales. Causal factors of environmental heterogeneity that presumably elicit adaptive changes are not readily guessed, let alone proved. Harper (1983) succinctly argues for long-term and ingenious field studies on genetic variation and patterns of life history together. And finally, continuous variability of most life-history traits as well as aggregate fitness parameters (e.g. mean lifetime reproductive value or Wrightian, $\overline{W}$) pose many uncertainties in the genetic analyses of evolutionary processes. As Grime points out, many species represent combinations of traits listed under C, S and R strategies, and not idealized 'pure' strategies.

What is an even more serious difficulty for an evolutionary biologist is the lack of adequate genetic knowledge of variation within species as well as almost total lack of information on the past evolutionary dynamics. We need research on the adaptive role of various life-history descriptors and their correlates in morphology or physiology. For example, a comparative study of an outbreeding species (*Limnanthes alba*) and the related inbred (*L. floccosa*) suggested seed dormancy, seed germination requirements, flowering time, and reproductive effort to have evolved into their 'risky' and 'safe' strategies, respectively. Based on the means and variances of various life-history traits, and some habitat and climatic variables, these species seem to fit Grime's competitive and stress tolerant classes (Ritland and Jain,

1984). Theoretical models on the simultaneous evolution of earliness and higher seed dormancy, as observed in *L. floccosa*, supported our strategy arguments, provided the maximization of population size is a valid optimality criterion.

Most strategy classifications treat seed dormancy as an important ecological trait, particularly with reference to annual plant communities. Various physiological and even some morphogenetic–embryological factors controlling variation in dormancy at the species level justify further evolutionary research. However, sources and level of intraspecific genetic and non-genetic variation are often difficult to disentangle. In two California grassland annuals, *Avena barbata* and *Trifolium hirtum*, interpopulation variation in the dormancy of freshly harvested seed and in its rate of loss during the summer was highly significant and may be adaptive in avoiding germination after irregular rainstorms of the typically arid summer months (Jain, 1982). Estimates of heritability based on the ratio of between- and within-families variation components (quite useful in predominant selfers) varied among species and populations. The next logical step involved selection of families with low versus high dormancy and transplant experiments to test their fate in natural populations. Survivorship of these families did not fit predictions from their dormancy levels. It appears that estimates of heritable variation in dormancy can vary widely depending upon the test environments as well as the precise seed processing procedures; founding colonies in nature can invoke many uncontrolled perturbations; and the key selective factor, summer precipitation, was not adequately simulated within the time span of our experiment. There are probably other difficulties in such tests of what appear to be rather simple propositions about natural selection (Endler, 1986, for similar discussions).

Besides the logistic queries in research on dormancy estimates, one might also ask: Is seed dormancy a well-defined trait from the viewpoint of population genetics? Can we identify simply inherited morphological or physiological attributes correlated with dormancy? This seems to be true in some cases. Can we develop experimental procedures to obtain better estimates of heritability (viz. specific environmental controls; use of known genetic stocks)? Perhaps. But as noted in an earlier section, genetic aspects relate largely to the input variation in such studies; analyses of fitness, multivariate shifts in form and function and search for environmental mechanisms pose difficult problems. For example, adaptive role of seed dormancy, no matter how elegantly assessed, makes sense only in combination with factors affecting reproductive output, dispersability, and fate of seedlings – in other words, the life historical approach to relative fitnesses. Reproductive systems enter into these studies in terms of parameters of seed output as well as genetic recombination, as noted earlier. Thus, in fact, many known examples of genetic variation in outbreeding, heterostyly, dioecy, sex ratio, or asexuality naturally combine the genetic and ecological viewpoints on variation and evolution (Richards, 1986; Ross and Gregorius, 1984). Recent upsurge in botanists' interests in sexual selection has also tended in this direction, and physiological ecology of resource allocation is equally pertinent to the evolution of life history and mating system (Bazzaz and Reekie, 1985; Crawley, 1986).

In this volume additional emphasis on genetics of variation in plant populations provides an important caveat. Life history and other quantitative ecological approaches to natural selection help answer ecological questions about species' distribution and abundance (Watkinson, 1986), whereas simply inherited traits would allow us to employ population and molecular genetic methods in evolutionary research. Detection and measurement of selective forces is the common goal. Wright (1978) concluded from an extensive survey of variation in natural populations: 'Conspicuous polymorphisms are especially favorable for the study of genetic variation. They are not, however, to be considered typical of the variation on which evolution depends.' According to Wright (1978), quantitative genetic variation and physiologically significant allozyme polymorphisms are most important. Crow (1985) and Milkman (1985) have attempted unified theory for these different forms of variation. Quantitative genetics of life history traits will hopefully provide an important test of these theories.

REFERENCES

Anderson, W. W. (1971) Genetic equilibrium and population growth under density-regulated selection. *Am. Natur.*, **105**, 489–98.

Bazzaz, F. A. and Reekie, E. G. (1985) The meaning and measurement of reproductive effort in plants. in *Studies on Plant Demography* (ed. J. White), Academic Press, London, pp. 373–87.

Begon, M., Harper, J. L. and Townsend, C. R. (1986) *Ecology: Individuals, Populations and Communities*. Sinauer, Sunderland.

Boyce, M. S. (1984) Restitution of r- and K-selection as a model of density-dependent natural selection. *Ann. Rev. Ecol. Syst.*, **15**, 427–48.

Charlesworth, B. (1980) *Evolution in Age-structured Populations*. Cambridge University Press, Cambridge.

Crawley, M. J. (1986) Life history and environment. in *Plant Ecology* (ed. M. J. Crawley), Blackwell, Oxford, pp. 253–90.

Crow, J. F. (1985) The neutrality-selection controversy in the history of evolution and population genetics. in *Population Genetics and Molecular Evolution* (eds T. Ohta and K. Aoki, Japan Scientific Society Press, Tokyo, pp. 1–18.

Endler, J. A. (1986) *Natural selection in the wild*. Princeton University Press, Princeton.

Harper, J. L. (1983) A Darwinian plant ecology. in *Evolution from Molecules to Men* (ed. D. S. Bendall), Cambridge University Press, Cambridge, pp. 323–46.

Jain, S. K. (1982) Variation and adaptive role of seed dormancy in some annual grassland species. *Bot. Gaz.*, **143**, 101–6.

Lewontin, R. C. (1979) Fitness, survival and optimality. in *Analysis of Ecological Systems* (eds. D. J. Horn, G. R. Stairs and R. D. Mitchell), Ohio State University Press, Columbus, pp. 3–21.

Milkman, R. (1985) Two elements of a unified thoery of population genetics and molecular evolution. in *Population Genetics and Molecular Evolution* (eds T. Ohta and K. Aoki), Japan Science Society Press, Tokyo, pp. 65–83.

Richards, A. J. (1986) *Plant Breeding Systems*. George Allen & Unwin, London.

Ritland, K. and Jain, S. K. (1984) The comparative life histories of two annual *Limnanthes* species in a temporally variable environment. *Am. Natur.*, **124**, 656–79.

Ross, M. D. and Gregorius, H. R. (1984) Outcrossing and sex function in herm-aphrodites: A resource allocation model. *Am. Natur.*, **121**, 204–22.

Stearns, S. C. (1977) The evolution of life history traits: A critique of the theory and a review of the data. *Ann. Rev. Ecol. Syst.*, **8**, 145–71.

Watkinson, A. R. (1986) Plant population dynamics. in *Plant Ecology*, (ed. M. J. Crawley), Blackwell, Oxford, pp. 136–84.

Wright, S. (1978) *Evolution and the Genetics of Populations*. Vol. 4. *Variability Within and Among Natural Populations*. University of Chicago Press, Chicago.

Epilogue

The contributed chapters and commentaries in this volume reveal that plant evolutionary biology has become a large and diverse field of inquiry. Research topics range from analyses of molecular variation and phylogenetic reconstruction to populational studies of life history traits, plus traditional subjects including systematics, genecology and physiological ecology. Attention is being given to more precise descriptions of morphological and physiological characters, their mode of inheritance and the developmental processes that result in different phenotypes. This will make it easier to assess the basis of fitness differences and, thereby, how variation within population responds to natural selection. Theoretical models that account for changes in genotypic and phenotypic composition are prompting improved studies of natural populations. Thus, research in the field reflects a basic duality with different themes emerging from natural history (*sensu* J. B. S. Haldane, G. E. Hutchinson, R. McArthur *et al.*) on the one hand and reductionist viewpoints on the other. We are convinced that evolutionary biology is now ready to carry out research that accepts both points of view because of the obvious potential to understand connections between mechanisms that produce variation at the individual level and those that sort out this variation at the population level.

Evolutionary changes are complex because their consequences are not limited to the level of the biological hierarchy (molecules, characters, individuals, populations, species) in which they initially occur. The complexity is interesting and, when taken into account in the design of research, we anticipate more complete explanations of evolutionary change. Such vertical studies would entail close genetic and phenotypic analysis of individuals coupled with studies of character expression and development. Insofar as characters are studied that affect survivorship and reproductive success of individuals in natural populations (and why not study such characters?), we can also learn about character variation and correlation to other traits, particularly in relation to concepts of optimality and other appropriately defined strategies. We expect, in fact, that these measures of fitness will find many close parallels in both morphogenetic and physiological descriptions of optimal form and function as well.

For example, several chapters in this book describe changes in breeding system from outcrossing to self-pollination. Explanations of the change are framed primarily in terms of the presumed influence of environmental factors (unavailability or reduced 'interest' of pollinators) that might favor selfing. We find it curious that selfing is often correlated with reduced floral organs and overall loss of attractiveness, defined as reduced amounts or quality of nectar and pollen resulting in resource reallocation and changes in petal pigmentation, and that these changes have evolved many times in unrelated lineages. Environmental factors may frequently be involved, but we wonder if the character correlations result from changes in developmental features, perhaps having to do with size and maturation of anthers with consequent changes in hormone levels (Lord, 1980; Minter and Lord, 1983). This particular hypothesis may not be universally valid; its significance is to point out that explanations utilizing genetics and development can also be formulated to explain phenotypic evolution at diverse biological levels.

To illustrate another new frontier in evolution, we anticipate important population studies following the recent cloning and sequencing of cDNAs encoded by the self-incompatibility locus in *Brassica oleracea* (Nasrallah *et al.*, 1985) and *Nicotiana alata* (Anderson *et al.*, 1986). Population biologists and population geneticists will want to know why this locus is hypervariable and whether there are qualitative differences between self-incompatibility within populations and incompatibility between species. The availability of the gene clones will facilitate studies of sequence variability within populations and help to define the level at which selection operates in this system, analogous to studies of alcohol dehydrogenase in *Drosophila melanogaster* (Kreitman, 1983; Aquadro *et al.*, 1986; Kreitman and Aguade, 1986). Gene sequences provide extensive information on population variability, consanguinity, and phylogenetic relationships, and we expect their availability to stimulate numerous population-level investigations.

Theoretical models will also be necessary to guide population research that asks 'What direction?' and 'How rapidly?'. The formal mathematical models that describe the dynamics of large systems with few parameters (gene frequency, outcrossing rate, dispersal rate, fecundity, etc.) have already been linked up with molecular evolution based upon gene sequence information (Ohta and Aoki, 1985). Similar connections, with particular relevance to plants, must now be made. The results will be critical in plants because, in the absence of dated fossils, sequence change provides the only estimator of rate of evolution.

For these reasons, this volume juxtaposes topics from molecular genetics/development to systematics to population biology, a juxtaposition essential to plant evolutionary biology and one practiced by Ledyard Stebbins.

REFERENCES

Anderson, M. A., Cornish, E. C., Mau, S. L. (1986) Cloning of cDNA for a stylar glycoprotein associated with expression of self-incompatibility in *Nicotiana alata*. *Nature*, **321**, 38–44.

Aquadro, C. F., Desse, S. F., Bland, M. M. (1986) Molecular population genetics of the alcohol dehydrogenase region of *Drosophila melanogaster*. *Genetics*, **114**, 1165–90.

Kreitman, M. (1983) Nucleotide polymorphism at the alcohol dehydrogenase locus of *Drosophila melanogaster*. *Nature*, **304**, 412–17.

Kreitman, M. and Aguade, M. (1986) Excess polymorphism at the Adh locus in *Drosophila melanogaster*. *Genetics*, **114**, 93–110.

Lord, E. M. (1980) Physiological controls on the production of cleistogamous and chasmogamous flowers in *Lamium amplexicaule* L. *Ann. Bot.*, **44**, 757–66.

Minter, T. C. and Lord, E. M. (1983) Effects of water stress, abscisic acid, and gibberellic acid on flower production and differentiation in the cleistogamous species *Collomia grandiflora* Dougl. ex Lindl. (Polemoniaceae). *Am. J. Bot.*, **70**, 618–24.

Nasrallah, J. B., Kao, T. H., Goldberg, M. L. and Nasrallah, M. E. (1985) A cDNA clone encoding an S-locus-specific glycoprotein from *Brassica oleracea*. *Nature*, **318**, 263–7.

Ohta, T. and Aoki, K. (1985) *Population Genetics and Molecular Evolution*. Japan Scientific Societies Press, Tokyo.

Index

405